ERGONOMIC DESIGN FOR PEOPLE AT WORK

VOLUME I

Suzanne H. Rodgers — Technical Editor

Elizabeth M. Eggleton, Editor — Commercial Publications, Eastman Kodak Company

Contributing Authors — Human Factors Section, Eastman Kodak Company

Stanley H. Caplan
Paul C. Champney
Kenneth G. Corl
Brian Crist
William H. Cushman
Harry L. Davis
Terrence W. Faulkner
Richard M. Little
Richard L. Lucas
Thomas J. Murphy
Waldo J. Nielsen
Richard E. Pugsley
Suzanne H. Rodgers
John A. Stevens

Designed by William Sabia

Figure drawings by John Edens

ERGONOMIC DESIGN FOR PEOPLE AT WORK

Volume I

Workplace, Equipment, and Environmental Design and Information Transfer

A Source Book for Human Factors Practitioners in Industry including safety, design, and industrial engineers, medical, industrial hygiene, and industrial relations personnel, and management.

by: The Human Factors Section
Health, Safety and Human Factors Laboratory
Eastman Kodak Company

VNR VAN NOSTRAND REINHOLD
New York

The Human Factors Section of Eastman Kodak Company acknowledges with gratitude permission received from the following companies and organizations to use material reprinted from their publications in the development of this book. Specific references are noted in the text, complete citations follow in the bibliography at the end of each chapter.

American Book Company, New York, N.Y.
American Institute of Physics, New York, N.Y.
ASHA Publications, Rockville, Md.
Butterworth Scientific Ltd., Guildford, Surrey, U.K.
CMP Publications, Inc., Manhasset, N.Y.
Flournoy Publishers, Inc., Chicago, Ill.
Journal Publishing Affiliates, Santa Barbara, Calif.
McGraw-Hill Book Company, New York, N.Y.
Pendell Publishing Company, Midland, MI
Prentice-Hall, Inc., Englewood Cliffs, N.J.
Taylor and Francis Ltd., London, England
The Human Factors Society, Publications Div.,
Santa Monica, Calif. 90406
University of California Press, Berkeley, Calif.
Western Electric Company, New York, N.Y.
John Wiley and Sons, Inc., New York, N.Y.
Van Nostrand Reinhold, New York, N.Y.

Printed in the United States of America

Van Nostrand Reinhold
115 Fifth Avenue
New York, New York 10003

Chapman & Hall
2-6 Boundary Row
London SE1 8HN, England

Thomas Nelson Australia
102 Dodds Street
South Melbourne, Victoria 3205, Australia

Nelson Canada
1120 Birchmount Road
Scarborough, Ontario M1K 5G4, Canada

16 15 14 13 12 11 10 9 8 7

Library of Congress Cataloging in Publication Data
Main entry under title:

Ergonomic Design for People at Work
Volume I

Bibliography: v. 1, p.
Includes index.
1. Human engineering—Handbooks, manuals, etc.
I. Eastman Kodak Company. Human Factors Section.
T59.7.E714 1983 620.8'2 83-719
ISBN 0-534-97962-9 (v. 1)
0-442-23972-6 (V. 1 ppr.)

NOTICE

CONTENTS

LIST OF ILLUSTRATIONS

LIST OF TABLES

PREFACE

Ergonomic Design for People at Work, Volumes I and II, are definitive works on industrial human factors/ergonomics. This book, the first in the series, is directed to a practical discussion of workplace, equipment, environmental design and of the transfer of information in the workplace. Its contents and purpose are further explained in Chapter I. The second volume will include guidelines for job design, manual materials handling, and shift work.

The application of human factors/ergonomics principles to the workplace has been of interest to Eastman Kodak Company for many years. This series summarizes current data, experience, and thoughts assembled from the published literature, internal research, and observation by the members of the Human Factors Section. The selection of material has been guided by the types of problems the section has been asked to address over the past 22 years. The guidelines and examples of approaches to design problems are most often drawn from case studies. The principles have been successfully applied in the workplace to reduce the potential for occupational injury, increase the number of people who can perform a job, and improve performance on the job, thereby increasing productivity and quality.

Members of the Human Factors Section have each brought their special expertise to the material included in these volumes. It is our hope that the experience we have gained from problem solving in an industrial setting with a group that includes many disciplines will be of value to others with fewer resources available to them, and that the material will be useful in the solution of human factors and ergonomics problems in industries in many countries.

Eastman Kodak Company

ACKNOWLEDGMENTS

Members of the Human Factors Section who contributed to the research and writing of each section of this book are listed opposite the title page and at the beginning of each chapter.

Lending support to the project were Harry L. Davis, Supervisor of the Human Factors Section; Kenneth T. Lassiter, Publications Director, Consumer/Professional and Finishing Markets; and Alexander Kugushev, Publisher, Lifetime Learning Publications.

The following people reviewed a draft of the manuscript and offered valuable suggestions for its improvement:

Thomas S. Ely, M.D., Assistant Director, Health, Safety and Human Factors Laboratory, Eastman Kodak Company
David Alexander, M.S., and Hart Kaudewitz, M.S., Human Factors Group, Tennessee Eastman Division, Eastman Kodak Company
Stover H. Snook, Ph.D., Liberty Mutual Insurance Company
Thomas Bernard, Ph.D., Westinghouse Corporation
Harry Snyder, Ph.D., Virginia Polytechnic Institute & State University
Thomas J. Armstrong, Ph.D., The University of Michigan School of Public Health
W. Monroe Keyserling, Ph.D., Harvard University School of Public Health

Eliezer Kamon, Ph.D., Pennsylvania State University, and David Kiser, Ph.D., Human Factors Section, Eastman Kodak Company, each reviewed the temperature and humidity section.

Carol McCreary, Human Factors Section, provided extensive technical assistance and traced many of the references for this project; she has our gratitude and admiration. Our appreciation also goes to Anne Wilkinson and Gerry Bommelje of the Photo Services Section, Research Division, Eastman Kodak Company, for their initial artwork. Elaine Villa and Shereta Harris are especially thanked for typing the manuscript. The photography was provided by Donald S. Buck of the Kodak Photographic Illustration Department and Allan Fink and Leslie Smades of the Kodak Park Industrial Studio.

Finally, we would like to thank McGraw-Hill, publishers of *Human Engineering Guide to Equipment Design* by Morgan, Cook, Chapanis, and Lund (1963), for permission to use much of the material in the equipment design chapter of this book.

The Human Factors Section
Health, Safety and Human Factors Laboratory

ERGONOMIC DESIGN FOR PEOPLE AT WORK

VOLUME I

Chapter I USING THIS BOOK FOR ERGONOMICS IN INDUSTRY: INTRODUCTION

Contributing Authors

Harry L. Davis,
B. S., Industrial Engineering
Supervisor, Human Factors Section

Suzanne H. Rodgers,
Ph.D., Physiology

Chapter Outline

1. Ergonomics/Human Factors in Industry: An Explanation and History of the Subject
   a. Explanation
   b. History of Ergonomics and Human Factors
2. Human Factors at Eastman Kodak Company
3. Scope and Purpose of the Book
4. Summary of the Contents
5. Criteria for Design: Who?

1. ERGONOMICS/HUMAN FACTORS IN INDUSTRY: AN EXPLANATION AND HISTORY OF THE SUBJECT

This chapter contains a short history of ergonomics/human factors and of the Human Factors Section of the Health, Safety, and Human Factors Laboratory at Eastman Kodak Company. The scope and purpose of the book, a summary of topics covered, and criteria for design relative to several human characteristics and capabilities are also described.

a. Explanation

Ergonomics/human factors is a multidisciplinary activity striving to assemble information on people's capacities and capabilities for use in designing jobs, products, workplaces, and equipment. In the United States the military and aerospace industries have generally accepted human factors principles, but most other industries have been less quick to understand the benefits. The probable benefits of well-designed jobs, equipment, and workplaces are improved productivity, safety, health, and increased satisfaction for the employees. Removing unnecessary effort from jobs or reducing demands by improving the way in which information is transferred between people or between product and people (inspection) allows for greater productivity and, ultimately, higher profitability.

As concerns about productivity, employee job satisfaction, and health and safety in the workplace have increased, interest in human factors/ergonomics has also increased. Many schools with an industrial engineering or a psychology department now include a course in human factors, and industrial hygienists are expected to know some ergonomics principles for certification. Medical professionals are recognizing the value of ergonomic analyses of jobs to assist them in the rehabilitation of people returning to work after illness. In addition, with increasing industrialization of developing countries, there is more demand for design guidelines that recognize the capabilities of people in manufacturing systems.

The terms *ergonomics* and *human factors* are often used synonymously. Both describe the interaction between the operator and the job demands, and both are concerned with trying to reduce unnecessary stress in the workplace. Ergonomics, however, has traditionally focused on *how work affects people.* This focus includes studies of their physiological responses to physically demanding work; environmental stressors such as heat, noise, and illumination; complex psychomotor assembly tasks; and visual-monitoring tasks. The emphasis has been on ways to *reduce fatigue* by designing tasks within people's work capacities. In contrast, human factors, as practiced in the United States, has traditionally been more interested in the man-machine interface, or human engineering. It has focused on people's behavior as they interact with equipment, workplaces, and their environment, as well as on human size and strength capabilities relative to workplace and equipment design. The emphasis of human factors is often on designs that *reduce* the potential for *human error*.

Since ergonomics/human factors requires two distinct focuses, we have addressed the subject in two volumes. This book, the first volume, includes the traditional human factors subjects of workplace and equipment design, information transfer, and physical environment. The second volume covers the traditional ergonomics subjects of job design, manual materials handling, and shift work.

b. History of Ergonomics and Human Factors

Healthful, safe, and comfortable working conditions have been the concern of many people since the beginning of the industrial revolution. Documented proof shows that a Polish educator and scientist, Wojciech Jastrzebowski, introduced the term *ergonomics* into the literature some one hundred twenty years ago. Ergonomics is from the Greek word *ergos,* meaning "work," and *nomos,* meaning "laws"—therefore, the laws of work. Jastrzebowski (1799–1882) was a professor of natural sciences at the Agronomical Institute in Warsaw-Marymount. This prominent educator, an expert on national physiography, especially its flora, played an enormous part in creating the basis for labor science, the study of work (Polish Ergonomics Society, 1979).

However, not much was heard of ergonomics until World War II. Complex and confusing aircraft, radar, and other equipment created adverse problems for operator performance and maintenance, which resulted in significant performance decrement (Chapanis, 1951). Since the problems were both engineering and behavioral, teams of psychologists, engineers, anthropologists, and physiologists were brought together to try to solve the design and training problems. Even though the multidisciplinary approach to solving workplace problems was first evidenced at this time, the concept was not identified as ergonomics until later. Terms like *human engineering* and *engineering psychology* were used to describe these early efforts.

The efforts that began as a result of World War II occurred simultaneously in the United States and Great Britain. Ergonomics work began to grow in most European countries at the same time. In the United States what came to be known as human factors work flourished in military programs and, subsequently, in the aerospace program. For example, since the early 1950s every system and aerospace design has been required to have a concurrent human factors design (Department of Defense, 1974).

In recent years human factors specialists, or ergonomists, have worked on such problems as the design of aircraft cockpits and manned space capsules. They have answered questions about the number of g's (acceleration of gravity) a pilot can endure without blacking out, and they have developed simulators to help train astronauts for space missions. The push-button telephone key arrangement was designed and evaluated by human factors personnel. Industrial problems, such as the way heat affects workers' well-being and productivity and the way the stress can be minimized, have also been approached from the ergonomics standpoint.

At home, work, or play new problems and questions must be resolved constantly. People come in all shapes and sizes, with varying capabilities and limitations in strength, speed, judgment, and skills; all of these factors need to be considered in the design function. Thus the scientific disciplines of physiology and psychology must be included with the engineering approach to solve design problems.

As interest in ergonomics grew, societies supporting the profession developed. The first known society that was formed to bring together the various disciplines concerned with people at work and to further the research effort was the Ergonomics Research Society of Great Britain (Edholm and Murrell, 1973). This society was formed in 1950, with the term *ergonomics* independently reinvented by Professor K. F. H. Murrell.

In the United States the formal concept for a human factors society was established in October 1956 at a joint meeting of the Aeromedical Engineering Association of Los Angeles and the San Diego Human Engineering Society. It was officially created in Tulsa, Oklahoma, on September 25, 1957.

At about the same time, people in Europe, Scandinavia, and the United States were seeking an umbrella association that would sponsor meetings and encourage the development of ergonomics worldwide. In 1961 the first international meeting of what was to become known as the International Ergonomics Association was held in Stockholm, Sweden. Today the association has at least fifteen affiliated societies including those in the United States, Great Britain, most European countries, Scandinavia, Japan, and Australia.

2. HUMAN FACTORS AT EASTMAN KODAK COMPANY

Although people have been applying human factors and ergonomics principles in the workplace for many years, few industries have formally recognized the field by establishing a group to study the full spectrum of people at work. E. I. du Pont de Nemours & Co., though, is one industry that has. Its Haskell Laboratory, headed by the late Dr. Lucien Brouha, is primarily concerned with heat stress problems and the capacities of people doing hard physical work (Brouha, 1970).

At Eastman Kodak Company Dr. Charles I. Miller and Mr. Harry L. Davis, having met with Dr. Brouha and having learned from him, began doing experiments in work physiology in early 1957. They formulated ideas and plans for a broad-spectrum human factors function within the company. By 1960 a small laboratory was developed, and a human factors group function formed. It was a joint effort of the Medical Department and the Industrial Engineering Division of the Kodak Park Division in Rochester, New York. The group specialized in workplace and job analysis and design and visual inspection tasks within a very large industrial complex that manufactures a diversity of photographic products, papers, chemicals, and hardware products. Because of the human being's complexity and the diversity of products manufactured under varied environmental conditions, it was determined that the group's makeup should be multidisciplinary. Today the group includes people with training in multiple scientific and engineering specialties: psychology, industrial engineering, physiology, medicine, electrical and mechanical engineering, and computer science.

Expansion of the group into a variety of disciplines has resulted in a corresponding increase in its activity. These activities include perceptual loading, analysis of physical effort, the effects of noise, illumination, and temperature, and visual capabilities. The area of product design has also been developed, and the group is now heavily involved in the design and evaluation of many company products. In 1972 the group became a member of the Kodak corporate staff when it was made a part of the newly created Health, Safety, and Human Factors Laboratory. It serves the entire corporation and interacts strongly with the manufacturing divisions and the Medical, Safety, Industrial Hygiene, Epidemiology, Industrial Relations, Design Engineering, Industrial Design, and Industrial Engineering staff groups to identify and resolve potential human factors problems. In addition to the corporate Human Factors Section in Rochester, there are persons performing human factors functions at plants of the Eastman Chemical Division in Kingsport, Tennessee; Columbia, South Carolina; and Longview, Texas; and at the Windsor, Colorado, and Toronto, Canada, plants.

3. SCOPE AND PURPOSE OF THE BOOK

This book is intended for use by practitioners of human factors in the design of workplaces, equipment, and the physical environment in the industrial setting. The guidelines in this volume are not specifically relevant to product design but may be applicable in many instances. Although many physiological and psychological data have been used to develop the guidelines, results are expressed in terms that engineering, safety, or medical personnel can easily transfer to the plant. Terms such as *reaches, heights,* and *comfort levels* are used throughout the book, wherever possible.

The art of applying human factors principles to the workplace depends on knowledge of the limitations of the data available. The information in this book is suitable for the design of new workplaces and equipment. The guidelines must be interpreted before being used to evaluate existing conditions.

What is optimal in new design is not essential for redesign in order to improve a workplace, a piece of production equipment, or an environment and to bring it within the capabilities of more people. Use of the guidelines for such evaluations is best shown through examples. A section on typical problems and use of the guidelines to evaluate them is included in Chapter VI, Appendix C. Once the problem has been identified, there are many possible solutions to it. The choice of solution will depend on factors that vary from location to location, so this book does not attempt to focus on solutions to existing design problems except in a very general sense.

Table I–1 summarizes the scope and goals of this book.

Table I–1: Scope and Purpose of the Book

IS	IS NOT
For industrial use by engineers, medical and safety personnel, supervision	Necessarily applicable to military industries
For workplace, equipment, and environmental design	Always applicable to product design
For the practitioner	For the basic researcher
A comprehensive reference source on the subject	A set of rules
Based on the U.S. population	Adjusted for differences in anthropometrics of non–U.S. populations such as Oriental, South American, etc. (see Chapter VI, Appendix A, for further comment)
Applicable to the large majority of the U.S. population (both sexes, all ages of the industrial population, varying fitness levels)	For the healthy young male only
For new design and evaluation of existing design	To be used to exclude people from jobs on the basis of their physical characteristics
Useful in identifying potential problems	Recommended primarily as solutions to specific problems
Subject to revision with improved information	The last word in industrial ergonomics/human factors

The audience, philosophy, and applicability of information in the book are identified in the first column. In the second column some of the limitations in applications of the information are noted.

4. SUMMARY OF THE CONTENTS

The organization of this book has been developed from an internal evaluation of an earlier edition and observations of the kind of questions most asked in the industrial setting. Published literature in the field has been drawn on heavily to provide a comprehensive source of information on the subject of human factors; the literature, research, and experience of the members of the Kodak Human Factors Section have also been utilized.

Published references are listed alphabetically at the end of each chapter. Examples are included in the text for illustrative purposes, and problems for the practitioner to solve are included in Chapter VI, Appendix C.

The major sections of the book are listed in the Contents. The reader will notice a large overlap between equipment and workplace design, environment, and information transfer. This redundancy has been reduced by cross-referencing wherever possible.

5. CRITERIA FOR DESIGN: WHO?

Until recently, much of the data available for use by human factors practitioners came from military studies on healthy young men (Damon, Stoudt, and McFarland, 1966; Hertzberg, Daniels, and Churchill, 1954; White, 1961). However, the industrial population is defined as men and women between the ages of 17 and 70 with the expected distribution of chronic illness and functional losses in capacity from congenital or developed disabilities. Designing workplaces, jobs, and equipment for healthy young men, therefore, could effectively result in nonoptimal design for about 75 percent of the potential industrial work force.

The data on which guidelines in this book are based include women and older populations where data are available. Where such information was not found, adjustments have been made to the data on men. This adjustment was done by using the relationships between men's and women's average capabilities for that measurement and estimating, from other measurements, the degree of overlap of the data. Revision of some of the guidelines may become necessary when better data are available on some of these capabilities for women and older people.

Designing workplaces, equipment, and the physical environment to fit the characteristics and capabilities of most people is a complex task. For example, clearances should always be designed for people with the largest dimensions (Hertzberg, Emmanuel, and Alexander, 1956; Thomson, Covner, Jacobs, and Orlansky, 1963). Other parameters require careful weighing of how the design will affect both extremes of the population. One always tries to design for a majority of the population, making trade-offs for both extremes. However, reasonable accommodations in the workplace may be easier to make for the smaller person than for the larger person. For example, consider workplace height. It may be relatively easy to provide a platform step for the smaller person, but in the same situation the larger person may have to stoop for long periods of time. Such considerations necessarily push the design criterion toward the upper side of the distribution.

Figure I–1 gives typical frequency distributions for an anthropometric variable. If these data represent stature, for instance, we can define clearance heights for standing workplaces or doorways that should accommodate at least 99 percent of the potential work force (at the far right of the male distribution). If the distributions represent forward functional reach, they can be used to assess the impact of a given reach requirement on men and women workers. Reaches should be designed to be within the capabilities of as many people as possible. A reach that accommodates people with the shortest arms may be very awkward for people with long arms. For example, if all reaches at a seated workplace are kept within 30 cm (13 in.) of the front of the body, people with long arms may not be able to rest their elbows while doing the task. This situation could contribute to shoulder muscle fatigue and reduced productivity. Examples of how to use anthropometric data to develop workplace and equipment design guidelines are given in Chapter VI, Appendix A. Use of the data to evaluate existing designs is demonstrated in Chapter VI, Appendix C.

Population capabilities for visually detecting defects, processing information, and working in the heat are less well defined. Determining the appropriate design levels in these areas is done more by trial-and-error rate analysis than by designing to a capability level. Tolerances to environmental factors such as heat and humidity at most industrial exposure levels appear to be quite individualized. These design guidelines are set by using observational data in conjunction with physiological information on body temperature control. Although the purpose of the guidelines is to accommodate the capacities of most people, designing to include people with high susceptibility to environmental stresses or very limited visual capability, for instance, would often not be feasible.

Figure I–1: Designing with Anthropometric Data

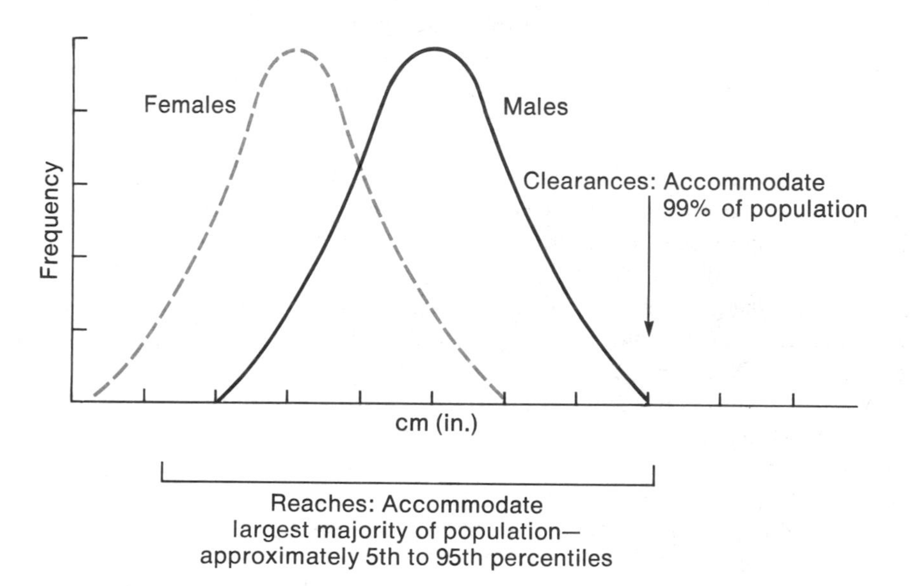

Examples of female and male frequency distributions for an anthropometric variable, such as stature or forward reach, are shown. Values for the variable, in centimeters (cm) and inches (in.), are indicated on the horizontal axis. When a sample population is measured, the number of people at each value is given on the vertical axis. In most size distributions the females' distribution curve falls to the left of that for the males, although there is considerable overlap of the populations. Dimensions that relate to body clearances—such as stature, shoulder breadth, hip breadth, and forearm circumference—should accommodate the largest (99th percentile) people. Workplaces that require reaches, such as forward functional or overhead reach, should consider both the large and small people. It is often necessary to compromise on a design that is appropriate for all but the extremes of the population.

REFERENCES FOR CHAPTER I

Brouha, L. 1970. *Physiology in Industry*. 2nd ed. New York: Pergamon Press, 164 pages.

Chapanis, A. 1951. "Theory and Methods of Analyzing Errors in Men-Machine Systems." *Annals of the New York Academy of Science, 51:* pp. 1179–1203.

Damon, A., H. W. Stoudt, and R. A. McFarland. 1966. *The Human Body in Equipment Design*. Cambridge, Mass.: Harvard Univ. Press, 360 pages.

Department of Defense. 1974. *Human Engineering Design Criteria for Military Systems, Equipment and Facilities*. MIL-STD 1472B, May 15, 1970, 239 pages, Washington, D.C.

Edholm, O. G., and K. F. H. Murrell. 1973. *The Ergonomics Research Society, A History 1949–1970*. London: Ergonomics Research Society, 39 pages.

Hertzberg, H. T. E., G. S. Daniels, and E. Churchill. 1954. *Anthropometry of Flying Personnel, 1950*. WADC Tech. Rpt. 52-321. Wright-Patterson AFB, Ohio: Wright Air Development Center.

Hertzberg, H. T. E., I. Emmanuel, and M. Alexander. 1956. *The Anthropometry of Working Positions. 1. A Preliminary Study*. WADC Tech. Rpt. 54-520, Wright-Patterson AFB, Ohio: Wright Air Development Center.

Morgan, C. T., J. S. Cook III, A. Chapanis, and M. W. Lund. 1963. *Human Engineering Guide to Equipment Design*. New York: McGraw-Hill, pp. 321–365.

Polish Ergonomics Society. 1979. *Ergonomia 2 (1)*. 7th International Ergonomic Association Congress. A comprehensive study of this subject was also published by J. Banka in the book *Narodzing filozofil nauky a pracy*, KiW, 1970.

Thomson, R. M., B. J. Covner, H. H. Jacobs, and J. Orlansky. 1963. "Arrangement of Groups of Men and Machines." Chapter 8 in Morgan et al. (1963), pp. 321–366.

White, R. M. 1961. *Anthropometry of Army Aviators*. TREP-150. Natick, Mass.: Environmental Protection Research Division, U.S. Army Quartermaster Research and Engineering Center, 109 pages.

Chapter II WORKPLACE DESIGN

Contributing Authors

William H. Cushman,
Ph.D., Psychology

Waldo J. Nielsen,
M.S., Industrial Statistics

Richard E. Pugsley,
M.S., Industrial Engineering

Chapter Outline

Section II A. Layout

1. Workplaces
   a. Sitting Workplaces
   b. Standing Workplaces
   c. Sit/Stand Workplaces
   d. Special Case: Design of VDU (Video Display Unit) Workplaces
   e. Special Case: Design of Chemical Hoods and Glove (Isolation) Boxes
2. Aisles and Corridors
3. Floors and Ramps
   a. Floors
   b. Ramps
4. Stairs and Ladders
   a. Stairs
   b. Ladders
5. Conveyors
6. Dimensions for Visual Work
   a. Viewing Angles
   b. Eye Height
   c. Eyeglasses
   d. Visual Work Area

Section II B. Adjustable Design Approaches

1. Adjusting the Workplace
   a. Shape
   b. Location: Height and Distance
   c. Orientation
2. Adjusting the Person Relative to the Workplace
   a. Chairs
   b. Support Stools, Swing-Bracket Stools, and Other Props
   c. Platforms, Step-Ups, and Mechanical Lifts
   d. Footrests
   e. Armrests
3. Adjusting the Workpiece or Product
   a. Jigs, Clamps, and Vises
   b. Circuit Board Assembly
   c. Parts Bins
   d. Lift Tables, Levelators, and Similar Equipment
4. Adjusting the Tool
   a. Design and Location of Tools
   b. Trays and Carriers

Section II C. Clearance Dimensions

Guidelines for the design of workplaces and work stations are presented in this chapter. Subjects discussed include layout, seating, clearances, and adjustments to accommodate individual differences in size and strength.

A workplace is a location where a person or persons perform tasks for a relatively long period of time. These periods may be interspersed with other activities that require the person to leave the workplace, such as procuring work supplies or disposing of the finished product. A work station is one of a series of workplaces that may be occupied or used by the same person sequentially when performing his or her job. The part or product is moved between stations either by equipment such as conveyors or by the operator. Work stations may also be locations where a person momentarily performs a task, such as monitoring or recording information from instrument panels. Additional information on controls and displays can be found in Chapter III, "Equipment Design."

Workplaces should be designed so that most people can safely and effectively perform the required tasks. Reaches, size, muscle strength, and visual capabilities have to be considered when developing design criteria (see Chapter I, the section titled "Criteria for Design: Who?"). Although reaches can be extended by stretching or leaning, and muscle strengths increased by provision of tools or other aids, designing workplaces to fit most people's capabilities helps to reduce unnecessary job stress and increase productive work.

SECTION II A. LAYOUT

The way a production or office area is laid out can have an effect on how efficiently people do their jobs. The design of a large production system, for instance, can determine the staffing needs of an operation. Extended travel distances and lack of space to store supplies or to inventory product can put excessive time pressure on an operator who is trying to keep a machine running.

Some general considerations when laying out a production workplace or office area are as follows:

- Services needed by several people should be placed in a central location.
- The communications needs of different operations should be evaluated, and people or workplaces should be located so as to maximize communication.
- Lines of sight and other visual requirements for operations should be kept clear. For example, it is important to be able to see from a control console to manufacturing equipment.
- Noisy, heat-producing, odor-producing, or visually distracting operations should be modified or located so as to minimize their effects on other operations. (See Chapter V, "Environment," for further discussion.)

- The work area should be arranged so that the product can flow through it, preferably in one direction, with minimal rehandling.
- Offices should be designed to permit people a minimum separation of 122 cm (48 in.), with 244 cm (96 in.) being more desirable (Hall, 1966).
- Postural flexibility and change should be provided. A person should not be restricted to a workplace in such a way that he or she cannot change posture during the shift.

In addition to these general design considerations, specific elements that should be considered when designing a workplace layout include the following:

- Sitting, standing, and sit/stand workplaces.
- Special workplaces such as computer terminals, chemical hoods, and glove (isolation) boxes.
- Aisles and corridors.
- Stairs, ramps, ladders, walkways, and floors.
- Conveyors.
- Workplaces where visual work predominates.

These elements are discussed in this section.

1. WORKPLACES

There are three major categories of workplace: sitting, standing, and sit/stand. Choice of the appropriate one depends on the task to be performed. Table IIA–1 indicates the recommended workplaces for combinations of tasks often found in industry.

Some general characteristics of workplaces in each of the three categories are summarized below. Sitting workplaces are best in the following situations:

- All items needed in the short-term task cycle can be easily supplied and handled within the seated work space.
- The items being handled do not require the hands to work at an average level of more than 15 cm (6 in.) above the work surface.
- No large forces are required, such as handling weights greater than 4.5 kg (10 lb) (adapted from Rehnlund, 1973). These large forces may be eliminated by using mechanical assists.
- Fine assembly or writing tasks are done for a majority of the shift.

Table IIA–1: Choice of Workplace by Task Variables (Developed from information in Ely, Thomson, and Orlansky, 1963; Murrell, 1965; Rehnlund, 1973, Woodson, 1981.)

Parameters	**Heavy Load and/or Forces**	**Intermittent Work**	**Extended Work Envelope**	**Variable Tasks**	**Variable Surface Height**	**Repetitive Movements**	**Visual Attention**	**Fine Manipulation**	**Duration > 4 Hours**
Heavy Load and/or Forces		ST	ST	ST	ST	S/ST	S/ST	S/ST	ST/C
Intermittent Work			ST	ST	ST	S / S/ST	S / S/ST	S / S/ST	S / S/ST
Extended Work Envelope				ST	ST	S/ST	S/ST	S/ST	ST/C
Variable Tasks					ST	S/ST	S/ST	S/ST	ST/C
Variable Surface Height						S	S	S	S
Repetitive Movements							S	S	S
Visual Attention								S	S
Fine Manipulation									S
Duration > 4 Hours									

Note: S = sitting; ST = standing; S/ST = sit/stand (an alternative to standing all day; a standing workplace could be used but would not be the preferred choice); ST/C = standing, with chair available.

Job and workplace characteristics are looked at, two at a time, in relation to the preferred workplace choice: sitting, standing, sit/stand, or standing with a chair provided. More than one type of workplace may be acceptable for these task combinations; the most appropriate choice is indicated.

Standing workplaces will be the best alternative in the following circumstances:

- If the workplace or work station does not have knee clearance for a seated operation.
- Objects weighing more than 4.5 kg (10 lb) are handled.
- High, low, or extended reaches, such as those in front of the body, are required frequently.
- Operations are physically separated and require frequent movement between work stations.
- Downward forces must be exerted, as in wrapping and packing operations.

Sit/stand workplaces should be considered in these instances:

- Repetitive operations are done with frequent reaches more than 41 cm (16 in.) forward and/or more than 15 cm (6 in.) above the work surface. Operations would be done at a sitting workplace if it were not for the reach requirements.
- Multiple tasks are performed, some best done sitting and others best done standing. Provision for both may not be feasible owing to space constraints.

In operations where a standing workplace is used for a majority of the shift, provision should be made for sitting down during machine time or other slack time. It is desirable to minimize static standing operations by having the operator move outside the immediate work area several times per hour (see "The Standing Work Area" section later). Such movement, however, should not be made a regularly occurring part of a short-duration, highly repetitive work cycle. Provision of floor mats at the workplace also reduces discomfort for people whose job requires them to stand all day. Where safety considerations prevent the use of floor mats, shoes with cushioned soles may increase a person's comfort in a standing workplace.

Since jobs may combine elements that favor each type of workplace, some priorities have to be established between or among the tasks. Some guidelines for this choice are as follows:

- The duration for each task should be assessed. Those that make up the majority of the work time should take precedence in establishing the type of workplace used.
- If critical visual tasks are involved, workplace choice should be geared to them, especially if they are a major part of the job.
- If extended reaches and exertion of forces are frequent on the job, the type of workplace chosen should optimize them.

In addition to general guidelines for the selection and design of sitting, standing, and sit/stand workplaces, this section gives information about the design of two specific types of workplaces: video display unit (VDU) workplaces and chemical hoods and glove boxes. VDUs may require extended visual work along with keyboard data entry. Chemical hoods and glove boxes require the operator to work behind a protective barrier or in gloves attached to arm ports. These special requirements should be considered in the design process.

a. Sitting Workplaces

(1) The Seated Work Area

Seated operators generally work in the space above the working surface. For the determination of where parts or controls may be located, it is necessary to visualize a three-dimensional space in front of the operator (Figure IIA–1).

The maximum forward reaches of a woman with short arms (5th percentile) are shown in Figure IIA–2. As the reach is located farther to the right of the body's centerline in the usual work posture, the forward reach capability is also reduced. Only the dimensions of the right arm's work area are shown. The left arm's workspace can be treated as a mirror image of the right's.

For example, if a workplace is used to pack small items into a kit, the distances from the center of the workplace to each supply bin should be designed to be within this seated, arm reach workspace. Suppose that 8 items had to be clustered around the kit assembly area, and at least a 25 × 25-cm (10 × 10-in.) work area was needed in front of the operator. Supply bins would then be more than 25 cm (10 in.) in front of the operator near the work surface. For the most efficient work motions the bins should be placed within 41 cm (16 in.) to the right or left of the center of the workplace and not more than 50 cm (20 in.) above the surface, preferably lower. To avoid fatiguing the shoulder muscles, one might decide to keep the procurement of items from supply bins 25 cm (10 in.) above the work surface. This technique would limit the comfortable reach distance to the left or right of the body's centerline to 41 cm (16 in.) and to about 36 cm (14 in.) in front of the operator. Although more extended reaches can be made, they should not be incorporated into a highly repetitive assembly task such as kit assembly.

Because these values represent people with less reach capability, it is advisable to add a 7.6-cm (3-in.) wide foldout extension to the front of the workplace or to provide a chair with adjustable armrests. These armrests permit people with long forearms to rest their elbows during repetitive assembly or inspection operations.

Any object that is to be frequently grasped or procured should be located within 15–36 cm (6–14 in.) of the front of the work surface. These ranges are the distances from which small objects can be procured without requiring the operator to bend forward. Large or heavy objects will need to be located closer to the front of the workplace. It is permissible to have an operator occasionally (a few times an hour) lean forward to procure something outside

Figure IIA–1: The Seated Workspace

The operator is seated behind a three-dimensional model that represents the reach capability of a person with short arms (5th percentile). The reach distances are shown for several different heights above the working surface. The operator's chair has been adjusted so that the work surface is at elbow height. The reaches are from the front of the workplace without leaning forward or stretching. See Figure IIA–2 for the reach distances.

the work area, but such reaches should not be made a regularly occurring part of a brief work cycle.

(2) Seated Workplace Height

The correct seated working height depends on the nature of the tasks being performed. A majority of manual tasks, such as writing and light assembly, are most easily performed if the work is at elbow height. If the job requires

Figure IIA–2: Forward Reach Capability of a Small Operator, Seated (Developed from data in Faulkner and Day, 1970.)

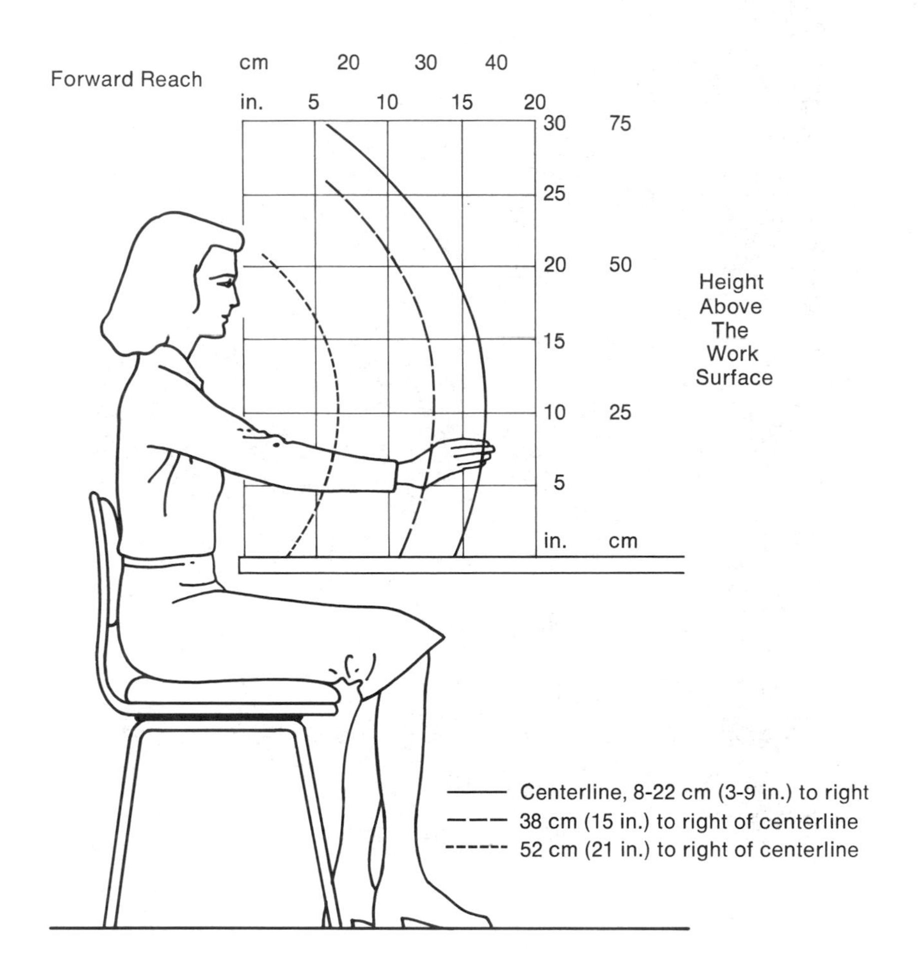

Three curves describe the seated reach workspace for a 5th percentile female's right hand. The view is from the side, similar to the angle in Figure IIA–1. The forward reach capability (horizontal axis) is affected by the height of the hands above the work surface (vertical axis) and by the arm's distance to the right of the body's centerline, as indicated by the three curves defined at the bottom of the figure. At 25 cm (10 in.) above the work surface, for example, the forward reach is 41 cm (16 in.) if the arm is within 23 cm (9 in.) of the centerline; if it is moved 53 cm (21 in.) to the right, forward reach falls to 18 cm (7 in.).

perception of fine detail, it may be necessary to raise the work to bring it closer to the eyes (see the section "Dimensions for Visual Work" later).

Seated workplaces should be provided with adjustable chairs and footrests. The recommended workplace dimensions for most seated tasks are shown in Figure IIA–3. If a workplace is used by only one person, as is a secretary's desk, the footrest can often be selected to fit the individual. Workplaces for people in wheelchairs should also follow these guidelines. Sitting wells should be at least 81 cm (32 in.) wide under the work surface (Mueller, 1979).

Figure IIA–3: Recommended Dimensions for a Seated Workplace with a Footrest (Adapted from P. C. Champney, 1975, Eastman Kodak Company; T. W. Faulkner, 1968, Eastman Kodak Company.)

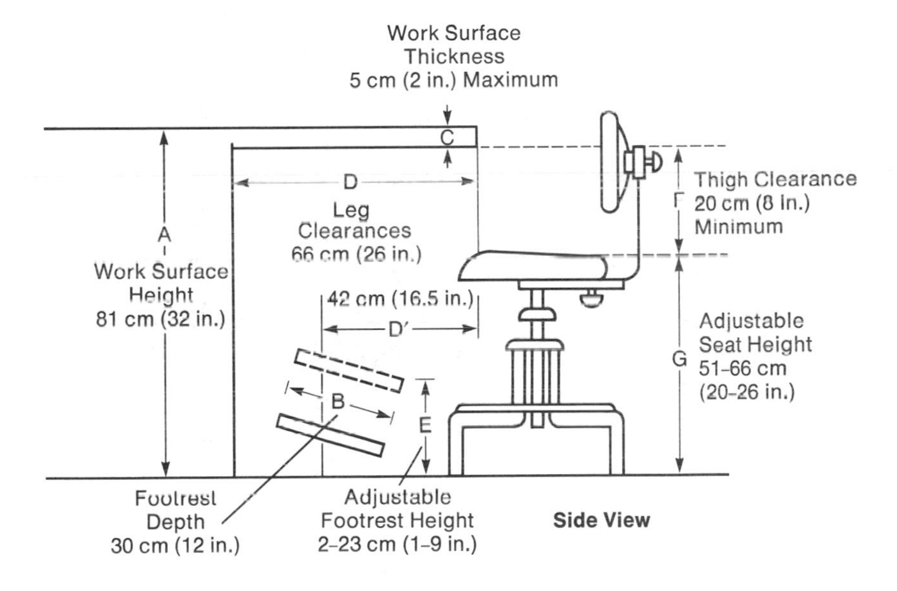

The heights, clearances, and work surface thickness of a seated workplace with a footrest and an adjustable chair are given. These design guidelines ensure that most people will be able to work comfortably at the workplace. The work surface height (A) given is the recommended level; chair (G) and footrest (E) heights should be adjustable to provide adequate thigh clearance (F) and leg comfort. When work surface height is 81 cm (32 in.), the table thickness (C) should not exceed 5 cm (2 in.). Two forward leg clearances are given: D is the recommended distance under the work surface, and D′ is the minimum distance from the edge of the work surface to about two-thirds of the depth (B) of the footrest. Seat and footrest adjustabilities are important in accommodating differences in size of people using a seated workplace.

If it is not possible to provide an adjustable footrest for a sitting workplace, or if a computer terminal or other equipment is being used at it, then the height of the work surface must be lowered. The dimensions for a sitting workplace without a footrest are shown in Figure IIA–4. This type of workplace requires additional forward leg clearance and a chair that adjusts to a lower height than that used at a workplace with a footrest.

Figure IIA–4: Recommended Dimensions for a Seated Workplace Without a Footrest (Adapted from P. C. Champney, 1975, Eastman Kodak Company; T. W. Faulkner, 1968, Eastman Kodak Company.)

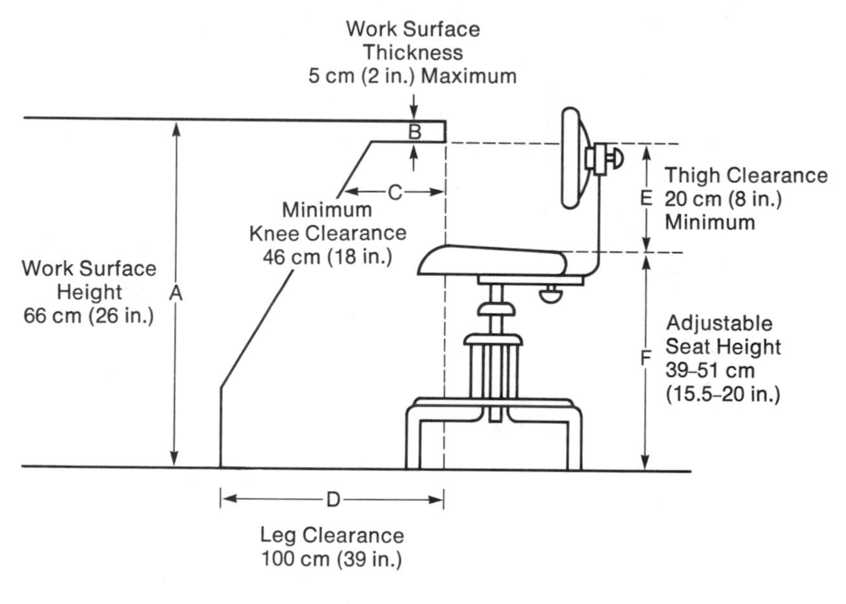

The heights, clearances, and work surface thickness of a seated workplace without a footrest are indicated. Because the feet must be able to rest on the floor, the chair has to adjust to lower levels (F), the forward leg clearance (D) must be greater, and the work surface height (A) has to be lower than in the case of the seated workplace with a footrest (Figure IIA–3). Work surface thickness (B) should not exceed 5 cm (2 in.) in order to ensure a thigh clearance (E) of at least 20 cm (8 in.). Knee clearance (C) should be at least 46 cm (18 in.) so that the operator can pull up to the edge of the work surface.

For specialized workplaces—for example, where manipulative tasks require only small arm, hand, and finger movements—the task should be located according to its visual requirements. It should be possible to raise the work surface or arm supports to function as elbow supports (see Section IIB, "Adjustable Design Approaches"). Sitting workplaces that are raised to provide arm support, extra storage space, or more convenience to the operator should not be raised more than 91 cm (36 in.) above the floor (P. C. Champney, 1975, Eastman Kodak Company; T. W. Faulkner, 1968, Eastman Kodak Company). A footrest must be provided.

b. Standing Workplaces

(1) The Standing Work Area

Standing operators frequently work in an area around a machine instead of at a given workplace. Even though the operator is free to move about, all handled items and controls should be positioned to eliminate excessive reaches, stooping and bending, twisting the body, and unnatural head positions because of visual requirements. A summary of standing reach guidelines is given in Chapter III, "Equipment Design." More specific information is provided below. Figures IIA–5 and IIA–6 illustrate the standing workplace area for forward reaches with one arm and both arms without bending the trunk forward.

The left arm's reach can be considered a mirror image of the right arm's pattern. Without stretching or leaning forward very much, most people can reach about 46 cm (18 in.) in front of the body as long as the object is 110 to 165 cm (43 to 65 in.) above the floor and not more than 46 cm (18 in.) to the side of the body's centerline. At farther distances to the side or heights less than or greater than the above range, forward reach capability falls off. The operator can achieve an extended reach only by leaning, stretching, stooping or crouching; these postures can all produce fatigue if they have to be assumed frequently or maintained for periods longer than a minute.

For tasks where two hands must be used, such as steadying and controlling an object or manipulating dual controls, the acceptable forward reaches are somewhat less than those for one-handed tasks. Because of restriction of arm movement across the body, the most extensive forward reaches (about 51 cm, or 20 in.) are within 15 cm (6 in.) of either side of the body's centerline. The farthest two-handed reaches to the side are only about 46 cm (18 in.) from the body's centerline; at this point only 36 cm (14 in.) of forward reach is possible without bending forward.

For occasional standing tasks where sustained activity is not required, such as activating a switch or marking a record, forward reach can be extended by bending forward over a work surface. If the bend can be made at the hips, an additional 36 cm (14 in.) of forward reach can be obtained. If the bend has to be made at the waist, as in leaning over a 89-cm (35-in.) high barrier, forward reach can be extended only 20 cm (8 in.) (B. M. Muller-Borer, 1981,

Figure IIA–5: The Standing Reach Area, One Arm (Adapted from B. M. Muller-Borer, 1981, Eastman Kodak Company.)

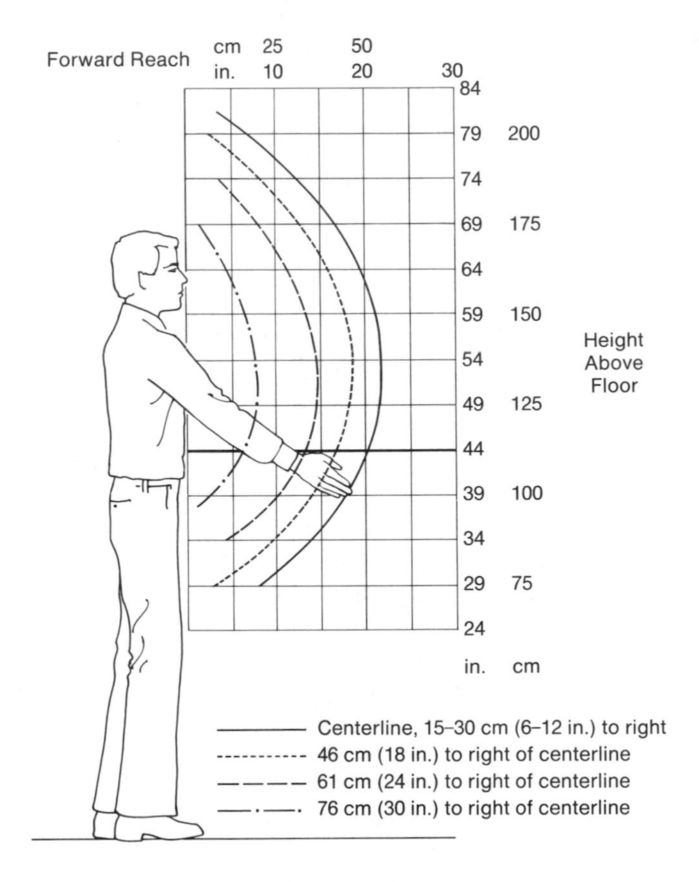

Four curves describe the forward reach from the front of body at different heights of the hand above the floor for a 5th percentile person (see Chapter VI, Appendix A). Distances are in centimeters (cm) and inches (in.). No leaning forward was permitted. The four curves represent forward reaches at different distances to the right of the body's centerline, as described at the bottom of the graph. The outermost curve shows the forward reach capability within 30 cm (12 in.) of the center of the body. Once the arm is positioned more than 30 cm (12 in.) to the right, there is a rapid reduction in forward reach capability at all heights above the floor. The dark line at 112 cm (44 in.) on the horizontal axis illustrates, by its intersection with the four curves; this reduction in forward reach with lateral arm movement; maximum forward reach falls from 51 cm (20 in.) to 15 cm (6 in.) as the arm moves 76 cm (30 in.) to the right. The left arm's forward reach capability can be considered to be the same, using distance to the left of the body's centerline to define the four curves.

Figure IIA–6: The Standing Reach Area, Two Arms (Adapted from B. M. Muller-Borer, 1981, Eastman Kodak Company.)

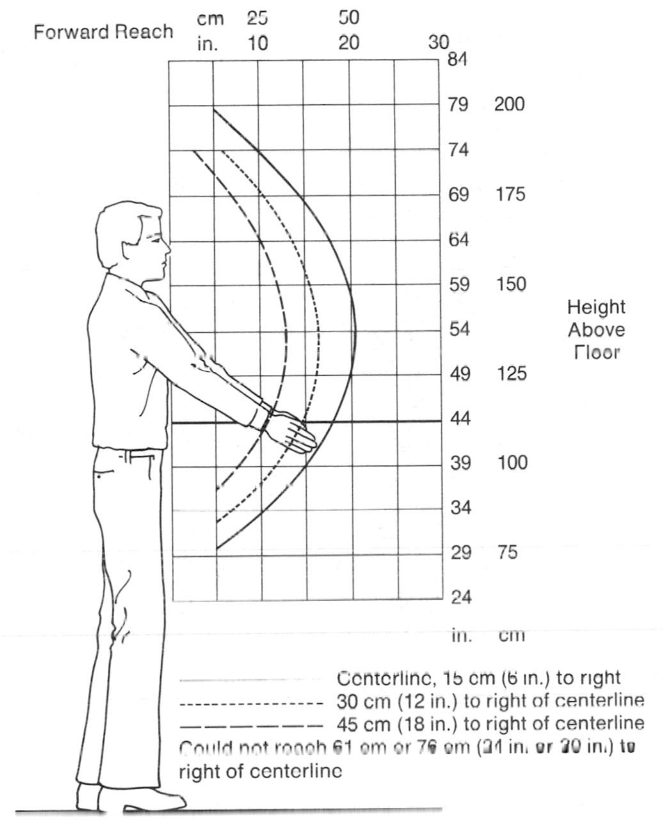

Three curves describe the forward reach from the front of the body for both arms together at different heights of the hands above the floor. Distances are in centimeters (cm) and inches (in.). No bending or leaning forward was permitted. The three curves describe forward reach capability at 15, 30, and 46 cm (6, 12, and 18 in.) to the right of the body's centerline. Reaches greater than 46 cm (18 in.) to the right could not be done because the left arm could not reach that far. The dark line at 112 cm (44 in.) illustrates, by its intersection with the three curves, how forward reach decreases as the arm moves laterally; it falls from 51 cm (20 in.) to 36 cm (14 in.) with a 46-cm (18-in.) move to the right of centerline. Forward reach is only marginally shorter for two-handed tasks than it is for one-handed tasks within the 46-cm (18-in.) lateral limit (see Figure IIA–5), except at the lowest and highest points above the floor.

Eastman Kodak Company). For more anthropometric data on reach capability, see Chapter VI Appendix A.

(2) Standing Workplace Height

Standing workplaces should be designed according to the dimensions indicated in Figure IIA–7. The optimal working height of the hands (A) is determined by compromise based on analysis of the total work sequence, as follows:

- For light assembly, writing, and packing tasks, the optimal working height of the hands (A) is 107 cm (42 in.).
- For tasks requiring large downward or sideward forces, such as casing operations and planing, the working height of the hands (A) should be at 91 cm (36 in.). Lower heights to about 76 cm (30 in.) may be appropriate in some very heavy force exertions.
- For tasks requiring large upward forces, as in clearing machine jams and removing components, the optimal working height of the hands (A) is 81 cm (32 in.).

The difference (B) between optimal working height (A) and the table surface height (C) is determined by the size of the objects being handled. Thus several values may result for B, each dependent on the particular item being handled and the optimal work method. Distance B should be adjustable to the height that allows the hands to be at the levels recommended for A above, most of the time. Bench cutouts or elevations should be provided to accommodate particular instruments that would be awkward to operate if placed on the bench surface.

Distance C is the height of the table and equals A minus B. Note that if different sizes of items are handled, several locations in the work area may be at different heights. If jobs requiring different sizes of items must be done at the same workplace, either an adjustable-height workbench should be used, or the height should be based on the most frequently used items (see Section IIB, "Adjustable Design Approaches").

c. Sit/Stand Workplaces

A sit/stand workplace should not be made by simply raising a seated workplace. Attention should be given to the variety of tasks that must be done and which type of workplace is best for each, as indicated below:

- A standing work surface for tasks that can be done while seated should be 102 cm (40 in.) above the floor. A seated task at a height of less than 102 cm (40 in.) will cause stooping and excessive static loading on the back and shoulder muscles for taller operators if the task is done from a standing posture. However, a sitting workplace at 102 cm (40 in.) is impractical for two reasons: (1) chairs raised to the best height for arm

Figure IIA–7: Recommended Standing Workplace Dimensions (Adapted from P. C. Champney, 1975, Eastman Kodak Company; T. W. Faulkner, 1970, Eastman Kodak Company; additional information from Kroemer, 1971.)

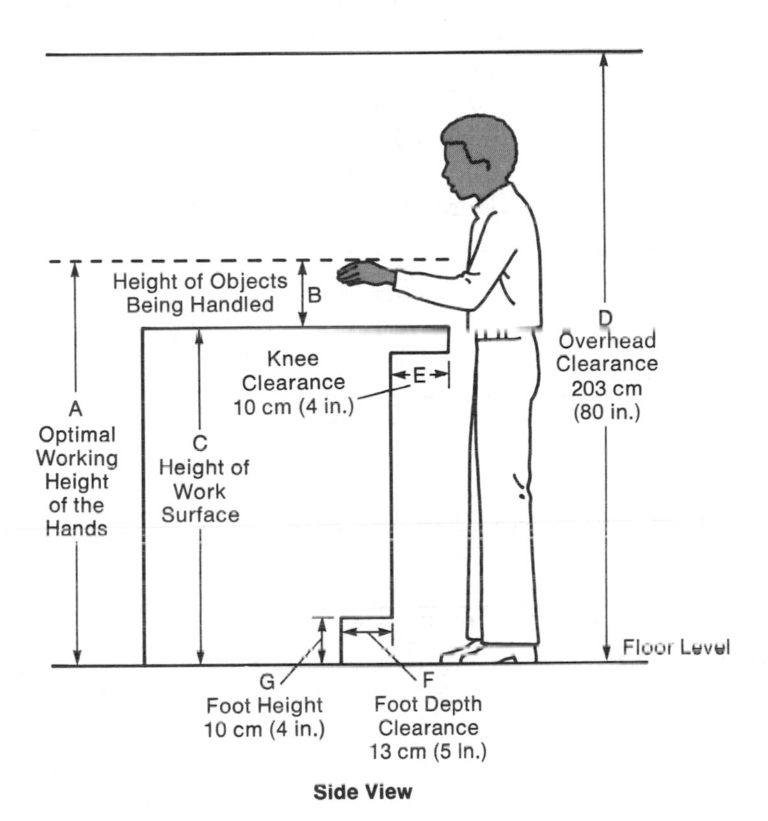

Side View

Workplace height (C) and overhead (D), knee (E), and foot (F and G) clearances are indicated for a typical standing workbench. The knee and foot clearances (E, F, and G) permit the operator to stand with knees bent and feet pointed straight ahead. Work height (C) varies according to the type of task being performed, the size of the objects worked on (B), and the location of the hands when doing it (A). Height A is the optimal working height of the hands, B is the typical height of the objects being assembled, packed, or repaired, and C is the height of the work surface without a product on it. Guidelines for determining the proper standing workplace height for a given task are further explained in the text.

work at this surface can be unsafe, and (2) it is difficult to reposition the chair once the operator is seated (P. C. Champney, 1975, Eastman Kodak Company).

- Operations that consist of several different tasks requiring different working heights (sitting and standing) can be done at a workplace 91 cm (36 in.) above the floor with an adjustable footrest. Workplace height for standing tasks can be raised by using 10–30-cm (4–12-in.) platforms on top of the work surface. At least 51 cm (20 in.) of forward leg clearance is required at a sit/stand workplace (see Figure IIA–3). Supports that can be used at sit/stand workplaces are discussed in Section IIB, "Adjustable Design Approaches."
- Work surfaces using large-size products or drawings, such as light tables used in photographic reproduction operations, are often 112 cm (44 in.) high. The work done on these tables lends itself to a sit/stand operation; provision of adequate leg clearance considerably reduces static loading on the legs and back of people as they lean over large drawings or negatives.

In most workplaces, whether seated, standing, or sit/stand in design, there is a tendency to use the space under the work surface for storage, thereby reducing the leg clearance. Cabinetry and storage shelves should be located so that they do not interfere with the clearances previously indicated in Figures IIA–3, IIA–4, and IIA–7.

d. Special Case: Design of VDU (Video Display Unit) Workplaces

Increasing use of video display units (cathode ray tubes, or CRTs, and keyboards) to communicate with computers has resulted in a need to optimize the design of these workplaces. In many console applications the operator must activate controls as well as enter data through the keyboard. Information covering acceptable reaches for seated and standing workplaces has been presented earlier. This section will address the location of the terminal, the viewing requirements, and the design approaches that minimize environmental distractions.

The VDU should be located in the production area or office so as to accomplish the following:

- Provide a reflection-free VDU screen (e.g., do not place it near a window where daylight may fall on the screen). See Section VC, "Illumination and Color," in Chapter V, "Environment," for further discussion.
- Permit the operator to see the production machinery affected by his or her data entry controls, that is, to see that one's action at the console has had the desired effect. If this technique is not feasible, feedback on the action should be provided at the console.

- Be accessible to all people who have to enter data or get status information from it.
- Be sufficiently out of the center of activity so that distractions are minimized and inadvertent activation of the keys cannot take place as can happen when a tool is dropped on the keyboard, for example.

The recommended dimensions for a seated VDU workplace without a footrest are those given in Figure IIA–4. Figure IIA–8 provides some additional dimensions for seated VDU workplaces. The keyboard and screen determine workplace height because of the motor and visual requirements of working at a VDU.

Figure IIA–8: Seated VDU Workplace Dimensions (Developed partly from information in Anon.–IBM, 1973.)

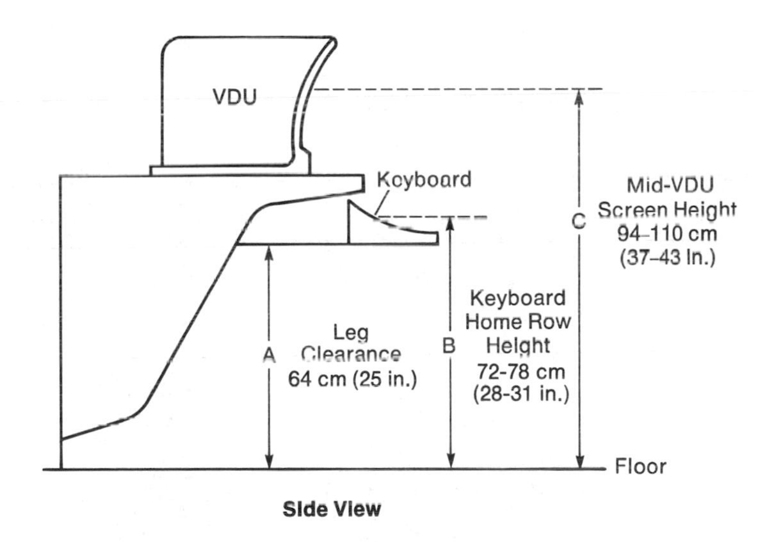

The workplace heights for leg clearance (A), keyboard home row location (B), and the center of a VDU screen (C) are given for a seated workplace. The other workplace dimensions are similar to those for a seated workplace without a footrest (see Figure IIA–4). A VDU with a keyboard separate from the display screen is shown since this arrangement permits optimal workplace design. If the VDU keyboard and screen are not separate, the table height should be in the 66–71-cm (26–28-in.) range. When frequent work at the VDU console is required, a seated operation is preferable.

Standing VDU workplaces are designed according to the guidelines in Figure IIA–7 with some alterations related to viewing and keying-in data. Figure IIA–9 summarizes design guidelines for these VDU stations.

In addition to the guidelines in Figures IIA–8 and IIA–9, some general principles of VDU workplace design should be considered. The following guidelines apply to both sitting and standing workplaces:

- A writing surface should be provided for data entry and monitoring operations. It should be at least 30 cm (12 in.) wide and 41 cm (16 in.) deep, preferably 76 cm (30 in.) in each direction (W. H. Cushman and B. Crist, 1979, Eastman Kodak Company).
- The VDU screen should be tilted back slightly (to a maximum of 15°) so that it is perpendicular to the operator's line of sight. If tilting results in increased reflections on the screen, improvements in this situation can be made by using a hood around the screen, by placing an antireflective filter directly on the screen, or by adjusting the overhead lighting (Ostberg, 1976). See Section VC, "Illumination and Color," in Chapter V, "Environment," for further information.
- Whenever possible, the VDU and keyboard should be movable so that an operator can select the most comfortable viewing and keying positions.
- Wearers of bifocal glasses often cannot see things easily at distances from 61 to 91 cm (24 to 36 in.) away because this range marks the transition between their near and far corrections. The operator should be able to adjust the viewing distance to fit individual visual needs. Preferably, both the seating and the screen should be adjustable. Individuals who wear glasses and work for long periods at a VDU should consider getting single-focal-length prescription lenses for the job.

e. Special Case: Design of Chemical Hoods and Glove (Isolation) Boxes.

The material in this section was developed from information in W. J. Nielsen (1981, Eastman Kodak Company) and Doxie and Ullom (1967).

Chemical hoods and glove boxes (also known as isolation or containment cabinets, clean boxes, and dry boxes) are used frequently in industrial, university, and governmental laboratories and in manufacturing processes in order to protect a person from direct contact with a chemical or to protect a product from environmental contamination. The performance of tasks in a hood or a glove box is subject to the constraints on arm movement and leaning imposed on the operator by plastic shields or armholes. There are also visual restrictions that may interact with the postural limitations to make an otherwise easy task very awkward to perform. In most chemical handling operations the operator will be wearing gloves, further reducing his or her dexterity. The type of work to be done in the hood or the glove box will determine how severely these factors affect the operator's ability to do the task.

There are numerous guidelines for the design of glove boxes and hoods that specify ventilation, seals, glove attachment, necessary services inside the

Figure IIA–9: Standing VDU Workplaces (Adapted from Anon.–IBM, 1973.)

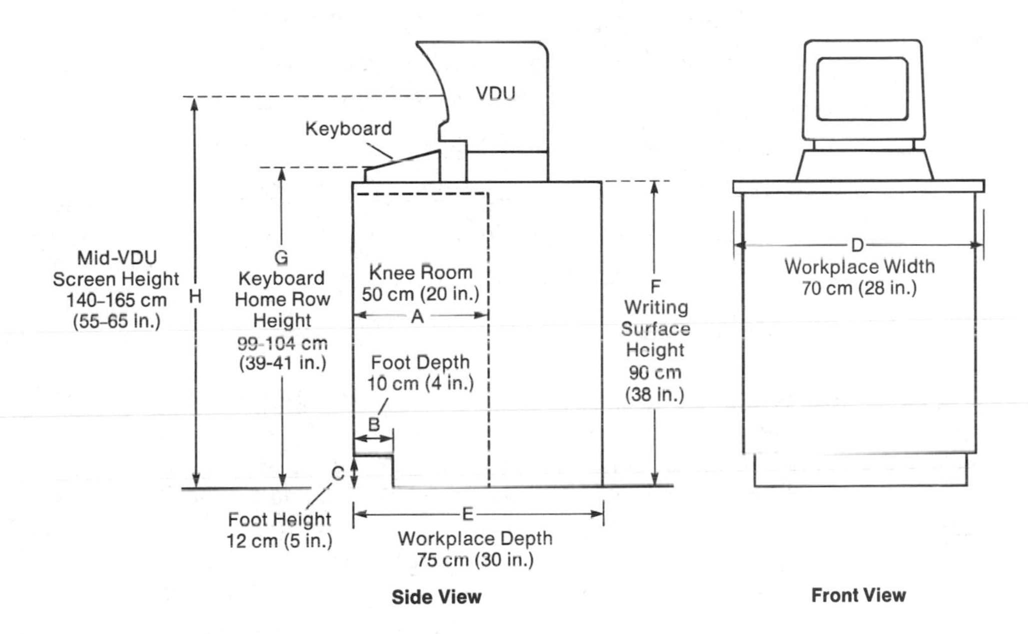

Recommended workplace heights and clearances are given for a typical standing VDU workplace where frequent interaction between the operator and the VDU occurs. Foot clearances (B and C) are the same as those shown in Figure IIA–7. More knee clearance (A) is allowed since the operator may wish to sit on a stool during some data entry operations. At least 70 cm (28 in.) of table width (D) is needed to have room for the papers from which the computer data entry is drawn. The table surface height (F) is determined by the recommended heights for the keyboard (G) and screen (H). For occasional lookup or data entry tasks on a VDU, these dimensions are less critical.

contained area, and cleaning and decontamination facilities. The human factors considerations are less well studied. Some of the factors that should be considered when designing a glove box are the following:

- Glove port, or opening, height (seated and standing).
- Glove port diameter.
- Separation between the glove ports.
- Reach limitations.
- Biomechanical constraints (very task-dependent).
- Visual constraints.
- Seat height and adjustability (for seated operations).
- Glove type and size.
- Location of controls and switches.
- Location of pass-through compartments into and out of the box.
- Access for cleaning, decontamination, or product changes.
- Design of tools, trays, and containers to be used inside the box.
- Probable task durations (time of continuous work in the glove box).

Most glove boxes are purchased from outside vendors. Two types are generally available: a low-profile glove box with a vertical glove port panel below the glass viewing panel (Figure IIA–10), and a high-profile glove box with glove ports set into the sloping glass front. The latter design offers advantages in visually demanding tasks. Glove ports are usually about 20 cm (8 in.) in diameter and 38 to 48 cm (15 to 19 in.) apart, center to center.

Chemical hoods include screens to prevent splashes to the face or eyes. The design of the screen can vary according to the tasks performed in the hood. Two-handed tasks require access to the hood either through glove ports or around both sides of the screen. If glove ports are not provided, the screen should be no more than 46 cm (18 in.) wide so that constant static loading of the shoulder muscles is not required during the job.

Selection of a sitting or standing workplace for glove boxes and hoods is determined by the nature of the task to be done. The same general principles that apply for other workplaces apply here. Reaches should be kept within 15–41 cm (6–16 in.) of the front of the work surface for seated operations and within 51 cm (20 in.) for standing operations (W. J. Nielsen, 1981, Eastman Kodak Company). Since chemical hoods are often manufactured with cabinets below the work surface, and with plumbing, gas lines, and other built-in services at extended reaches from the front surface, most are used as standing workplaces. Hoods meeting the guidelines for standing arm reach and working surface height given earlier in this section should be selected wherever possible.

Figure IIA–10: Example of a Glove Box

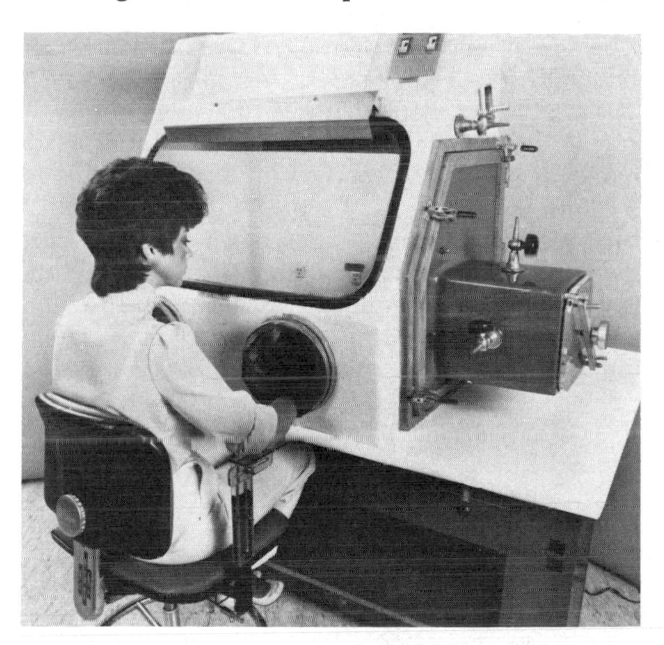

A low-profile glove box is illustrated. The operator puts her arms in the glove ports and manipulates small vials of chemicals in the enclosed box. The chair and table are both adjustable. The chair's adjusting lever has been modified so that it can be adjusted with the forearm; the operator does not have to remove her hands from the gloves to change seated height. The right side of the glove box contains a "pass-port" that can be isolated from the box through a double set of port covers.

For glove box operations, seated workplaces are preferred, and an easily adjusted chair, such as a pneumatic chair, should be provided. Leg clearances of at least 66 cm (26 in.) are recommended. Some means should be provided to enable the operator to adjust the chair to a comfortable height for the arms and shoulders without having to remove his or her hands from the gloves. Some tasks are easier to do with the chair adjusted to its highest level, while others are best done with the chair at its lowest level. An example of the former is a task in which the force is exerted downward; an example of the latter is a task involving extended reaches. Providing height adjustment by foot pedal, forearm switch, or a similar method allows the operator to move the chair to suit the task. Time is not lost in manual adjustments, then, and muscle fatigue from awkward working postures is less probable. Chair adjustability should be in a range of 15 cm (6 in.), and there should be adequate clearance to permit these adjustments to be made without wedging the thighs between the underside of the glove box floor and the seat pan of the chair.

If a standing glove box or a plastic screen in a hood is needed, the center of the arm ports should be 132 cm (52 in.) above the floor. This height should be comfortable for taller people; shorter operators will need a retractable step stool or platform with standing levels at 8 and 15 cm (3 and 6 in.) above the floor (W. J. Nielsen, 1981, Eastman Kodak Company). Since there is less flexibility in adjusting these platforms, every effort should be made to adapt the work to a seated operation, especially if the glove box is used regularly and for a majority of the shift.

The biomechanical problems of working through arm ports come from the restriction of body posture options; it is often not possible to utilize the most powerful muscle groups during lifting and force-exerting tasks. The shoulder and elbow ranges of motion are restricted, putting more stress on the hands and wrists. The options for moving objects to the side are reduced since the body cannot pivot with the arms in glove ports; thus passing an object through a side pass-through port can be very awkward. Consequently, techniques that reduce the need to handle objects in the box are desirable. Equipment to accomplish this reduction includes the following:

- Conveyors to supply and remove product (use of conveyors may not be possible if contamination is a problem). Leg clearance should not be compromised by the conveyor mechanism.
- Air pumps to remove liquids from stock bottles so that the liquids need not be poured manually.
- Trays or containers to permit parts or product to be moved into, out of, and around the inside of the box conveniently (especially if small parts are involved).
- Small platforms of different heights inside the box. The platforms should be movable so that the operator can adjust the height of an operation to a comfortable working posture.

2. AISLES AND CORRIDORS

The accumulation of equipment, supplies, and product in workplace aisles is a common problem in some operations. Aisles and corridors should be designed to meet minimum clearance guidelines when the system is running at full capacity. Figure IIA–11 provides guidelines for minimum clearances in aisles and corridors.

Some parts of corridors and aisles may be designated as *marshalling areas* where trucks, carts, and products on pallets are stored prior to being taken to the workplace or warehouse. Space requirements for these areas should be determined not just by the size of the items stored but also by the needs for maneuvering handling equipment used there. This maneuvering room can add as much as 25 percent more to the space requirements in some operations.

For aisles in warehouses or storage areas where high-stacking fork trucks

Figure IIA–11: Minimum Clearances for Aisles and Corridors (Adapted from Thomson, Jacobs, Covner, and Orlansky, 1963.)

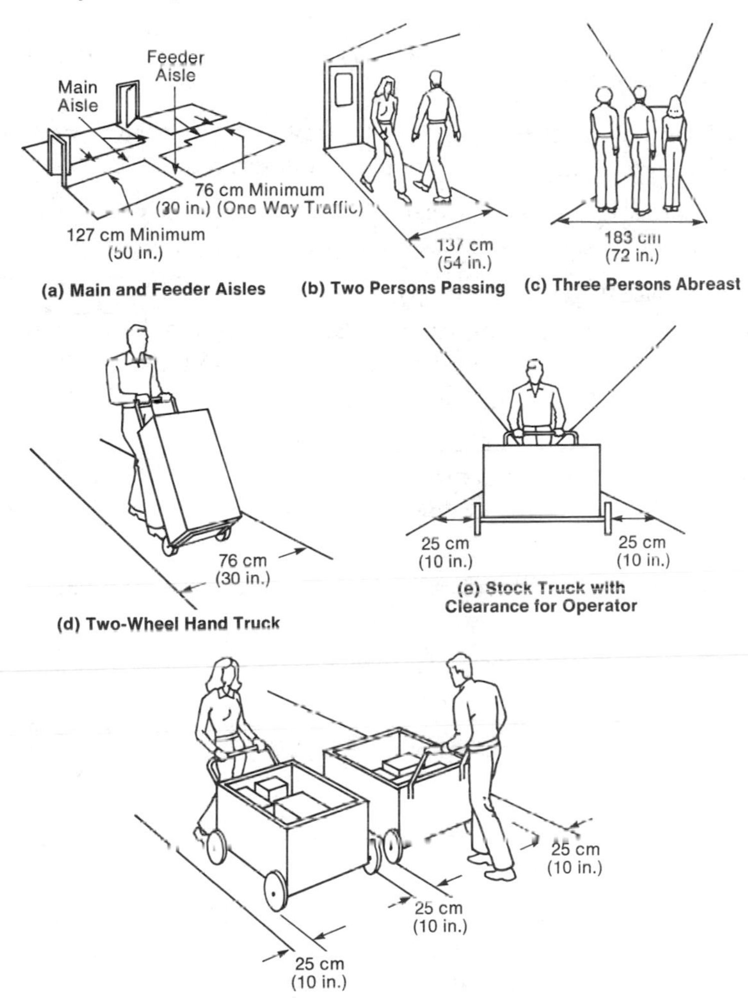

(a) Main and Feeder Aisles **(b) Two Persons Passing** **(c) Three Persons Abreast**

(d) Two-Wheel Hand Truck **(e) Stock Truck with Clearance for Operator**

(f) Two Stock Trucks with Clearance for Operators

The aisle widths shown in these illustrations are the minimum values for traffic and for handling trucks in production areas. Main aisles should be wider than feeder aisles (see part a). Both widths should be determined by the traffic needs (see parts b, c, and d) but should not be less than the values given in part a. Where trucks and carts are used, there should be 25 cm (10 in.) of clearance on either side of them (see part e) and between them if there is two-way traffic in the corridor (see part f). Minimum aisle width will be set by truck width, its clearance needs, and the traffic pattern in the production area.

are used, aisle width should be about 5–10 percent greater than the values given in Figure IIA–11. The visual demands of judging distance in the high lifts may require the truck to be positioned farther from the shelves than would be the case for lower lifts (Drury, 1974).

Some additional guidelines for the design of aisles and corridors are given below (Thomson, 1972):

- Avoid blind corners. Arrange machinery and workplaces to allow visibility around corners. Use a mirror, if necessary.
- Locate paths for minimum distances, using flow charts or diagrams to indicate where the densest traffic will be.
- Mark traffic guides (aisle limits, arrows) on floors, walls, or ceilings.
- Design aisles, machinery, and workplaces to avoid the possibility of employees brushing up against equipment and inadvertently activating switches or knocking unattached objects to the floor.
- Avoid having doors opening into corridors. Occasionally, there is no alternative, such as in the case of fire doors or a small utility closet in an aisle. In these instances use folding, sliding, or recessed doors, if possible.
- Keep aisles clear. Do not allow structural support columns or production equipment to protrude into the aisle space.
- Where possible, avoid locating an aisle against a wall, because this location permits access only from one side.
- Avoid one-way traffic restrictions in aisles. They are practically unenforceable.

3. FLOORS AND RAMPS

a. Floors

The characteristic of floors in the workplace can determine the potential for slip-and-trip incidents, the forces required to move carts, trucks, or products manually, and the comfort of the people working at standing workplaces. The choice of floor material, how it is maintained, and what footwear the people working in the area are wearing will influence worker safety and performance.

(1) Floor Material

The following factors are usually considered when a floor material is chosen:

- cost
- architectural considerations (load-bearing characteristics)
- aesthetics (appearance)
- durability
- nature of work being done
- maintenance needs

In addition to these factors, attention should be paid to the slipperiness of the floor surface both when dry and when wet from spilled materials or cleaning operations. Slip-resistant coatings, such as paint with sand in it, have been successfully used to reduce the potential for slip-and-trip incidents in areas where wet floors are common, such as liquid-chemical preparation areas or cleaning stations (Archea, Collins, and Stahl, 1979; Anon., 1980). Too much slip-resistant material on a floor, however, may make walking or handling of carts and trucks more difficult. The difficulty arises from the increased frictional resistance to the sliding that occurs naturally as part of walking or maneuvering a vehicle manually.

Grates are sometimes used over floors in order to raise the operators during handling operations or to permit them to remain relatively dry during cleaning operations. Floors with grates can significantly increase the force requirements for the operator manually handling trucks and carts, though. Larger casters and wider treads are often needed on the carts to compensate for this increased resistance to motion (Lippert, 1954).

Rugs or mats can be used in the workplace to improve the comfort of standing operations, reduce ambient noise levels, reduce breakage in some assembly operations, or improve the appearance of a work area. While they are most often considered beneficial, rugs and mats have increased resistance to rolling motion and, therefore, make the manual movement of carts, trucks, or other objects across them difficult. The mat or rug should contrast with the floor color so that its edges are clear and do not produce a trip hazard. The edges should be tapered down to the floor to make the transition between floor and rug smoother (Anon., 1981). Areas where reduced light levels make visual identification of floor coverings difficult should have full floor coverage, up to entrance and exit doors, or other architectural cues to mark where the floor-to-mat or floor-to-rug transition occurs. These cues may include support columns, panels, or workplace delimiters.

Entrances to buildings from the outdoors should have mats to reduce tracking in of water, snow, or mud in inclement weather. Ideally, the mats should be long enough to permit about ten steps to be taken (about 6 m, or 20 ft) before the regular floor surface is stepped on. In many buildings it is not possible to provide this much space since a stair flight is present just inside the door. Use of stair mats or a roughened portion on the stair step to improve traction can reduce the potential for slip-and-trip incidents (Asher, 1977).

(2) Floor Maintenance

The maintenance of floors can be divided into three categories:

- Housekeeping, such as cleaning up spills and removing dropped objects.
- Cleaning, such as waxing, vacuuming, and scrubbing.
- Repairing, such as filling cracks, repainting, and replacing rugs or mats.

Although the latter two categories are important to ensure that the floors are properly cared for, it is the housekeeping of floors that frequently is improperly done and contributes to slips and trips (Doering, 1974, 1981). Choice of floor according to the type of work being done can help reduce housekeeping problems, as indicated in the following list:

- Rubber or cocoa fiber mats can be used at entrances to buildings from the outdoors. These mats help remove dirt, water, or snow from people's shoes so that less is carried into the work area.
- Gratings or open plastic mats can be used on floors in areas where water is commonly present, as in cleaning stations or chemical preparation workplaces. By raising the worker above pools of water, gratings or mats reduce the amount of water tracked to other workplaces in the area.
- Rugs or mats can be placed at an assembly work station where parts may be dropped on the floor. Rug, mat, and floor colors should be chosen to contrast with the items that could be spilled on them. Thus the parts would be visible and more likely to be cleaned or picked up. This scheme would be feasible in workplaces where only a few operations are done or where few items are involved.

Attention to three other workplace characteristics can also reduce housekeeping problems in a work area:

- A small catch-trough, a depression near the front of the work surface to catch parts before they fall to the floor, may reduce housekeeping problems on floors. Figure IIA–12 illustrates a catch-trough at a seated workplace.
- Cracks, depressions, or other irregularities in a floor's surface can require much greater forces from operators moving hand trucks or carts manually. The effort needed to dislodge a truck from a floor crack, for instance, may result in product spilling from the truck. This spill could present a slip or a trip hazard if it is not quickly cleaned up. Floor surfaces should be kept in repair, especially in areas where trucks or equipment are moved.
- Regular cleaning to reduce the accumulation of dirt, excess wax, or other materials is also needed, because these substances make the surface uneven and may contribute to handling or slip-and-trip incidents.

3. Footwear

The material in this section was developed from information in R. A. Day, and W. J. Nielsen, (1978, Eastman Kodak Company).

Figure IIA–12: Workplace Catch-Trough

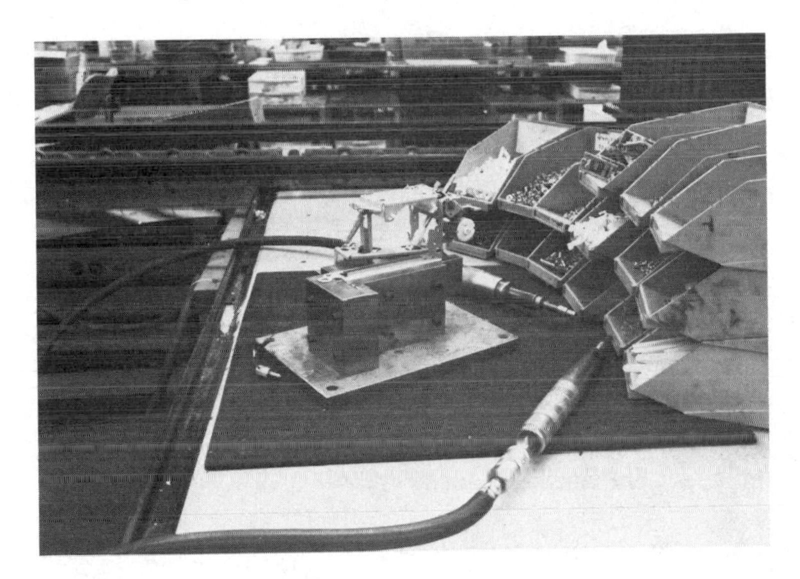

The front surface of the workplace includes a small indentation that will capture small parts before they roll off the edge onto the floor. The trough should be sloped gradually, without sharp edges, so it does not represent a pressure point for the operator's forearms during the assembly task. Also illustrated is an aluminum block fixture used to orient several small parts that are being fastened to a product in this assembly operation. The fixture may be fastened to the work surface or be movable. The powered hex nut drivers on either side of the fixture are used for the fastening tasks.

The operator's footwear should be selected on the basis of the floor surface, the standing requirements of the job, the nature of the work being done, and the potential for exposure to environmental hazard, such as electric shock. Cleaning operations where wax strippers are used on linoleum floors, for instance, result in very slippery floors. Leather-soled shoes increase the slip hazard, whereas neoprene soles with a pattern that includes ridges and spaces (see Figure IIA–13) reduce the hazard.

When a worker has to exert large forces—>222 N (newtons), or 50 lbf (pounds force)—to move an object, the slippage of the shoe soles on the floor may determine how much force can be exerted. Provision of a rigid support in the floor, which will not be a trip hazard but against which the operator can push to avoid slipping, makes force exertion easier (Kroemer, 1969; Kroemer and Robinson, 1971).

Aside from slip-and-trip considerations, footwear can be chosen to improve comfort in standing operations. Shoes with well-cushioned insteps and

Figure IIA–13: Shoe Sole Design to Reduce Slippage

The frictional resistance of the illustrated shoe soles, when wet, increases from left to right in the picture. Leather soles (far left) are smooth and offer the least resistance for slippage on wet floors. When the sole is patterned with ridges and spaces (center), and when rubberlike materials are used (right), the soles have higher coefficients of friction on wet floors, thereby reducing the opportunity for slipping accidents.

soles can be worn in areas where floor mats cannot be used. In some special circumstances, such as in areas where equipment design makes high reaches very difficult for short people, shoes with very thick soles (elevator shoes) have been worn by operators to increase their reach height (D. A. Alexander, 1980, Tennessee Eastman Company). This approach is not generally recommended since walking on such shoes requires added work and may result in lower leg muscle and ankle joint soreness.

b. Ramps

Ramps are found in access ways to buildings and are used to join two areas with different floor heights (adjoining two buildings, or adjoining a special purpose room, such as a computer facility, with its neighboring rooms). The slope of the ramp should be kept below 7° (12 percent grade) so that wheelchair users can negotiate it without excessive effort (Tica and Shaw, 1974). Where there is not enough space to provide a low-sloped ramp, powered equipment is recommended. For truck and cart handling a ramp slope of 15° (27 percent grade) marks the point where powered equipment is recommended over manual handling (Corlett, Hutcheson, DeLugan, and Rogozenski, 1972).

Because it is more difficult to walk up ramps than to walk up stairs, ramps should have a nonskid surface and handrails on each side. Figure IIA–14 illustrates a ramp for pedestrian and vehicle use.

Many times, a ramp leads to a door. That door often moves outward (toward the ramp) because it is the fire exit from the building or room to the outside. Pulling a door toward oneself while resisting the rolling motion of a cart, truck, or wheelchair down the ramp can lead to awkward postures and the increased potential for accidents. Manually handled equipment should be provided with a brake to assist the operator in negotiating doors on ramps.

Figure IIA–14: Ramp Design for Pedestrian and Vehicular Traffic (Adapted from Thomson, Jacobs, Covner, and Orlansky, 1963.)

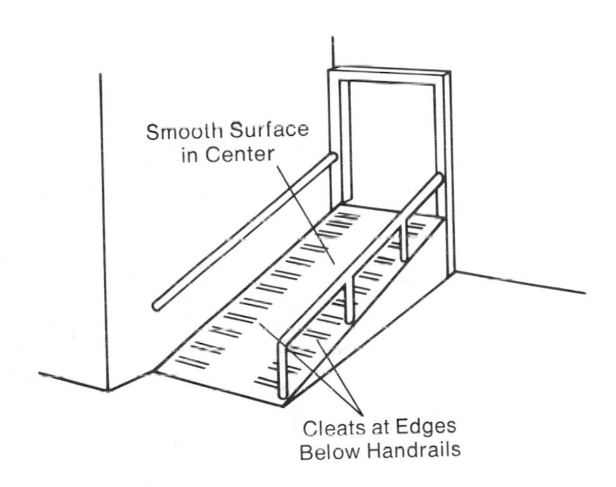

The ramp illustrated can be used by both pedestrians (at the sides) and trucks (in the center). Because the ramp must accommodate truck traffic, its center is smooth. Steps have been formed near the handrails by ridges or cleats to provide stability for the pedestrian's feet when ascending and descending the ramp. The width of these steps should not be less than 61 cm (24 in.).

Ramps should not be located directly in line with floor openings (pits) or stairwells. If a piece of equipment rolls away from a handler on a ramp, it should be able to come to a gradual stop on a flat surface without endangering people working in the area or damaging itself or other equipment.

4. STAIRS AND LADDERS

Falls from ladders or on stairs are one of the leading causes of injury and death in the United States (National Safety Council, 1980). Attention to the design of stairs and ladders cannot be expected to eliminate all of these incidents because many are related to inattention or risk-taking behavior (Templar, Mullet, and Archea, 1976). Good design can, however, reduce the potential for misstepping or provide the person about to fall with a way to retrieve balance.

a. Stairs

Detailed recommendations for stair design have been developed by federal, state, and local building code and safety organizations (Anon., 1978a; Archea,

Collins, and Stahl, 1979; Carson, Archea, Margulis, and Carson, 1978). These recommendations should be referred to when a specific design has been developed in order to be certain that it meets existing regulations.

Stair designs have become integrated so completely into the aesthetics of architecture that some new human factors problems have been created relative to anticipated stair riser height and tread width. Designs that are pleasing to the eye may be hazardous because they do not take into account the normal walking gait or expected step height (Templar, Mullet, and Archea, 1976). A stairway that is not difficult to ascend may be very difficult to descend. In this section stair and handrail dimensions are discussed.

(1) Stair Dimensions

Figure IIA–15 illustrates a typical fixed stairway section with recommended dimensions for riser height, tread depth, and tread width.

The slope of a staircase should be about 30–35° from the floor (Thomson, Jacobs, Covner, and Orlansky, 1963). The maximum range is 20°–50°. Because stairway slope will affect the mechanics of walking on stairs, tread depth and riser height will have to be adjusted accordingly. Table IIA–2 shows this relationship. Figure IIA–16 shows how stair slope is measured.

Figure IIA–15: Fixed Stairway Design (Adapted from Thomson, Jacobs, Covner, and Orlansky, 1963.)

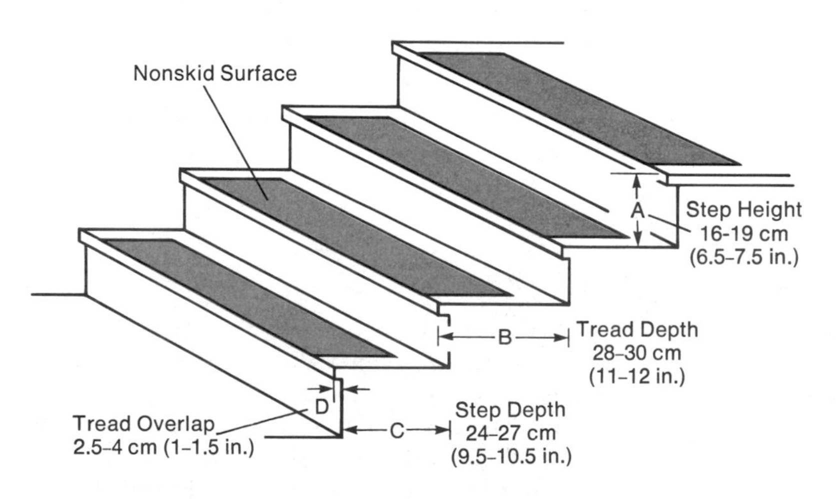

A fixed stairway should follow the guidelines for riser, or step, height (A) and step (C) and tread (B) depth shown here. The tread should overlap the riser horizontally (D) by 2.5 to 4 cm (1 to 1.5 in.). A nonskid surface is recommended for the front surface of each tread.

Table IIA–2: Effect of Fixed Stair Slope on Recommended Riser Height and Tread Depth (Anon., 1978a.)

Slope (°)	Riser Height		Tread Depth	
	Cm	In.	Cm	In.
30	16	6.5	28	11.0
35	18	7.2	26	10.2
40	20	8.0	24	9.5
45	22	8.8	22	8.8
50	24	9.5	20	8.0

Column 1 presents a selection of slopes, in degrees, that cover the common range (30°–50°) for a fixed stair. Columns 2 and 3 indicate the combinations of riser height and tread depth that would be needed to accommodate each slope. At higher slopes the riser and tread designs become less optimal.

(2) Stair Surfaces

So that the opportunity for slipping on stairs is reduced, a nonskid surface is often placed on the leading edge of each tread. This surface can be a strip of metal, hard rubber, or synthetic material, or a special paint that resists sliding of the foot and increases the stair user's stability. These nonskid surfaces should be maintained regularly, especially in areas of heavy traffic.

Outdoor stairways or stairways in work areas where water is frequently on the walk surfaces, as in chemical-making areas, should have a means to direct the water away from the treads. This feature helps to prevent the accumulation of water, slush, or snow on the stairs, all of which could result in increased slip hazard for stair users.

(3) Visual Considerations in Stair Design

Stair safety problems are frequently associated with misstepping and catching a heel on the edge of a step. Visual distractions in the stairway can contribute to stair user inattention and, thereby, to increased potential for misstepping. The following factors should be considered to improve stairway visibility:

- Use a color or hue on the edge of the tread (the nonskid material) that contrasts with the rest of the tread.
- Use matte, not high-gloss, finishes on the steps so that overhead lighting or daylight does not create sources of glare for the stair user (see Section VC, "Illumination and Color," in Chapter V, "Environment").

Figure IIA–16: Fixed Stair Slope Range (Adapted from Thomson, Jacobs, Covner, and Orlansky, 1958.)

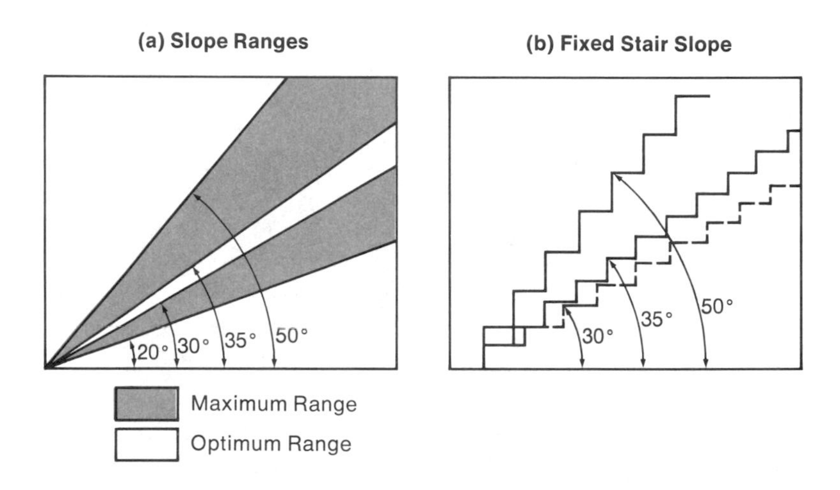

The maximal range of a fixed stair slope is from 20° to 50°. The optimal range is shown as 30°–35°. These slopes are shown in part a. The optimal and maximum slopes are further illustrated, as stairways, in part b. The extremes of this range result in awkward and more strenuous stepping requirements, because the tread and riser design do not match most people's normal gait (see Table IIA–2). Slopes below 20° are for ramps; slopes above 50° are for stair ladders.

- Do not use carpeting patterns that are visually distracting and blend to disguise differences in depth, such as narrow stripes of strongly contrasting colors.
- Use a handrail that contrasts with wall and stair colors. A handrail of contrasting color is an easier target to focus on when descending the stairs, providing a sensation of improved stability for some people.

(4) Handrails

The design of handrails is often affected by architectural as well as functional considerations. A handrail should be graceful enough to add to the aesthetics of the staircase, but it must be functional enough to allow it to be grasped in the event of a slip or to be used routinely in ascending or descending the stairs. Figure IIA–17 presents guidelines for the height and grasp characteristics of handrails.

Figure IIA–17: Handrail Design Guidelines (Adapted from Thomson, Jacobs, Covner, and Orlansky, 1963.)

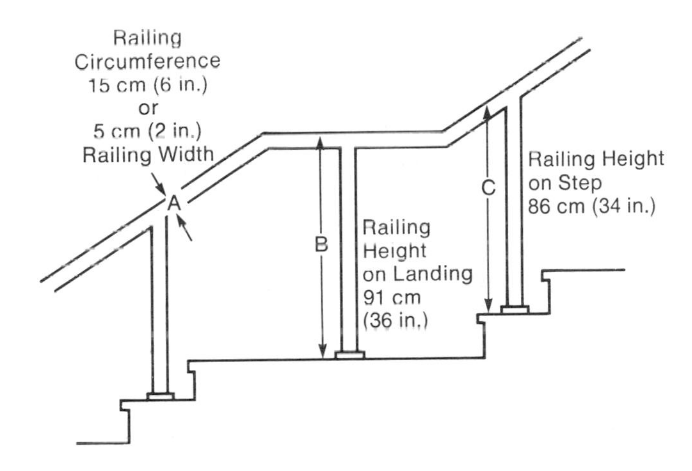

Guidelines for handrail design on fixed stairs are given. The railing height on a landing (B) is 5 cm (2 in.) more than its height on a step (C). Railing circumference (A) should not exceed 15 cm (6 in.). For rectangular, rather than circular, railings, the width of the grasping surface should be about 5 cm (2 in.). Indentations allow easier grasping of the railing in the event that a person loses balance when descending or ascending the stairway.

In addition to a handrail, open stairways should incorporate a guardrail at about 38 cm (15 in.) above the stair surface so that a person who slips cannot fall off the side of the staircase.

b. Ladders

A ladder is usually thought of as portable and is generally used to move, vertically, up slopes in excess of 75° above the floor. Stair ladders also exist, which are fixed ladders, usually with a slope between 50° and 75° (Thomson, Jacobs, Covner, and Orlansky, 1963). Stair ladders are frequently found in workplaces where large processing equipment, such as reactors or extruding machines, requires operators to move between several levels on an occasional basis. Fixed ladders usually have handrails, whereas movable ones may not. As was the case for stairways, the slope of the stair ladder will determine the appropriate riser height and tread depth. The more vertical it is, the more shallow is the tread and the higher is the recommended riser height. Figure IIA–18 summarizes the recommended dimensions for stair ladder and ladder design and selection.

Figure IIA–18: Design of Stair Ladders and Ladders (Developed from information in Anon., 1978b; Thomson, Jacobs, Covner, and Orlansky, 1963; Woodson, 1981.)

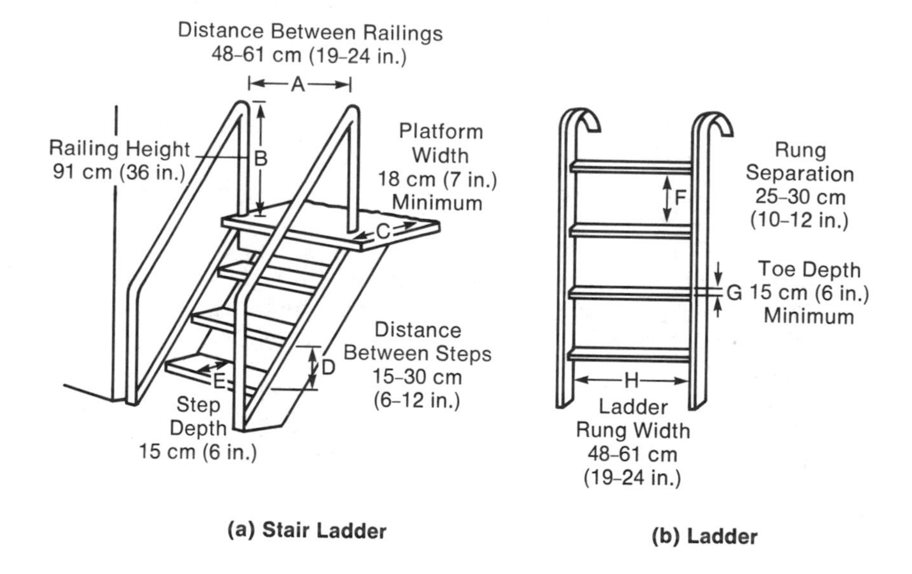

The recommended step depth (E), distance between steps (D), distance between the railings (A), railing height (B), and minimum platform width (C) are shown for stair ladders (part a). Stair ladders have a slope greater than 50° and usually less than 75°. The rung width (H) and separation (F) and minimum toe depth (G) for a vertical ladder are shown in part b. The rungs should be flattened on the top to provide stable footing. Stair ladders and ladders are often found on large pieces of production equipment or in areas where platforms are provided to improve access to a loading station.

There are a large number of step stools and portable stairs used in the workplace to help access high shelves or parts of production machinery. Figure IIA–19 illustrates a step stool and one of these shorter ladders. The ladder should be designed or selected according to the guidelines given above for stair ladders. Stairs with retractable casters (which become very stable as soon as a person stands on them) have the added advantage of being very easy to move around the workplace. Thus they may be used more frequently than stairs that have to be carried or dragged into position.

Figure IIA–19: Portable Stairs and a Step Stool

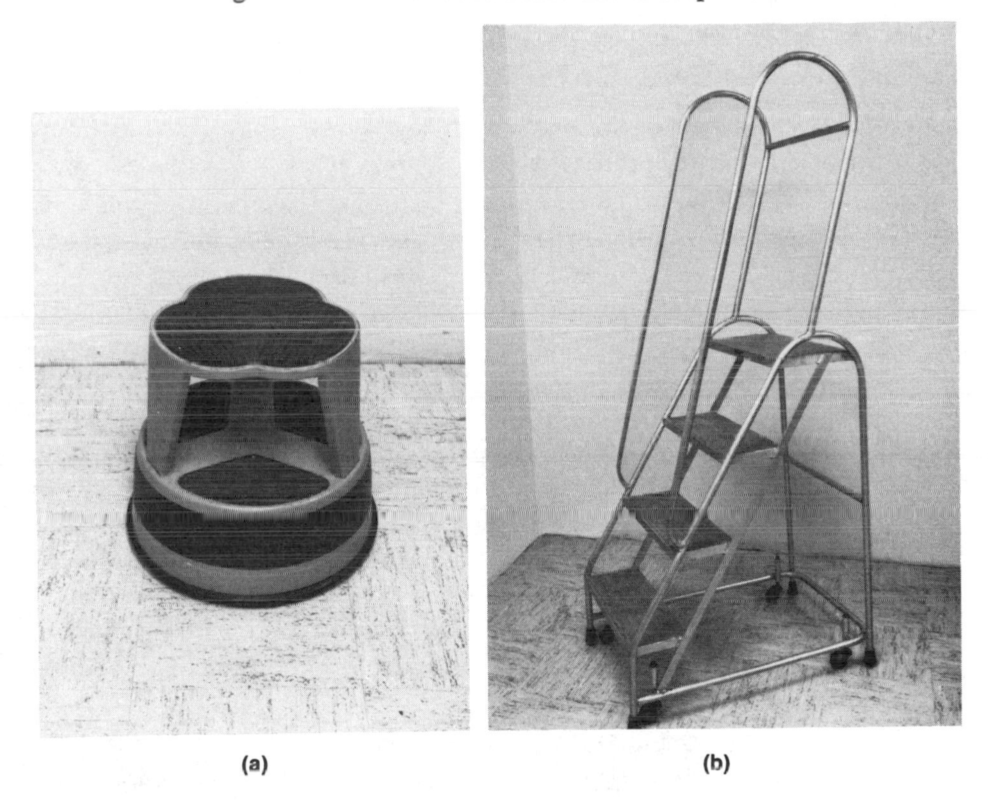

(a) (b)

Reaching items above shoulder height or lifting materials up to shelves above 127 cm (50 in.) is made easier with portable step stools or stairs. These devices may be two steps about 30 cm (12 in.) high each, as seen in part a, or a small stairway with railings, as seen in part b. The small stairway has retractable casters and four fixed supports. The casters permit the stairs to be moved around the workplace easily; the fixed supports provide a secure base once a person has stepped on the stairway.

5. CONVEYORS

Conveyors are often used to link workplaces in a manufacturing system. Products and supplies are moved in and out of the workplace on them in assembly, storage, transportation, and supply operations. For some situations assembly work is done directly on the conveyor. The type, location, height, width, and pace of conveyors can all influence the way a person works by determining the postures and strengths required and the amount of time pressure involved. Large manufacturing systems incorporate automatic assembly with hand assembly operations, making the impact of conveyor design on people greater with regard both to the workplace dimensions and pacing.

The following information should be used when installing conveyors in manufacturing and service areas:

- Conveyors should be accessible from both sides, especially in locations where large, heavy products are transported and where jams can occur.
- Crossovers or gates in conveyors should be provided where people need to move in and out of workplaces or where supplies are handled by hand pallets, trucks, or hoists. The gates should be counterweighted and easy to raise and should lock into place when down or fully raised.
- Conveyor height and width for a given operation should be determined by the size of the units handled on it and by hand location when working on the product (Figure IIA–20). The guidelines for work height of the hands in standing workplaces also apply to conveyors in standing tasks (see Figure IIA–7). Conveyor heights of 69 to 79 cm (27 to 31 in.) are often used in casing operations or other finishing areas. The seated workplace dimensions given in Figures IIA–3 and IIA–4 should be used when assembly work is done on conveyors. Large drums, 208 L (liters), or 55 gal (gallons), on a filling line are best transferred on conveyors close to the floor, so they can be chimed (rolled on edge) on and off the conveyor to pallets.
- Conveyors in sequential assembly workplaces should be located within the sitting or standing arm reach areas shown earlier in this chapter. Leg and knee clearance should be adequate for seated work. Whenever possible, the operator should be able to slide, rather than lift, the part or tray on and off the conveyor.
- In work areas where the conveyors carry the assembly task and are run either continuously or at a preset rate—as in the case of pulse conveyors or computer-controlled assembly workplaces—the conveyor rate should be set as a compromise between the most- and least-skilled operators. At conveyor speeds greater than 10 m/min (32 ft/min), susceptible people may develop conveyor sickness symptoms such as nausea and dizziness (T. G. and R. L., 1975).

Figure IIA–20: Standing Workplace at a Conveyor

The handler is labeling cases on the conveyor prior to putting them on pallets (lower part of the picture). The roller conveyor is adjusted to about 76 cm (30 in.) above the floor, making the labeling area of the most frequent case size about 102 cm (40 in.) high. This height is convenient for most handlers in applying labels and visually checking them. This height is also good for transferring the case from the conveyor to the pallet.

- Since unloading conveyors has been shown to be three times as likely to result in overexertion as loading them (Cohen, 1979), a space should be provided that can be used to temporarily accumulate parts or trays after sliding them from the conveyor. This arrangement removes some of the pace pressure from the operator and permits more careful handling from and to the conveyor.
- When bulk materials and cases in shipping or receiving operations are handled, snake conveyors for truck and railroad car loading or similar conveyors that permit some flexibility in location should be used. Wherever possible, the manufacturing process and the shipping operations

should be linked to minimize the need to rehandle product. Continuous conveyors should be designed to move product to the shipping area without interfering with other activities on the work floor.

6. DIMENSIONS FOR VISUAL WORK

We get much of the information about how to do our work from our eyes. For many tasks the visual requirement is a primary one. If the workplace is being designed for close visual tasks, the objects being viewed should be 15 to 25 cm (6 to 10 in.) above the recommended working surface height (P. C. Champney, 1975, Eastman Kodak Company). If magnifiers are used, they should be designed or located to avoid stretching the neck. If magnification is 8× or less, fiber optic magnifiers should be considered for inspecting flat products.

A common viewing distance for reading or monitoring information is 36 to 46 cm (14 to 18 in.). The maximum and minimum distances depend on the size of the object being viewed. See Section III B, "Displays," in Chapter III, "Equipment Design," for further discussion.

a. Viewing Angles

Because the size of an area that an operator can easily observe depends on the distance from which it is viewed, dimensions for visual work are expressed in terms of the operator's viewing angle. The values for preferred viewing angles are shown in Figure IIA–21.

The normal vertical angle of downward rotation of the eyes from the horizontal is 15°. This angle can be maintained for long periods where constant viewing is required. If the downward angle is increased to 45° or more, continuous viewing time will be decreased because of postural muscle fatigue (Chaffin, 1969). Constantly viewing material that is above the horizontal line of sight results in rapid fatigue of neck and shoulder muscles. Bifocal wearers who work at VDUs may have this problem and so are advised to get special prescription glasses for that work.

These angles should be kept in mind especially when designing visually demanding tasks that are highly repetitive. It is often possible for a person to rotate the body, as in a swivel chair, to see a signal or defect that is outside these guidelines. So that detection is more probable in monitoring or inspection tasks, however, the above guidelines should be followed.

b. Eye Height

The standing eye height measurement is from the floor to the eye. It ranges from 147 to 175 cm (58 to 69 in.) in most industrial populations. The sitting measurement is from the seat to the eye and ranges from about 66 to 79 cm (26 to 31 in.) (NASA, 1978). These dimensions represent the range of eye heights that must be accommodated and not the acceptable design range. Both sets of measurements assume normal clothing, including shoes, and normal postural slump.

Figure IIA–21: Preferred Viewing Angles (Developed from data in Crouch and Buttolph, 1973; Ely, Thomson, and Orlansky, 1963; Lehmann and Stier, 1961.)

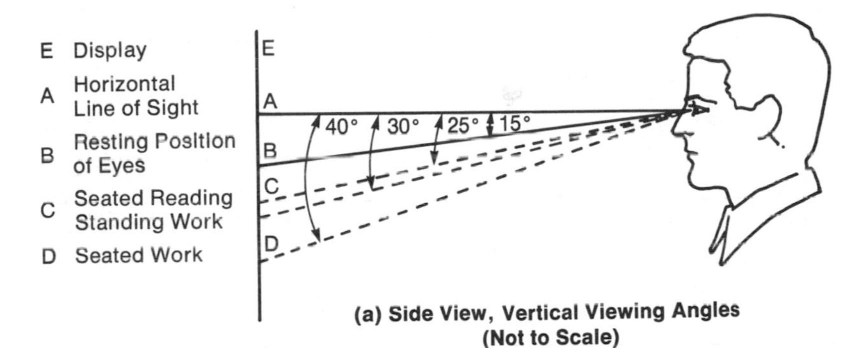

(a) Side View, Vertical Viewing Angles (Not to Scale)

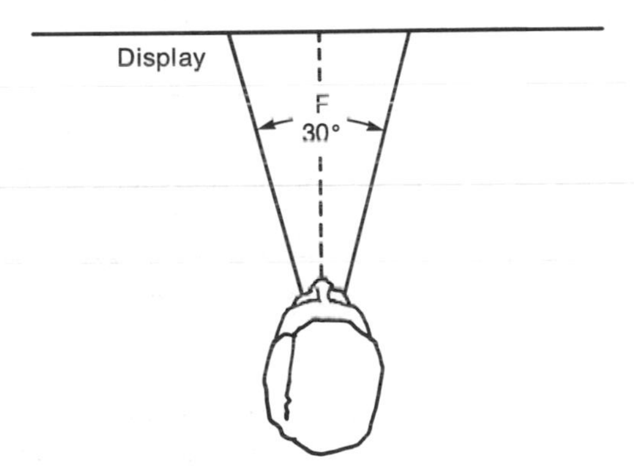

(b) Top View, Lateral Viewing Angles

Side view (part a) and top view (part b) of preferred viewing angles are shown. The vertical viewing angle is the angle between the operator's horizontal line of sight (A) and the actual line of sight (B, C, or D). The most comfortable angle is about 15° downward (B) and straight ahead. The preferred lateral viewing angles are within 15° of either side of the centerline (F). Observations of people at work have indicated that visual angles up to 40° downward are common for visual tasks (C, D). It is possible to keep most objects within these preferred viewing angles by moving the head and eyes.

c. Eyeglasses

About 50 percent of an industrial population wears some form of eye-correction lens. More than half of this group wear bifocals or some other multifocal lens. Bifocal wearers cannot easily focus on signs or dials that are from 61 to 91 cm (24 to 36 in.) in front of them. The bifocal wearer also cannot easily observe any close object (less than 18 cm, or 7 in., in front of the body) that is above eye level or near the floor. Workplace designers should recognize these limitations and try to provide workplaces that avoid these problem areas.

d. Visual Work Area

When a workplace is used for monitoring tasks or for activities where visual control is essential, special attention must be paid to the location of the displays and controls relative to the operator's normal work area. Figures IIA–22 and IIA–23 summarize visual work dimensions for sitting and standing workplaces. These dimensions are designed to accommodate both large and small individuals by considering the postural implications as well as the visual needs of the tasks. These recommended workplace dimensions will permit a console operator or other equipment-monitoring person to perform efficiently and without fatigue buildup from awkward head postures.

SECTION II B. ADJUSTABLE DESIGN APPROACHES

People vary in size and strength. Thus no one design can be optimal for all people. Adjustable workplaces or pieces of equipment that accommodate individual differences are, therefore, very desirable. Appendix A in Chapter VI includes information on the anthropometric characteristics of industrial and military populations. These data are used to assess the impact of a proposed or existing design on the potential work force. Most workplaces require attention to more than one anthropometric characteristic. The design of seated glove boxes, for instance, has to consider forward reach, shoulder breadth, visual angles, hip-to-thigh length, thigh breadth, and upper arm circumference. If one designed it for a person with 5th percentile (short) reaches and 99th percentile (very large) clearances, it would probably be unsuitable for most people (McConville and Churchill, 1976).

A good fit between the person and the task can be obtained by making the workplace adjustable. The needed adjustment can be achieved in one or more ways: the work surface can be raised or lowered; the person can change his or her position; a tool can be moved or used to extend a reach; or the product or object being worked on can be relocated or reoriented.

Although adjustable features are provided in the workplace, they may not be utilized. Use depends on how much time and effort are needed to make the changes and on the perceived benefits to the operator. Not every person will need to employ the adjustment. But if the adjustment is there, more people will be comfortable in the workplace. An aid that is shared between widely spaced workplaces, such as a drum truck or stacker truck, may not be

Figure IIA–22: Visual Work Dimensions for Seated Workplaces (Developed from data in Kennedy and Bates, 1965; Woodson, Ranc, and Conover, 1972.)

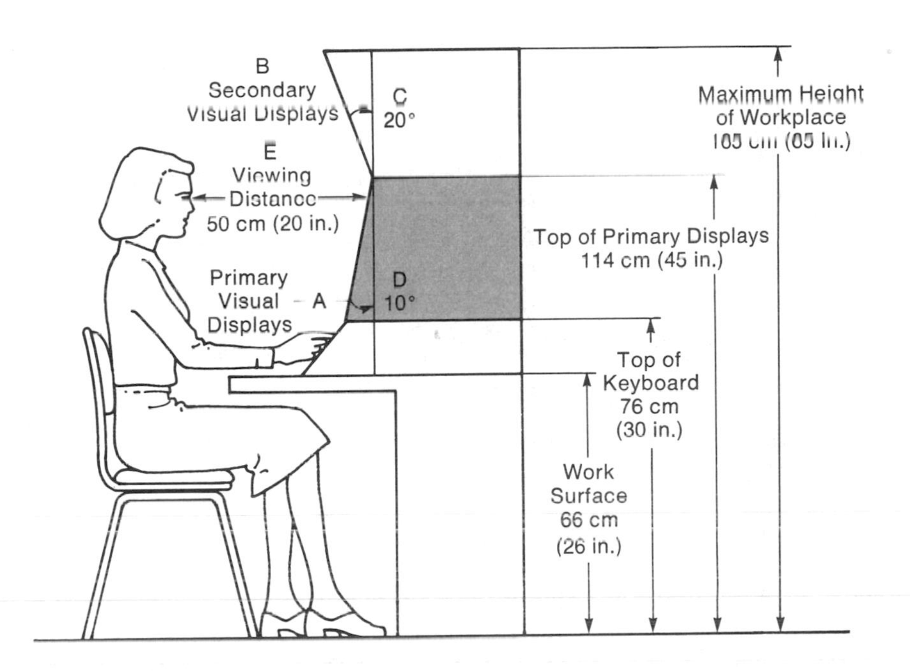

The preferred heights of primary (A) and secondary (B) visual displays for a seated console are shown. Primary displays are those most frequently monitored; secondary displays are less critical to the operation, but they give the operator information about the status of equipment, as needed. The console is at a 20° angle from the vertical plane at heights above 114 cm (45 in.) (C) and at a 10° angle below that height (D) in order to keep the display within 50 cm (20 in.) of the operator's eyes (E) in the resting posture. These suggested heights for primary displays will keep the visual angle within the ranges given in Figure IIA–21 for most people.

Figure IIA–23: Visual Work Dimensions for Standing Workplaces (Developed from information in Woodson, Ranc and Conover, 1972.)

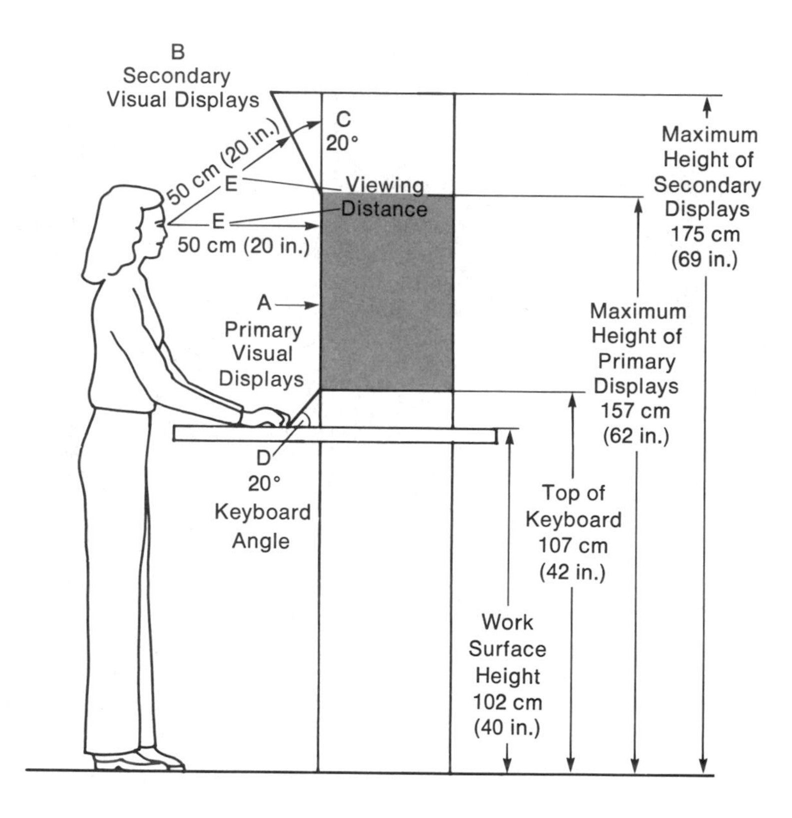

The recommended heights of primary (A) and secondary (B) visual displays for a standing console are given. Primary displays are those most critical to the operation. Secondary displays are consulted less frequently and usually give the operator information about the status of equipment or processes. The front surface of the console is angled at heights above 157 cm (62 in.) (C) to keep the distance to the operator's eyes at 50 cm (20 in.) (E). The keyboard is also at a 20° angle (D) from the work surface top. These guidelines for console design should reduce the potential for neck muscle fatigue by keeping the visual displays within the visual angles recommended in Figure IIA–21. Visual displays should also be kept within 51 cm (20 in.) of either side (lateral) of the primary work area.

taken advantage of by an operator because it will take too long to procure. Availability and accessibility of the aid, such as an air hoist or levelator, will also determine its use. Examples of three levels of adjustability are given in Table IIB–1.

Table IIB–1: Levels of Adjustment

Level	Characteristics	Examples
High	Instantaneous (<5 sec) Continuous Powered or mechanical assist	Pneumatic chair Air hoist Lowerators, levelators
Moderate	Takes 5 to 30 sec Incremental adjustment Manual effort	Chain hoist Pallet truck foot pedal Foldout steps Adjustable lighting fixtures Mechanically adjusted chairs
Low	Takes more than 30 sec Only 2 levels of adjustment Manual effort: pushing or lifting	Sliding, wing-nut chair or footrest adjustment Unpowered platform movement

Three levels of adjustment (column 1) are given. Their adjustment characteristics, including time to adjust, way of responding (continuous or incremental), and method of making the adjustment are identified in column 2. Examples of industrial equipment illustrating each level of adjustment are given in column 3. The higher the adjustability, the more likely it is that this feature will be used.

Flexibility is not the same as adjustability, although it is a desirable workplace feature. A flexible workplace is one that can be readily changed or modified to accommodate a product or task change. Once the change is made, the workplace remains fixed. Flexibility is particularly useful in an area where frequent product changes occur.

1. ADJUSTING THE WORKPLACE

The shape, location, and orientation of the work surface are determined by the overall layout of the production line. How the person interfaces with the workplace can be influenced by these factors.

a. Shape

Where forward reaches in excess of 41 cm (16 in.) are required, such as in disposing of a product to a conveyor more than 51 cm (20 in.) directly before the front surface of the workplace, a semicircular cutout can be made in the workplace to bring the operator closer to the reach point (see Figure IIB–1). The cutout can only be made if the requirements for workspace in front of the operator are small. An additional advantage of this cutout is that aisle space behind the operator is increased when the chair is pulled into the workplace.

Figure IIB–1: Seated Workplace with a Cutout to Reduce Reach Requirements

The front surface of this workplace has been cut out to permit the operator to move closer to the conveyor, which runs along the far side. Semicircular cutouts like this one may be suitable in assembly work stations; storage bins of parts can be clustered around the assembly area within the seated reach space (see Figure IIA–2).

b. Location: Height and Distance

A drafting board or an adjustable-height table is especially useful at a standing workplace, because the angle or height or both can be varied to accommodate people of different sizes. To be effective, the adjustment mechanism has to be easily found and activated. Reaches can be reduced by tilting the work surface toward the operator in some tasks. All job requirements should be evaluated, however, to ensure that none would be affected negatively by use of the tilt capability.

c. Orientation

Positioning of the workplace in relation to a conveyor line or other flow of materials can affect the reaches and strengths required to procure and dispose of product. Orienting the work surface at a 45° or 90° angle to the conveyor has been used to reduce the reach when the worker is lifting heavy objects. The suitability of such an orientation has to be evaluated in terms of all the tasks performed; visual and communication needs should also be considered.

2. ADJUSTING THE PERSON RELATIVE TO THE WORKPLACE

Height adjustment of the work surface is often not feasible, because many services to the workplace, such as pipes, vents, and conduits, are rigidly attached to it. In this event people can be located on an adjustable chair or platform or given footrests, armrests or other aids to improve their interaction with a nonadjustable work surface.

a. Chairs

Vertical adjustment can be achieved by changing chair or seat height. Some horizontal adjustment can be obtained with chair casters and swivel seats, which extend a person's reach by extending his or her range. Guidelines for seat height and back rest adjustability were given earlier. It is important for a chair to provide correct posture and comfort and features compatible with the workplace and task. Poor seating can lead to fatigue, poor performance, and interference with work. The dimensions shown in Figure IIB–2 summarize characteristics that should be looked for in selecting a chair for a workplace.

The operator in a production workplace should be provided with a chair that has the following characteristics (T. W. Faulkner, 1967, 1968, 1970, Eastman Kodak Company):

- An easily adjusted seat, such as is found in a pneumatic chair.
- An easily adjusted backrest giving lumbar support with up-and-down as well as in-and-out movement.
- A narrow enough backrest so that an operator's arms and rib cage do not strike it if the torso is rotated during a work cycle.
- A seat upholstered in woven fabric to improve comfort and reduce heat transfer problems (resulting in sweating) in the warmer months.

Chairs with casters are suitable for seated workplaces without footrests. Seat height should not be adjusted to more than 51 cm (20 in.) above the floor in order to maintain stability.

Recent studies (Mandal, 1981) have explored the use of tilted seat pads on chairs to reduce back curvature in seated workplaces. A 20° tilt forward of the seat can reduce hip joint flexion and may improve seated work posture, especially in jobs where people remain seated most of the shift.

Figure IIB–2: Recommended Chair Characteristics (Developed from information in Akerblom, 1954; Anon., 1974; T. W. Faulkner, 1967, 1968, 1970, Eastman Kodak Company; Floyd and Roberts, 1958; Grandjean, 1980.)

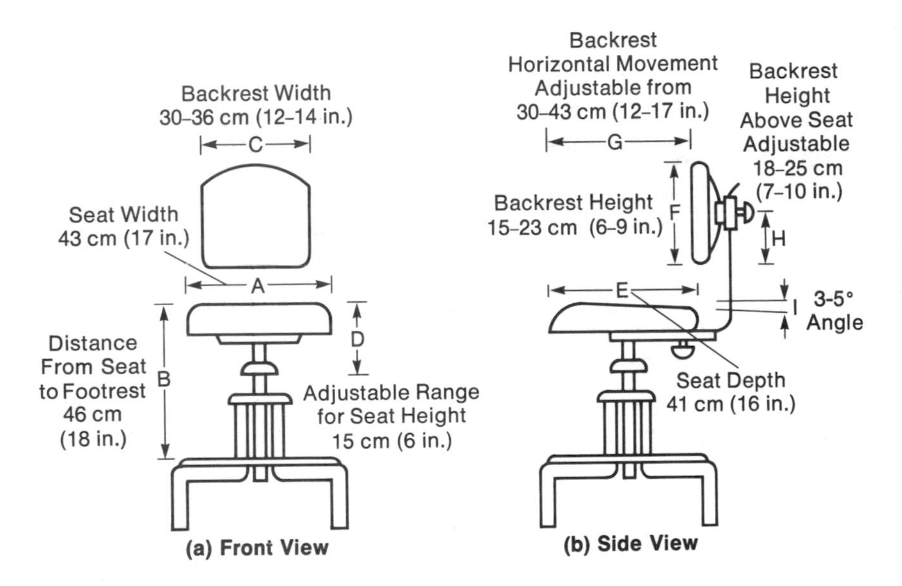

Dimensions for chair seat width (A), depth (E), vertical adjustability (D), and angle (I) and for backrest width (C), height (F), and vertical (H) and horizontal (G) adjustability relative to the chair seat are given, using both front (part a) and side (part b) views. The angle of the backrest should be adjustable horizontally from 30–43 cm (12–17 in.), by either a slide-adjust or a spring, and vertically from 18–25 cm (7–10 in.). This adjustability is needed to provide back support during different types of seated work. The seat should be adjustable within at least a 15-cm (6-in.) range. The height above the floor of the chair seat with this adjustment range will be determined by the type of workplace, with or without a footrest (see Figures IIA–3 and IIA–4). The 46-cm (18-in.) distance between the chair seat and the footrest should be maintained by having the footrest move vertically 15 cm (6 in.) with the seat.

b. Support Stools, Swing-Bracket Stools, and Other Props

A chair or stool used at a sit/stand workplace must be very stable; for instance, five legs are preferred to four, with a wide base. Consideration should also be given to providing an adjustable-height, swing-out or swing-bracket stool or chair (see Figure IIB–3, parts a and b). This type is stable and safe and can be

Figure IIB–3: Examples of Support Stools

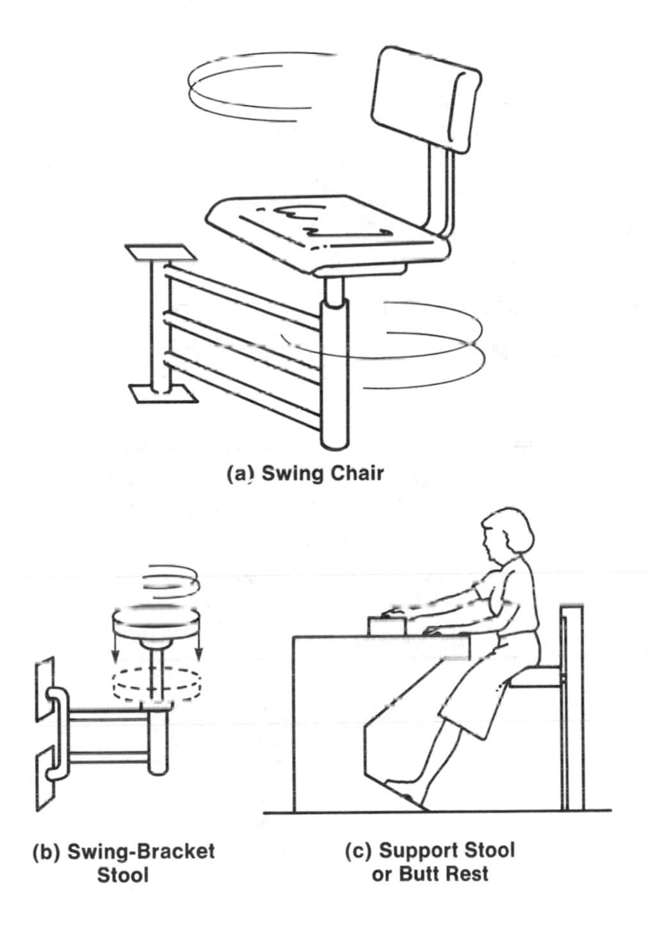

(a) Swing Chair

(b) Swing-Bracket Stool

(c) Support Stool or Butt Rest

Three types of support stool or chair are illustrated. The swing chair (part a) and swing-bracket stool (part b) are each stored under a table or a bench or against a wall. They can be swung out and latched when needed. The stool's height can be adjusted, but the chair often cannot be. The support stool, or butt rest (part c), can be adjusted in height and angle and can be swung away from the workplace for standing operations. The base of the support stool is located in the work area, and it may hinder the movement of trucks and materials into the workplace.

recessed when not in use. When a job includes a support stool or other prop, an angled foot plate and cutout in the equipment should be provided (as shown in Figure IIB–3, part c) to permit the operator to move as close as possible to the work surface.

A swing-bracket stool can also be used as a support stool or prop; the operator leans against the seat rather than sitting on it, thereby getting postural relief during an extended standing operation. It is important that the prop be located so that the operator may continue performing the job. Props like this can be helpful in extending monitoring activities. Figure IIB–4 illustrates a number of props that can be used in industrial workplaces. These include a foot rail, padded arm rail, jump seat, and a prop stool.

c. Platforms, Step-Ups, and Mechanical Lifts

Another means of adjusting a person to the workplace is to use a platform or step-up stool. These aids do not provide a change in posture, as the props discussed above do, and they may present a tripping hazard to people unfamiliar with a work area. Ideally, such a platform or step-up stool should be retractable; that is, designed to be moved out of the way when not in use. In areas where low light levels predominate, full floor platforms present less of a tripping hazard, but these devices may not be acceptable if materials-handling equipment such as a pallet truck has to be used in the area.

For areas where it is necessary to work above head height for extended periods, a mechanical lift or elevating platform can raise the person to reduce arm and shoulder fatigue. Riding trucks with mechanical lifts are often used in warehousing and construction or maintenance activities.

d. Footrests

For accommodation of all sizes of operators, a workplace should include an adjustable footrest. An adjustable chair, by itself, is insufficient, because achieving the best height for working at the work surface may leave the feet unsupported. This puts pressure on the underside of the thighs, which is uncomfortable.

Some of the types of footrests available are (see Figure IIB–5) a portable footrest or platform, a chair foot ring, and a footrest built into a workbench. For the workplace where a low chair is used and the feet are close to the floor, a portable, angled footrest can be used. With the chair adjusted to the correct working height, people with short legs can use the footrest to reduce discomfort on the underside of their thighs.

Whatever type of footrest is provided, it must be easily adjustable. Rings on a chair may not be satisfactory because they are often close to the floor and fixed. If the seat is raised, a person with short legs may not be able to reach the foot ring. Some chairs are manufactured with foot rings that move with the seat as it is adjusted, and others have foot rings that can be adjusted independently. Because foot rings are generally close to the center post of the chair, the operator has to position the legs backward to use them and cannot operate foot pedal controls from them.

Figure IIB–4: Other Props for Operators in Standing Workplaces

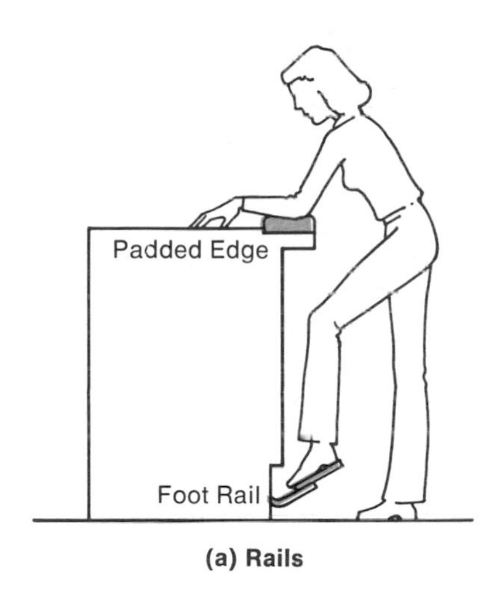

(a) Rails

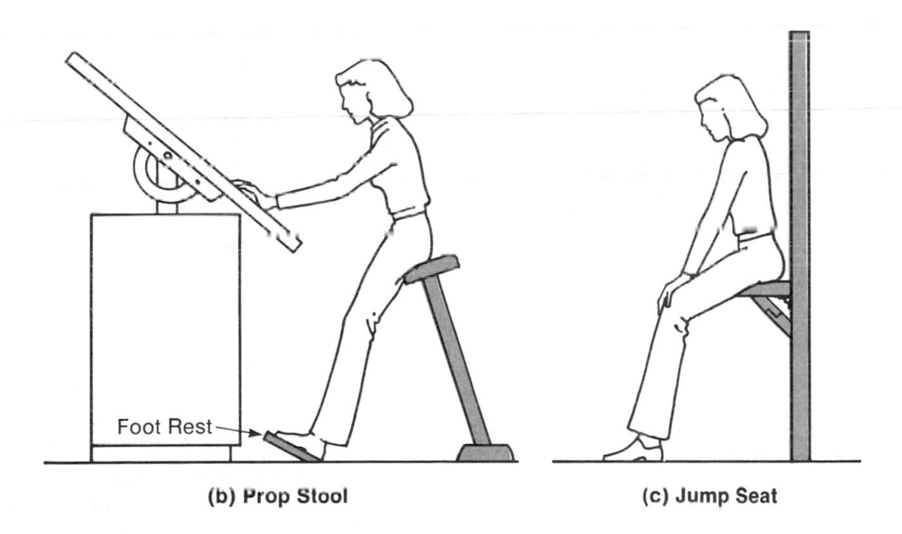

(b) Prop Stool **(c) Jump Seat**

Four types of props that can be used to aid operators in predominantly standing workplaces are shown. The foot and arm rails (part a) provide ways to rest the legs, one at a time, and the arms, by cushioning the elbow. The prop stool (part b) is similar to the one shown in Figure IIB–3 (part c), but it is portable and can be moved out of the area if necessary. It is also less stable, and it is not adjustable. The jump seat (part c) is also not adjustable, but it provides a temporary seat for an operator during a break in the work cycle or for a short monitoring activity.

Figure IIB–5: Examples of Footrests for Seated Workplaces

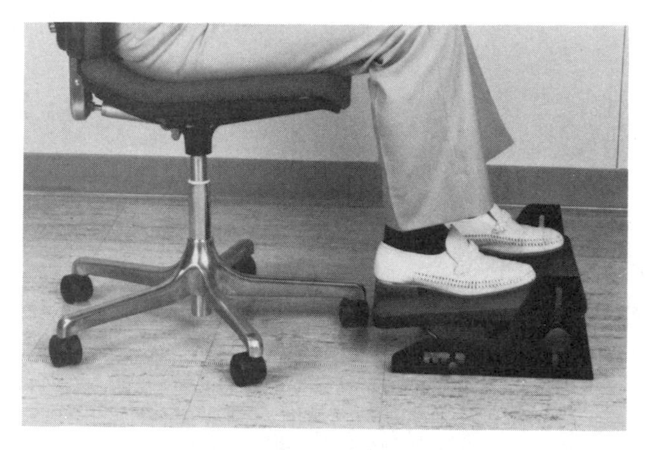

(a)

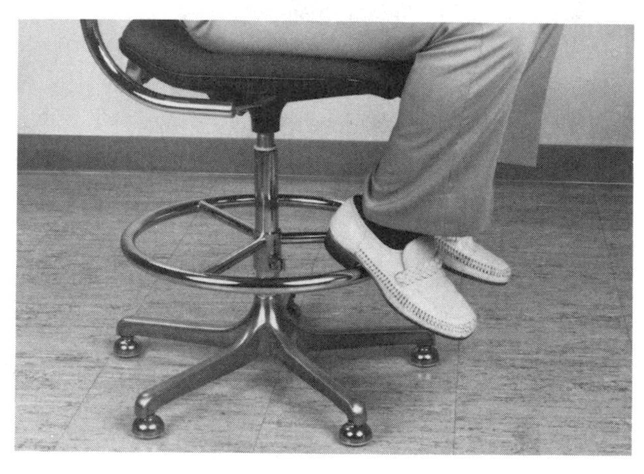

(b)

Two types of footrests are illustrated. Part a shows an adjustable platform that is set on the floor under a seated workplace; it can be moved to the most comfortable location by the operator. A footrest on an adjustable chair is shown in part b. These footrests are often not easily adjustable, making them less suitable for people with short legs whose work requires them to use their chair at the upper range of its adjustability.

Portable footrests must be large enough to support the soles of both feet. A surface of 30 by 41 cm (12 by 16 in.) should be adequate. If the footrest is built into the workplace, it should be 30 cm (12 in.) deep and wide enough to reach across the width of the seat well. The footrest inclination should not exceed 30° (Roebuck, Kroemer, and Thomson, 1975). Its top should be covered with a nonskid material to reduce slippage. Bars, brackets, or narrow strips are not adequate footrests.

It is usually best to build the footrest into the workplace. A board whose height can be varied in 5-cm (2-in.) increments (like a bookshelf or refrigerator shelf) is satisfactory for most situations.

e. Armrests

Armrests should be provided in assembly or repair tasks when the arm has to be held away from the body and is not moved extensively during the work cycle. A soft foam or plastic cushion on the armrest, covered with a non-soiling fabric, will permit easy movement of the forearm and avoid discomfort from hard edges (Kellerman, van Wely, and Willems, 1963). The armrests should be located near the front surface of the workplace, but should be easily moved to fit the variety of tasks an operator may have to do. They should tilt without having to be readjusted manually. Wrist supports can also be useful in delicate assembly work to steady the hands.

3. ADJUSTING THE WORKPIECE OR PRODUCT

Adjusting or repositioning the workpiece or product enables the operator to maintain a comfortable working posture while continuing a series of tasks. The workpiece can be adjusted or held in a fixture, parts can be supplied in a revolving supply bin, or the product can be adjusted on a leveling device such as a lift table.

a. Jigs, Clamps, and Vises

It is often necessary to hold a workpiece still while an operation is done on it. If one hand is primarily needed for holding, use of a fixture can improve the efficiency of an assembly operation by reducing static effort. Jigs, clamps, and vises are fixtures that can be used to hold a workpiece. When rotation is added, with a swivel ball and joint, for example, and motion along a track is allowed for translational movement, fixtures can become an indispensable tool to an assembler. Location of the fixture in the workplace should not require awkward reaches. These can best be avoided by making the location adjustable, for example, by mounting the fixture on a sliding track. Figure IIA-12 illustrates a fixture used to hold a component during an assembly task.

b. Circuit Board Assembly

Boards or holders are available on which larger parts can be mounted. These boards permit a wide range of motions so that the operator does not have to use awkward hand and wrist motions to complete parts of the assembly. They are often used in electrical circuit board assembly operations (Figure IIB–6). For other applications a tilting-easel workplace may be useful.

Figure IIB-6: Circuit Board Assembly Aid

The boards are attached to a ball-and-socket fixture that allows them to be rotated to, and fixed in, a multitude of positions. This feature permits the operator to move from one side of the boards to the other without having to repeatedly remove and replace the units. The boards are held in position by the fixture, so the operator has two hands available for the assembly task.

c. Parts Bins

In tasks where a large number of parts are used, such as electronic assembly operations, a revolving bin is sometimes useful to improve accessibility of parts (Anon., 1959a). In workplaces where parts storage space is limited, a multitiered set of bins can reduce the need for extended and awkward reaches. It can also utilize otherwise inaccessible space. Parts bins that tip forward for easier access are also available. Figure IIB–7 shows parts bins used in an assembly task work station.

Figure IIB–7: Parts Bins for Small Parts Assembly

Small parts are stored in individual bins at the workplace. They are located directly in front of the operator within the seated, arm reach space (see Figure IIA–2). Incoming and outgoing product is stored to the sides of the assembly area. Such bins are especially recommended where confusion between parts can occur and where many parts are used. An overhead support for a powered screwdriver is also shown. This support permits the assembler to bring the screwdriver down to the work as needed; the tension reel (at the top of the photograph) lifts the tool out of the way when it is not in use.

d. Lift Tables, Levelators, and Similar Equipment

In pallet loading and some packing operations, the product height can be adjusted by using a lift table, as shown in Figure IIB–8 (Anon., 1959b). A levelator, lowerator, stacker, or forklift truck can also be used. Where powered equipment cannot be used or justified economically, use of a wooden platform or two or three stacked pallets to provide increased height often adjusts the product sufficiently.

Figure IIB–8: Lift Table for Adjusting the Height of a Pallet

A scissors-type lift table that can be adjusted vertically from 25 to 102 cm (10 to 40 in.) above the floor is shown at a palletizing station. A pallet is placed on the scissors table; its height is then adjusted to permit the operator to transfer product horizontally or downward to it from the conveyor line or workplace.

4. ADJUSTING THE TOOL

a. Design and Location of Tools

There are many tasks where tools are used to perform operations that the hand cannot do, to add strength to the hand, or to increase the arm's reach capability. Power wrenches or screwdrivers, for instance, are used in many assembly operations. The weight of the tool is enough to recommend that it be supported from above; but the way in which that support is given can force the hand into an awkward position during assembly tasks. Whenever possible, the tool should be supported so that it has several degrees of motion; for example, it should be on a track or spring with a low-tension reel that allows the operator to move it through 180°–270° without having to fight the power cord. Figure IIB–7 illustrates a power screwdriver support suspended above the workplace.

A tool can be designed to reach an otherwise inaccessible part of a product or workpiece or to make an awkward task more comfortable. An example of the latter is the design of a curved pair of pliers (see Figure IIID–11 in Section III D) for a specific electronic assembly wiring operation. Biomechanically designed tools are usually specific to a task and are not necessarily appropriate for all types of applications. Chapter III on equipment design contains more information about tool design and selection.

b. Trays and Carriers

A tray or carrier can be used as an extension of the arm. It can be used to move a piece of equipment or a part from one position to another when the reaches are beyond the guidelines given previously or where awkward postures are required. For example, a tray or carrier can be used to help move an item through a passage in the side of a glove box (a glove box is used to reduce contact with chemicals in some filling operations or to reduce contamination of the product in clean operations; see Figure IIA–10). The carrier's length allows it to be moved through the port without requiring the operator to reach the full distance to the box's side.

SECTION II C. CLEARANCE DIMENSIONS

Clearances should be dimensioned according to the measurements of the largest worker. It should be possible for larger workers to fit easily into a workspace. Some especially critical clearances, such as head clearances at doorways, should accommodate the 99th percentile worker and have a safety margin to spare. Minimum clearances for several activities are shown in Figures IIC–1 through IIC–6 and in Table IIC–1.

Figure IIC–1: Minimum Clearances for Walking (Adapted from Thomson, Jacobs, Covner, and Orlansky, 1963.)

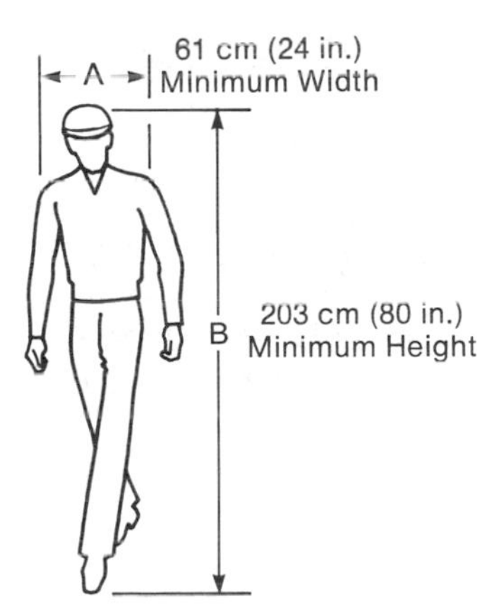

The minimum amount of space needed to permit a person to walk normally is shown. The minimum width (A) includes about 5 cm (2 in.) of clearance on either side of the shoulders of a very broad-shouldered person. The aisle clearance recommendation of 76 cm (30 in.), given in Figure IIA–11, should be used for clearances whenever possible instead of this minimum value. Vertical clearance (B, minimum height) should be measured from the floor or working surface; the dimension given will accommodate a very tall person wearing thick-soled shoes and a hard hat.

Figure IIC–2: Work Area Clearances, Horizontal (Developed from information in Croney, 1971; Hertzberg, Emanuel, and Alexander, 1956; Rigby, Cooper, and Spickard, 1961.)

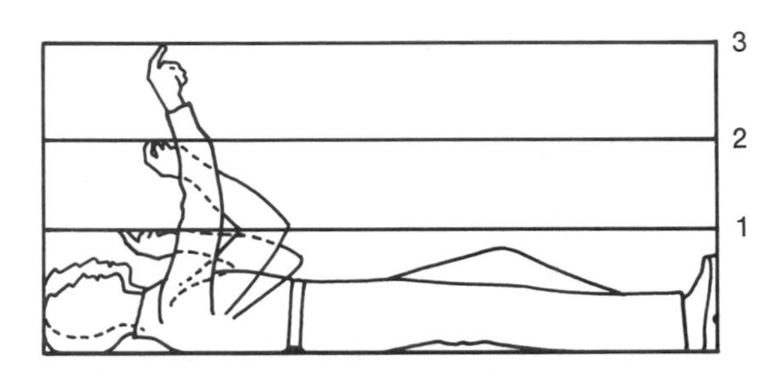

Position	Minimum Dimensions	
	Vertical	Horizontal
1. Lying for inspection	46 cm (18 in.)	193 cm (76 in.)
2. Restricted space for small tools and minor adjustments; power from elbow extension not possible	61 cm (24 in.)	193 cm (76 in.)
3. Space for reasonable arm extension; 152–203-mm (6–8-in.) length power tools could be used	81 cm (32 in.)	193 cm (76 in.)

The minimum horizontal space for a work area when a person has to lie in a supine position (on the back) is shown; in the third column of the accompanying table, this dimension is given as 193 cm (76 in.). This space will accommodate most people comfortably. The vertical clearance dimensions (1, 2, and 3) vary with the task to be performed; more space is needed if the arms have to exert force or use tools (2 and 3).

Figure IIC–3: Work Area Clearances, Upright and Prone (Adapted from Alexander and Clauser, 1965; Croney, 1971; Hertzberg, Emanuel, and Alexander, 1956; Rigby, Cooper, and Spickard, 1961.)

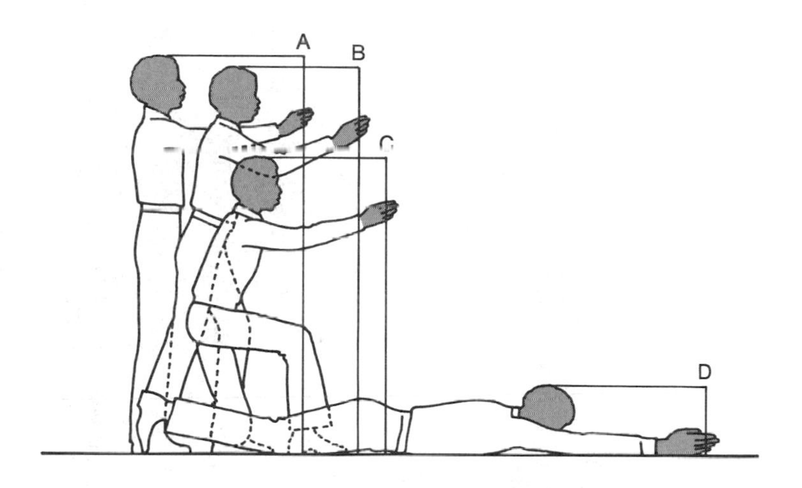

Position	Minimum Dimensions: Vertical	Minimum Dimensions: Horizontal
1. Standing	203 cm (80 in.)	76 cm (30 in.)
2. Standing, Legs Braced	203 cm (80 in.)	102 cm (40 in.)
3. Kneeling	122 cm (48 in.)	117 cm (46 in.)
4. Prone Arm Reach	46 cm (18 in.)	243 cm (96 in.)

The vertical dimension (column 2 of the accompanying table) and horizontal dimension (column 3) required for a person doing tasks while standing erect (1), with a foot braced (2), kneeling (3), and lying prone with the arms outstretched (4) are given. The breadth should be at least 61 cm (24 in.), as given in Figure IIC–1. Horizontal distances are measured from the back of the rear foot to the outstretched hand's knuckles.

Figure IIC–4: Clearances for Entering Open-Top Vessels (Adapted from R. E. Pugsley, 1975, Eastman Kodak Company.)

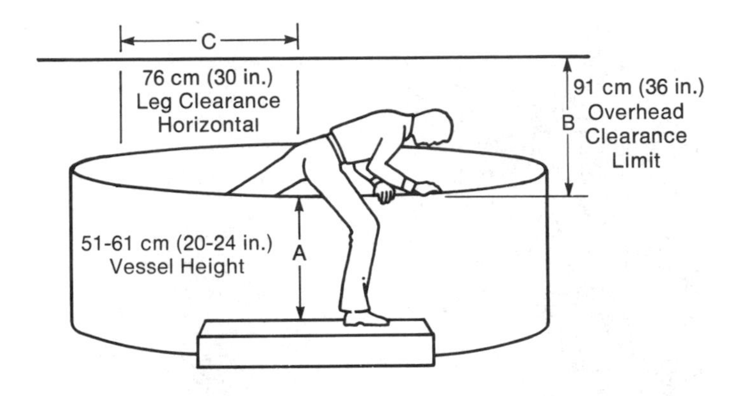

The dimensions of the work area around an open-top vessel, such as a chemical reactor or a tank, are shown. The distance between the walking surface (often a platform on the floor) and the top of the vessel (A) can be about 10 cm (4 in.) less and still accommodate most workers. The clearance above the vessel to any overhead obstruction (B), such as pipes or an overhead hoist, is needed to minimize the operator's risk of bumping his or her head and shoulders. The horizontal distance from the point where the operator enters the vessel to the nearest vertical barrier (C) should be at least 76 cm (30 in.) to permit leg extension.

Figure IIC–5: Some Selected Clearances for Arms and Hands (Adapted from Kennedy and Filler, 1966; Woodson and Conover, 1964.)

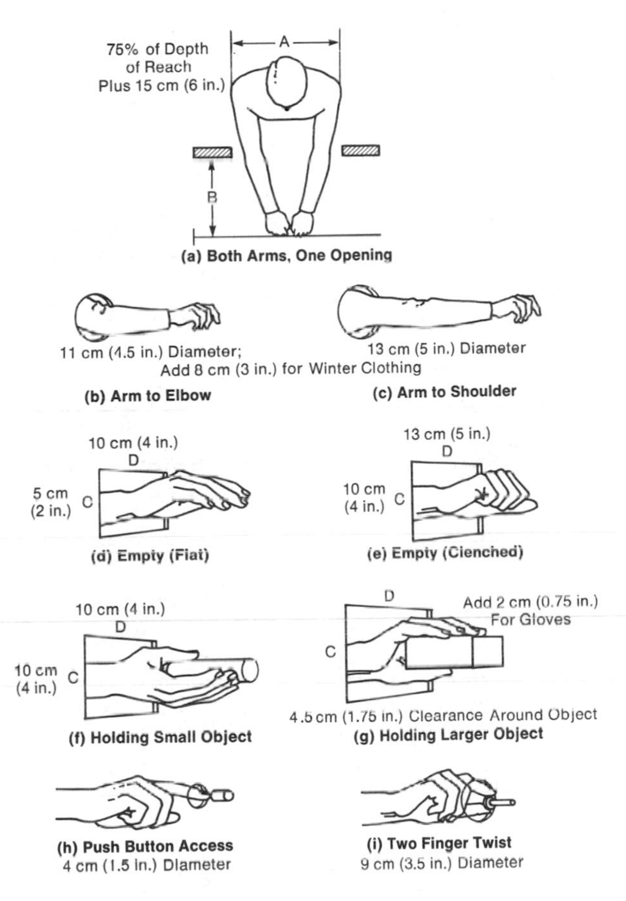

The dimensions for access ports in equipment that will permit the finger, hand, arm, or both arms to enter are given. If both arms must enter (part a), a minimum of 61 cm (24 in.) of horizontal clearance (A) is needed to provide a 61-cm (24-in.) forward reach (B). The port diameters for arm-to-elbow (part b) and arm-to-shoulder (part c) access must be increased if the operation is done under conditions where heavy clothing is worn. Height (C) and width (D) clearances for the hand when empty or holding an object are given in parts d, e, f, and g. These values should be increased by 2 cm (0.75 in.) if work gloves are worn. The access diameters shown in parts h and i are for one- or two-finger access. The size of the part being adjusted will determine the proper diameter of the two-finger access port; the larger the part, the larger is the opening that is needed to access it.

In Figure IIC–6 minimum access port dimensions are 51 by 61 cm (20 by 24 in.). These dimensions just allow a large person to move through the port. A 76-cm (30-in.) diameter port (e.g., a manhole) permits arm and leg bending as well.

Figure IIC–6: Minimum Full-Body Access Port Clearances (Adapted from U.S. Army, 1978.)

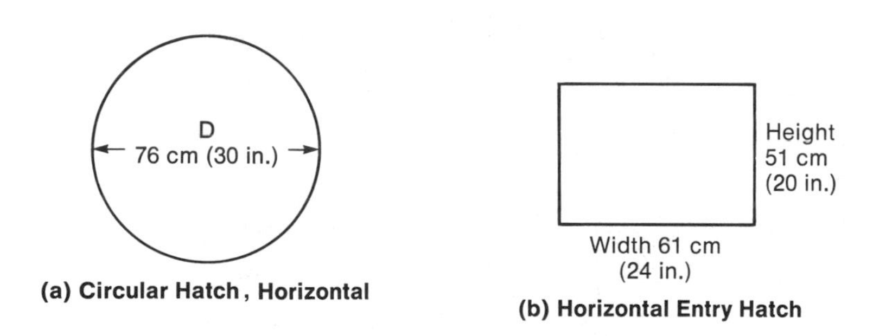

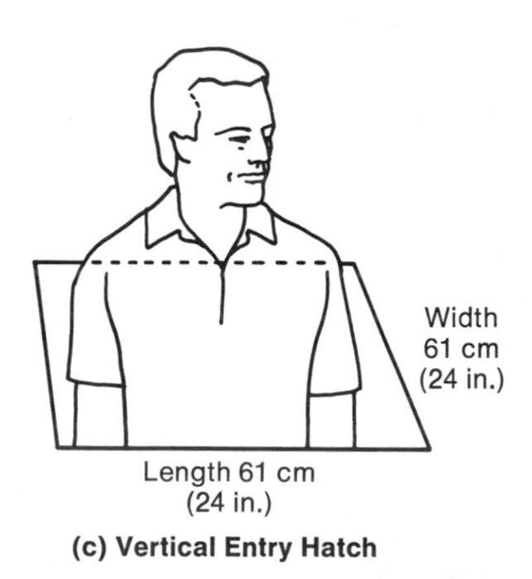

(c) Vertical Entry Hatch

Minimum dimensions for three full-body access ports are shown: a horizontal, circular hatch (part a), such as a pipe, with diameter D of 76 cm (30 in.); a rectangular, horizontal entry hatch (part b), with a height of 51 cm (20 in.) and a width of 61 cm (24 in.); and a 61-cm (24-in.) square, vertical port (part c). People wearing heavy clothing need 10–20 cm (4–8 in.) more clearance than shown here.

Table IIC–1: Minimum Clearances for the Working Hand (Adapted from Baker, McKendry, and Grant, 1960.)

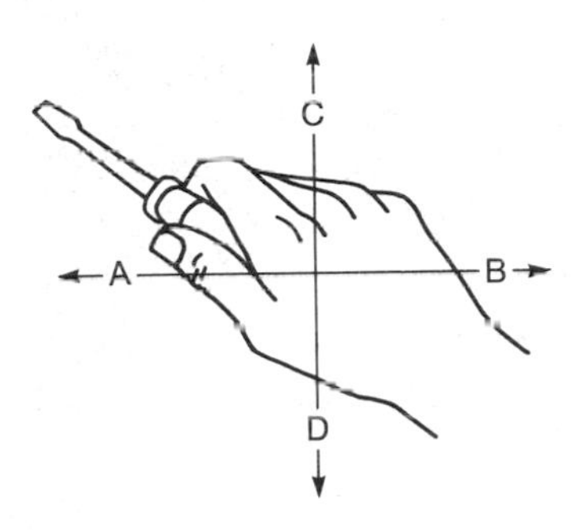

Hand Action	Minimum Dimensions For Hand Clearance			
	A, To Left	B, To Right	C, Up	D, Down
Turning Screwdriver [20-cm (8-in.) Length] or Spinate Wrench [15-cm (6-in.) Length]	38 mm (1.5 in.)	62 mm (2.4 in.)	62 mm (2.4 in.)	42 mm (1.7 in.)
Grasping, Turning, and Cutting with Needle-Nosed Pliers [14-cm (5-in.) Length] or Wire Cutters [13-cm (5-in.) Length]	53 mm (2.1 in.)	73 mm (2.9 in.)	46 mm (1.8 in.)	68 mm (2.7 in.)
Turning Socket Wrench [10-mm ($\frac{3}{8}$-in.) Base, 7-cm (3.2 in.) Shaft]	54 mm (2.1 in.)	83 mm (3.2 in.)	83 mm (3.2 in.)	73 mm (2.9 in.)
Turning Allen Wrench [5-cm (2-in.) Length]	36 mm (1.4 in.)	86 mm (3.4 in.)	94 mm (3.7 in.)	64 mm (2.5 in.)

The minimum space needed to work with common tools (identified in column 1) in maintenance or adjustment tasks is given. Measurements of tool length are given in column 1; the clearances assume that multiple rotations are needed, even where space is tight. The four directions of clearance are radial deviation (A), usually to the left; ulnar deviation (B), usually to the right; extension (C), or up; and flexion (D), or down. More space is desirable, if possible.

REFERENCES FOR CHAPTER II

Akerblom, B. 1954. "Chairs and Sitting." In *Human Factors in Equipment Design,* edited by W. F. Floyd and A. T. Welford, pp. 29–35. London: H. K. Lewis.

Alexander, M., and C. E. Clauser. 1965. *Anthropometry of Common Working Positions.* Tech. Rpt. 65–73. Wright-Patterson AFB, Ohio: Aerospace Medical Research Labs.

Anon. 1959a. "Part III: Small Parts Containers for the Bench." *Modern Materials Handling, 14:* pp. 105–107.

Anon. 1959b. "Part IV: Tables as Positioning Devices." *Modern Materials Handling, 15:* pp. 100–103.

Anon.–IBM. 1973. *IBM 3270 Information Display System: A Human Factors Study of Work Station Design.* IBM Rpt. No. GA27-2759-0, 12 pages.

Anon. 1974. "Seating in Industry." Chapter 7 in *Applied Ergonomics Handbook,* pp. 53–59. Surrey, England: IPC Science and Technology Press, Ltd.

Anon. 1978a. "Part 1910.24, Fixed Industrial Stairs." Subpart D, Title 29, Chapter XVII, in *OSHA Safety and Health Standards* (29 CFR 1910). Washington, D.C.: U.S. Department of Labor and OSHA (OSHA 2206).

Anon. 1978b. "Part 1910.27, Fixed Ladders." Subpart D, Title 29, Chapter XVII, in *OSHA Safety and Health Standards* (29 CFR 1910). Washington, D.C.: U.S. Department of Labor and OSHA (OSHA 2206).

Anon. 1980. "Answers to the Problem of Unsafe Surfaces Include Plates, Mats, Coatings and Housekeeping." *National Safety News, 122:* pp. 56–59.

Anon. 1981. "Floor Mats and Runners, National Safety Council Data Sheet 1-595-81." *National Safety News, 123:* pp. 59–62.

Archea, J. C., B. L. Collins, and F. I. Stahl. 1979. *Guidelines for Stair Safety.* NBS Building Science Series 120. Washington, D.C.: National Bureau of Standards, U.S. Department of Commerce, 119 pages.

Asher, J. K. 1977. "Towards a Safer Design for Stairs." *Job, Safety and Health, 5:* pp. 27–32.

Baker, P., J. M. McKendry, and G. Grant. 1960. *Supplement III—Anthropometry of One-Handed Maintenance Actions.* Tech. Rpt. NAVTRADEVCEN 330-1-3. Port Washington, New York: U.S. Naval Training Device Center.

Burlin, L. W. 1960. "Props for Operators." In *Human Engineering, Product Engineering Reprints,* pp. 28–30. New York: McGraw-Hill.

Carson, D. H., J. C. Archea, S. T. Margulis, and F. E. Carson. 1978. *Safety on Stairs.* NBS Building Science Series 108. Washington, D.C.: National Bureau of Standards, U.S. Department of Commerce, 120 pages.

Chaffin, D. B. 1969. "Physical Fatigue: What Is It?—How Is It Predicted?" *Journal of Methods—Time Management, 14:* pp. 20–28.

Cohen, H. 1979. *Conveyor Safety*. Washington, D.C.: U.S. Department of Health, Education, and Welfare, NIOSH.

Corlett, E. N., C. Hutcheson, M. A. DeLugan, and J. Rogozenski. 1972. "Ramps or Stairs." *Applied Ergonomics, 3 (4):* pp. 195–201.

Croney, J. 1971. *Anthropometrics for Designers*. New York: Van Nostrand Reinhold Co., 176 pages.

Crouch, C., and L. Buttolph. 1973. "Visual Relationships in Office Tasks." *Lighting Design and Application, 35:* pp. 23–25. Cited in Konz (1979).

Doering, R. D. 1974. "Defining a Safe Walking Surface." *National Safety News, 116 (8):* pp. 53–58.

Doering, R. D. 1981. "A System Approach—Slips and Falls—An Accident Analysis." *National Safety News, 123 (2):* pp. 47–50.

Doxie, F. I., and K. J. Ullom. 1967. "Human Factors in Designing Controlled Ambient Systems." *Western Electric Engineer, XI (1):* pp. 24–32.

Drury, C. G. 1974. "Depth Perception and Materials Handling." *Ergonomics, 17 (5):* pp. 677–690.

Ely, J. H., R. M. Thomson, and J. Orlansky. 1963a. "Design of Controls." Chapter 6 in Morgan et al. (1963), pp. 247–280.

Ely, J. H., R. M. Thomson, and J. Orlansky. 1963b. "Layout of Workplaces." Chapter 7 in Morgan et al. (1963) pp. 281–320.

Faulkner, T. W., and R. A. Day. 1970. "The Maximum Functional Reach for the Female Operator." *AIIE Transactions, 2:* pp. 126–131.

Floyd, W. F., and D. F. Roberts. 1958. *Anatomical, Physiological and Anthropometric Principles in the Design of Office Chairs and Tables*. BS3044. London: British Standards Institute.

Grandjean, E. 1980. *Fitting the Task to the Man—An Ergonomic Approach*. New York: International Publications Service, 379 pages.

Hall, E. T. 1966. *The Hidden Dimension*. Garden City, New York: Doubleday and Co. Cited in J. Marcotti, "How Close Is Your Neighbor?" *Industrial Engineering*, October 1969, pp. 14–18.

Hertzberg, H. T. E., I. Emanuel, and M. Alexander. 1956. *The Anthropometry of Working Positions I. A Preliminary Study*. WADC Tech. Rpt. 54-520 (AD110-573). Wright-Patterson AFB, Ohio: Wright Air Development Center.

Kellerman, F. T., P. A. Van Wely, and P. J. Willems. 1963. *Vademecum—Ergonomics in Industry*. Eindhoven, Netherlands: Phillips Technical Library, 102 pages.

Kennedy, K. W., and C. Bates. 1965. *Development of Design Standards for Group Support Consoles*. AMRL-TR-65-163 (AD630 639). Wright-Patterson AFB, Ohio: Aerospace Medical Research Labs. Cited in Roebuck, Kroemer, and Thomson (1975).

Kennedy, K. W., and B. E. Filler. 1966. *Aperture Sizes and Depths of Reach for One-Handed and Two-Handed Tasks.* AMRL-TR-66-27. Wright-Patterson AFB, Ohio: Aerospace Medical Research Labs.

Konz, S. 1979. Work Design. Columbus, Ohio: Grid Publishing Co., 592 pages.

Kroemer, K. H. E. 1969. *Push Forces Exerted in Sixty-Five Common Working Positions.* AMRL-TR-68-143. Wright-Patterson AFB, Ohio: Aerospace Medical Research Labs.

Kroemer, K. H. E. 1971. "Seating in Plant and Office." *American Industrial Hygiene Association Journal, 32 (10):* pp. 633–652.

Kroemer, K. H. E., and D. Robinson. 1971. *Horizontal Static Forces Exerted by Men in Common Working Positions on Surfaces of Various Tractions.* AMRL-TR-70-114. Wright-Patterson AFB, Ohio: Aerospace Medical Research Labs.

Lehmann, G., and F. Stier. 1961. "Mensch and Gerät." In *Handbuch der Gesamten Arbeitsmedzin 1,* pp. 718–788. Berlin: Urban and Schwarzenberg. Cited in Grandjean (1980).

Lippert, R. R. 1954. "Rolling Resistance." *Modern Materials Handling, 9* (June 1954): pp. 85–89.

Mandal, A. C. 1981. "The Seated Man (Homo Sedens), the Seated Work Position—Theory and Practice." *Applied Ergonomics 12 (1):* pp. 19–26.

McConville, J. T., and E. Churchill. 1976. *Statistical Concepts in Design.* AMRL-TR-76-29. Wright-Patterson AFB, Ohio: Aerospace Medical Research Labs, 57 pages.

Morgan, C. T., J. S. Cook III, A. Chapanis, and M. W. Lund. 1963. *Human Engineering Guide to Equipment Design.* New York: McGraw-Hill, 609 pages.

Mueller, J. 1979. *Designing for Functional Limitations.* Washington, D.C.: George Washington University, Rehabilitation Research and Training Center, Job Development Laboratory, 79 pages.

Murrell, K. F. H. 1965. *Human Performance in Industry.* New York: Reinhold, 478 pages.

NASA. 1978. *Anthropometric Source Book. Volume II: A Handbook of Anthropometric Data* (Reference Publication 1024). Yellow Springs, Ohio: NASA, 424 pages.

National Safety Council. 1980. *Accident Facts,* 1980 Edition. Chicago: National Safety Council, 94 pages.

Ostberg, O. 1976. "Office Computerization in Sweden: Workers Participation, Workplace Design Considerations, and the Reduction of Visual Strain." In *Proceedings of the NATO Advanced Study Institute on Man-Computer Interactions, Mati, Greece, September 5–18, 1976.* NATO, 28 pages.

Rehnlund, S. 1973. *Ergonomi.* Translated from the Swedish by C. Soderstrom. A. B. Volvo Bildungskoncern, 87 pages.

Rigby, L. V., J. I. Cooper, and W. A. Spickard. 1961. *Guide to Integrated System Design for Maintainability.* Tech. Rpt. 61-424, Aeronautical Systems Div., Wright-Patterson AFB, Ohio: U.S. Air Force.

Roebuck, J. A., Jr., K. H. E. Kroemer, and W. G. Thomson. 1975. *Engineering Anthropometry Methods.* New York: Wiley, 459 pages.

Templar, J. A., G. M. Mullet, and J. Archea. 1976. *An Analysis of the Behavior of Stair Users.* Washington, D.C.: Directorate for Engineering and Science, Consumer Product Safety Commission.

T. G. and R. L. 1975. "Conveyor Belt Sickness." *National Safety News,* 117: p. 37.

Thomson, R. M. 1972. "Design of Multi-Man-Machine Work Areas." Chapter 10 in Van Cott and Kinkade (1972), pp. 419–466.

Thomson, R. M., H. H. Jacobs, B. J. Covner, and J. Orlansky. 1958. *Arrangement of Groups of Men and Machines.* ONR Rpt. No. ACR33. Washington, D.C.: Office of Naval Research.

Thomson, R. M., H. H. Jacobs, B. J. Covner, and J. Orlansky. 1963. "Arrangement of Groups of Men and Machines." Chapter 8 in Morgan et al. (1963), pp. 321–366.

Tica, P. L., and J. A. Shaw. 1974. *Barrier Free Design, Accessibility for the Handicapped.* Publication No. 74-3. New York: City University of New York, Institute for Research and Development in Occupational Education, 31 pages.

U.S. Army. 1978. *Human Engineering Design Data Digest.* MIL-STD-1472B Redstone Arsenal, Alabama: U.S. Army Missile Research and Development Command, Human Engineering Lab Detachment, 149 pages.

Van Cott, H. P., and R. G. Kinkade, 1972. *Human Engineering Guide to Equipment Design.* Rev. ed. Washington, D.C.: American Institutes for Research, 752 pages.

Woodson, W. E. 1981. *Human Factors Design Handbook.* New York: McGraw-Hill, 1072 pages.

Woodson, W. E., and D. W. Conover. 1964. *Human Engineering Guide for Equipment Designers.* 2nd ed. Berkeley: University of California Press.

Woodson, W. E., M. P. Ranc, Jr., and D. W. Conover. 1972. "Design of Individual Workplaces." Chapter 9 in Van Cott and Kinkade (1972), pp. 381–418.

Chapter III EQUIPMENT DESIGN

Contributing Authors

William H. Cushman,
Ph.D., Psychology

Richard M. Little,
M. S., Electrical Engineering

Richard L. Lucas,
Ph.D., Psychology

Richard E. Pugsley,
M.S., Industrial Engineering

John A. Stevens,
B. S., Mechanical Engineering

Chapter Outline

Section III A. Design of Production Machinery: General Principles

1. Reaches and Clearances
2. Maintainability
   a. Areas to Consider When Planning Maintainability Needs
   b. Other Maintenance Design Features
3. Environment and Safety

Section III B. Displays

1. Signal Detection
2. Mode of Display
   a. Auditory Presentation
   b. Visual Presentation for Detection of a Signal
3. Visual Displays for Information Transfer
   a. Dial Design
   b. Viewing Distance
   c. Installation of Displays: Design and Environment
   d. Electronic Display Design and Installation
4. Video Display Unit (VDU) or Cathode Ray Tube (CRT) Displays
   a. Hardware Selection Characteristics
   b. Software Enhancement of VDU Displays

Section III C. Controls and Keysets

1. Population Stereotypes
2. Design, Selection, and Location of Controls
   a. Location
   b. Spacing
   c. Shape Coding
   d. Control Resistance
   e. Types of Controls
3. Keyboards for Data Entry
   a. The Nature of the Task
   b. Operator Proficiency
   c. Keyboard Layout
   d. Keyboard Parameters
   e. Locating the Keyboard in the Workplace

Section III D. Hand Tool Selection and Design

1. Factors of Concern in Hand Tool Design
   a. Static Muscle Loading: Fatigue and Soreness
   b. Awkward Hand Positions
   c. Pressure on Tissues or Joints
   d. Vibration and Noise
   e. Pinch Points
   f. Other Factors
2. Design and Selection Recommendations
   a. Handle Length
   b. Handle Diameter
   c. Handle Span
   d. Other Features of Handle Design
   e. Switches and Stops
   f. Tool Weight and Suspension
   g. Special-Purpose Tools

The design of equipment can have profound effects on the safety and performance of people at work. In this chapter four aspects of equipment design are covered: general principles, including reaches, clearances, maintainability, and safety; displays, including dials and video display units; controls and keysets; and hand tools. Factors to consider when selecting or designing this hardware are discussed.

The rapid changes in equipment and manufacturing systems in today's technological expansion present new problems for designers. Systems are becoming complex, causing the consequences of design error to be magnified. Many workplaces and machines are being designed that simply have no past history—nothing like them has been designed before. Equipment that has been in use for some time often does not receive needed modification. Certain designs may perpetuate system deficiencies that need drastic, methodical restructuring. The profitability of such equipment, however, often precludes substantial redesign. This situation is particularly true of reliable production equipment that, though ergonomically deficient, has made product at a profit for many years. When a company is installing machines to meet expanding production needs, copies of existing machines may be used to save the extra costs of engineering and drafting a new design.

For a system to operate successfully, the efficient operation of machines is not enough; the machines must be usable within the constraints of the system operator—the person. Further, not only should the system be usable, but it should also demand enough attention and skill to be interesting and rewarding. The former concerns are within the realm of human factors. The latter issues are very important, but they are usually within the province of job enrichment and organizational development experts.

The time to think about how a person will interact with the equipment is at the very beginning of the project. Human factors principles cannot be tacked on to an existing design as effectively as they can be incorporated into the equipment's basic design. A sign-off role for human factors on equipment or facility design can only identify deficiencies. If the people in the system are considered from the beginning, though, human capabilities can be incorporated into the design, thereby reducing the opportunity for an operator to limit the equipment's productivity.

The human is the most flexible part of the system, and the person will fill the gap in order to overcome deficiencies in the equipment. This effort in gap filling, however, usually means that less time is available for other activities that are more within the human realm, thereby underutilizing the human resource. Designing the equipment and selecting the controls, keysets, and tools to be within the strength, endurance, reach, and sensory- and information-processing capabilities of a large number of people are desirable, if not always possible. Factors that make it difficult to achieve these goals are time pressure stress and a tendency to order components off the shelf instead of designing them in-house. These factors can be addressed in the following ways:

- By training engineers to use human factors information.
- By providing designers and engineers with a source book of human factors information.
- By inserting the human factors evaluation of a given design into an earlier stage of the capital project.
- By providing human factors specifications as part of the requisition when ordering parts of production systems off the shelf (displays, controls, tools, and indicators, for instance).
- By providing evaluations of existing production machinery from an ergonomic standpoint.

SECTION III A. DESIGN OF PRODUCTION MACHINERY: GENERAL PRINCIPLES

As vital elements in the manufacturing system, production machines can be designed to improve the performance of the human system element in two ways: first, by assigning people to tasks that they perform best; second, by designing equipment within the capabilities of people to optimize their performance. Table IIIA–1 summarizes the relative strengths of people and machines.

Once a task has been allocated to either a machine or a person, the designer should consider the interface, or points of interaction, between the machine and its operator. These interactions include being able to reach and access parts of the equipment, planning for convenient maintenance, and providing safe and comfortable conditions for the operators in the workplace. These factors are discussed below.

1. REACHES AND CLEARANCES

Some of the questions a systems designer should be asking when designing the human interface with the machine are as follows:

- What does the person have to do to operate the machine?
- How high can a supply hopper be?
- How much room is there for the maintenance person to work in?
- How easy is it to reach equipment locations where jams are likely to occur?

The data used to determine the answers to these questions permit one to evaluate the trade-offs in a number of design options.

Chapter II on workplace design provided guidelines for reaches and clearances in the workplace; these guidelines can also be used for equipment design. A short summary of the salient guidelines follows:

Table IIIA–1: Humans Versus Machines (Adapted from Woodson and Conover, 1964.)

Humans Excel In
• Sensitivity to a wide variety of stimuli
• Ability to react to unexpected, low-probability events
• Ability to exercise judgment where events cannot be completely defined
• Perception of patterns and making generalizations about them

Machines Excel In
• Performing routine, repetitive, or very precise operations
• Exerting great force, smoothly and with precision
• Operating in environments that are hostile to humans or beyond human tolerance
• Being insensitive to extraneous factors
• Repeating operations very rapidly, continuously, and precisely the same way over a long period
• Doing many different things at one time

The capabilities of humans are compared to those of machines in order to identify whether an industrial task is best handled by a person or by automated equipment. Humans are best at tasks that require judgment and integration of information, whereas machines are best at routine tasks that have to be done precisely, rapidly, and continuously in nonoptimal environments.

- Forward reaches should be kept within 46 cm (18 in.) of the front of the body (measured at the ankles), wherever possible.
- Objects weighing more than 10 kg (22 lb) should be handled between 25 and 100 cm (10 and 39 in.) above the floor.
- The higher the lift, and the farther in front of the body it is, the less is the weight that can be handled. Provide either automatic equipment (such as air conveyors) or aids to operators (such as platforms or hoists) when production machine supplies must be loaded into hoppers that are more than 100 cm (39 in.) above the floor. Similar approaches are desirable if the load must be supported more than 36 cm (14 in.) in front of the body.
- A horizontal clearance of at least 117 cm (46 in.) should be provided beside a piece of equipment that requires on-site maintenance.

- A vertical clearance of 203 cm (80 in.) should be provided above any piece of equipment that requires overhead maintenance.
- Space should be provided around components that may have to be removed. A minimum of 4.5 cm (1.8 in.) around each side of the component to be grasped is recommended.
- At seated workplaces (e.g., machine consoles), 100 cm (39 in.) of forward leg room is recommended. Work surface height should be about 65 cm (26 in.) above the floor; an adjustable-height chair should be provided. Overhead reaches should not be more than 60 cm (24 in.) above the chair seat.

The alternative to designing production equipment within the reach and strength capacities of the industrial population is to select the people whose anthropometric characteristics make them suitable for the operation of a given piece of machinery. Because selection of people to fit job demands requires special testing and validation of the selection criteria for each job (EEOC, 1978), proper job design is the preferred approach. Figure IIIA–1 illustrates the difficulties of selecting operators to operate a modern industrial lathe. The lathe controls are located so that the ideal operator for this job would be 1.4 m (4.5 ft) tall, have a shoulder breadth of 0.6 m (2 ft), and have an arm span of 2.4 m (8 ft). Individuals possessing these characteristics may have the required reaches and strengths but lack communicative skills and have an unbridled passion for bananas.

2. MAINTAINABILITY

The material in this section was developed from information in Chapanis, Cook, Folley, and Altman (1963) and Crawford and Altman (1972).

Production equipment should be designed from the start with maintainability in mind. As systems become more complex and interdependent, evaluation of the maintenance needs and provision of aids to troubleshoot problems in a timely manner become even more important. The best source of information about maintenance needs are the maintenance mechanics, who are also most affected by poor design. In the planning of a system the following questions should be dealt with to ensure an effective maintenance program:

- What must the system do, and how reliably must it do it?
- What routine and nonroutine kinds of service are needed? What are the criteria for total overhaul or replacement?
- Where will maintenance be done? On the machine? In a shop? By contract or in-house mechanics?
- How much time will be available for completing maintenance activities? Will the mechanic be working under time stress?

Figure IIIA–1: Industrial Lathe Design: Human Interface (Adapted from Singleton, 1962. *Ergonomics for Industry,* Dept. of Scientific and Industrial Research, London, England.)

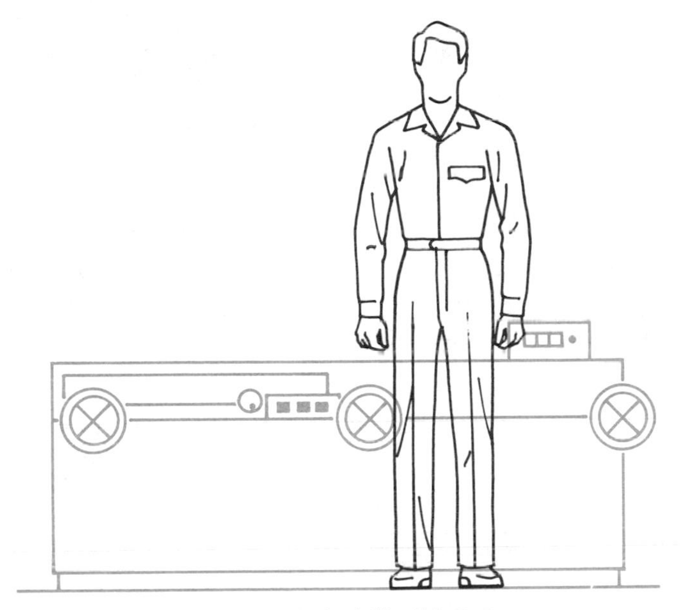

(a) Most people look like this, but . . .

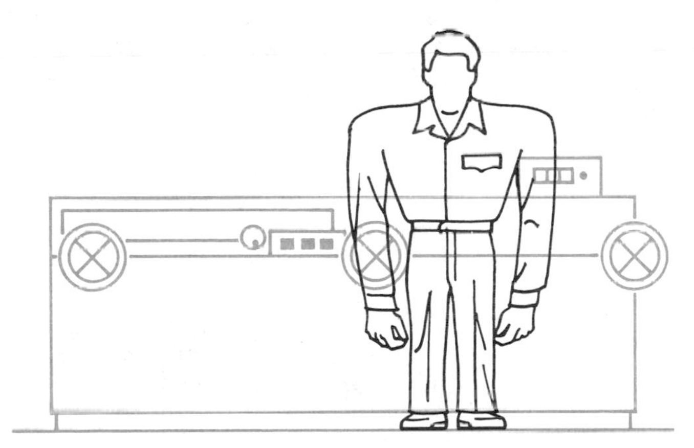

(b) Some designers think that people are shaped like this

Part a shows the location of controls for an industrial lathe in relation to a typical person's body size. Many of the controls are below waist height and at more than arm's reach from the center of the workplace. Part b depicts what a person would have to look like in order to possess the reach and visual control capabilities needed to comfortably operate this lathe.

- What information is needed to permit the mechanic or machine operator to make trade-offs among factors such as cost, speed, reliability, labor, and flexibility?
- Has a maintainability concept document been established for each piece of production equipment? This document should include the following:
  A review of the maintenance program.
  Information from other concerned areas in the organization.
  Maintenance criteria for the designers and developers of the system.

a. Areas to Consider When Planning Maintainability Needs

There are five general areas to consider when planning the maintainability needs of a production system. These areas are presented below with general guidelines for each.

(1) Prime Equipment

- Use modular or unit packaging. When warranted, use throwaway packaging.
- Design replaceable units that are independent and interchangeable. Removing and replacing one unit should not require extensive adjustment of other units.
- Provide easy access to test points and internal parts of the equipment.
- Provide self-checking features or test points for checking by auxiliary equipment.

(2) Installation

Design equipment so that it can be serviced in the place where it is finally installed. Use the guidelines for reaches, clearances, and strengths included in Chapter II, "Workplace Design."

(3) Test Equipment

Design production machinery systems so that they can be checked with readily available, standard test equipment. If this cannot be done, design and build special test equipment that will be available when the prime equipment is ready to use.

(4) Maintenance Manuals

- Write up the maintenance procedure with aid from the system designers and experienced service personnel.
- Provide the maintenance manual at the time the equipment is ready for use.
- List all of the steps necessary to maintain the equipment. Use illustrations, descriptive material, checklists, and diagrams.

- Keep the manual up to date.

(5) Tools

- Use standard tools wherever possible.
- Minimize the number of different tools needed to repair or service the equipment.

b. Other Maintenance Design Features

Several other design features of production equipment will influence its ease of maintenance: accessibility, connectors and couplings, and labeling. The guidelines in the following sections pertain to these features (U.S. Army, 1975).

(1) Accessibility

- Provide access to components that will need maintenance, preferably through openings large enough to accommodate both hands and to permit visual access as well. See Section IIC, "Clearance Dimensions," in Chapter II for dimensions of access ports.
- Consider what the maintenance tasks require in terms of tool use, exertion of force, and depth of reach when determining the dimensions of access ports. A diameter of 20 cm (8 in.) is needed for one-handed tasks requiring force exertion.
- Locate access ports so that they do not expose the maintenance operator to hot surfaces, electrical current, or sharp edges.
- Locate access ports so that the maintenance operator can see the appropriate displays when making adjustments. This often means providing access ports on the front, rather than on the back, of the equipment.

(2) Connectors and Couplings

- Provide access port covers that are easy to remove and, if possible, that are hinged. When open, the covers should not block other components that may have to be manipulated or seen.
- Provide fasteners on access covers that are easy to operate with gloved hands; a tongue-and-slot design is recommended (see Figure IIIA–2a).
- Keep to a minimum the number of turns needed to remove or replace a component (usually less than ten turns).
- Use a hex (six-sided) bolt with a slot as a screw fastener so that it may be removed by either a screwdriver or a wrench (see Figure IIIA–2b).
- Provide electrical connectors with easily detached self-locking connectors that can be actuated with one hand.

Figure IIIA–2: Examples of Fasteners (Adapted from Woodson, 1981.)

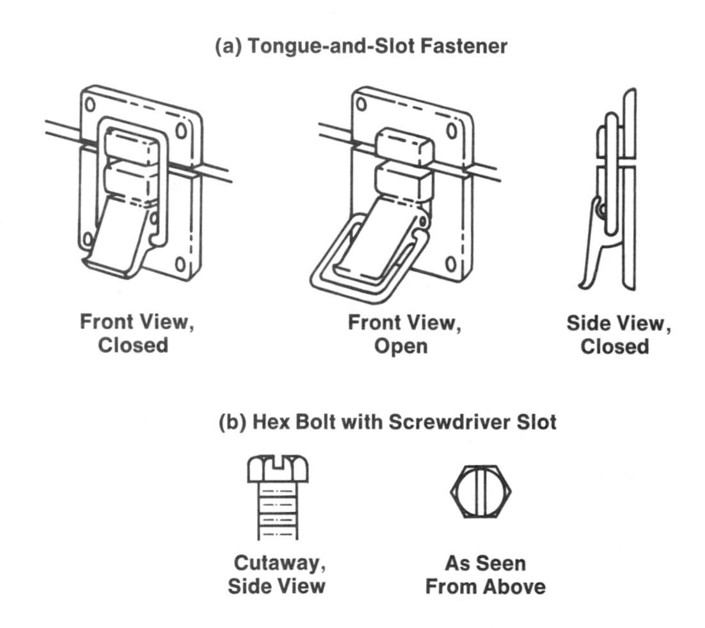

Two types of fasteners used to secure covers on equipment or to keep parts in place are illustrated. The tongue-and-slot fastener (part a) is shown in two front views, closed and open, and one side view. This type of fastener provides a tight seal and can be opened and closed with one hand, even if gloves are worn. The hex bolt with a screwdriver slot (part b) is shown in a cutaway side view and as seen from above. It can be fastened or loosened with either a wrench or a screwdriver.

- Keep the replaceable seals for couplings between pipes visible to ensure that they are replaced during assembly or repair.
- Design the fasteners for covers over components so that they are easily accessible and visible from the maintenance operator's usual work posture.

(3) Labeling

- Use labels to identify potential hazards (hot surfaces, electrical current); make them very apparent to the casual operator or maintenance worker.
- Use labels to identify test points and to present critical information for specific maintenance procedures. Keep the message short and clear.

- Place the labels where they will not be obliterated by dirt or oil.
- Follow the stereotypes for label placement relative to the controls or test points (see Section III C on controls later in this chapter).
- Label access ports with information about what components can be reached through them.
- If any fasteners are not familiar or do not follow the usual movement stereotypes, label them to indicate how they should be operated.

3. ENVIRONMENT AND SAFETY

Accidents are often ascribed to human error. Poor equipment design may often be a major contributor to that human error by requiring the operator to work to the limits of his or her capabilities for information handling, perception, or exertion of strength. Awkward lifting and twisting and similar overexertions may be produced by systems designed with excessive reaches or inadequate clearances. If human factors principles are incorporated into the design of production equipment, the system should be easier to operate and, thereby, safer and more effective.

The following suggestions should improve the safety of machinery and remove from the operator the burden of being constantly aware of possible hazards:

- Provide handles on components weighing more than 4.5 kg (10 lb).
- Avoid or guard against pinch points.
- Provide protection against accidental activation of control switches. This protection can be achieved both by shielding the control so that it cannot be activated if struck by another part of the body or by equipment moving through the area and by locating the controls in the workplace so that accidental activation is unlikely.
- Provide lockouts on machine controls to ensure that others cannot start the machine when it is being maintained or cleaned.
- Provide lock-ins on ladders, stands, and telescoping extensions to prevent their inadvertent collapse.
- Round off sharp edges and corners to reduce impact injuries.
- Provide guard rails around platforms used for monitoring or maintenance activities.
- Keep machine parts out of aisles so that tripping hazards are reduced.
- Provide aids (color coding, lighting, standard location) for readily accessing, identifying, and activating emergency equipment.
- Provide an environment with controlled levels of heat, humidity, noise, illumination, and chemical and physical substances so that the operator

can perform the job without undue risk. See Chapter V on environment for further information.

- Design reaches and clearances within the guidelines presented in Chapter II, "Workplace Design." Try to keep lifts between 25 and 130 cm (10 and 51 in.) above the floor. Design supply stations so that bulk materials can be slid or automatically conveyed instead of requiring lifting.

SECTION III B. DISPLAYS

The purpose of a display in a production system is to give information to the operator about the functional condition of the equipment or the process. The information can be categorized as follows:

- need to know
- nice to know
- historical

Information is sometimes displayed in a confusing format, with less critical data obscuring the presence of information on which action must be taken. The guidelines included in this section are meant to improve the transmission of information from a display to an operator in order to improve operator efficiency and reduce error opportunities.

1. SIGNAL DETECTION

Since the purpose of a display is to notify the operator of a situation, two prime concerns in designing the display are to make it easily detectable and to have it indicate clearly any required actions. The operator may experience a decrement in monitoring and detection performance over the work shift related to the repetitiveness of the task and the frequency of appearance of signals to be detected. Table IIIB–1 indicates factors that contribute to signal detectability.

Table IIIB–1: Task Conditions Affecting Signal Detectability During Extended Monitoring (Adapted from Van Cott and Warrick, 1972)

To increase the probability of detecting a signal
• Use simultaneous presentation of signals (audio and visual)
• Provide two operators for monitoring; allow them to communicate freely
• Provide 10 minutes of rest or alternate activity for every 30 minutes of monitoring
• Introduce artificial signals that must be responded to. These signals should be the same as real signals. Provide feedback to the operator on detection of the artificial signals

Factors that decrease the probability of signal detection
• Too many or too few signals to be detected and responded to
• Introduction of a secondary display-monitoring task
• Introduction of artificial signals for which a response is not required
• Instructions to the operator to report only signals of which there is no doubt

Conditions that make it easier to detect a signal, such as a defect in a product, are listed. Factors that make detection more difficult are also given. To improve inspection performance, one should implement the appropriate suggestions in the first part of the table and avoid the situations in the last part.

2. MODE OF DISPLAY

Information from displays is generally brought to the operator's attention by visual or auditory cues. Tactile cues may be used in situations where ambient noise levels interfere with auditory signals or where vision is obscured or reduced, such as in reduced ambient illumination. Because the majority of displays are visual and auditory, they are discussed below. Table IIIB–2 indicates the relative advantages of these two types of displays and indicators in several task situations.

Table IIIB–2: Visual Versus Auditory Presentation of Signals (Adapted from Deatherage, 1972.)

Use Visual Presentation if
• The person's job allows him or her to remain in one position
• The message does not call for immediate action
• The message is complex
• The message is long
• The message will be referred to later
• The auditory system of the person is overburdened
• The message deals with location in space
• The receiving location is too noisy

Use Auditory Presentation if
• The person's job requires him or her to move about continually
• The message calls for immediate action
• The message is simple
• The message is short
• The message will not be referred to later
• The visual system of the person is overburdened
• The message deals with events in time
• The receiving location is too bright or if preservation of dark adaptation is necessary

The upper part summarizes conditions in the workplace, or conditions related to the information to be communicated, that make visual presentation the preferred method. The lower part provides a comparable list for auditory presentation of information. Visual presentation is preferred for complex messages in noisy environments where response time is not critical. Auditory presentation is preferred for simple messages in areas where people move around frequently and where response time must be rapid.

a. Auditory Presentation

In most instances auditory presentation is used as a supplement to visual presentation, as in warning signals. The following guidelines for auditory presentation of warning signals should be noted:

- Signals should be audible, that is, discriminable from and about 10 dB above the ambient noise level (see the section on noise in Chapter V, "Environment").
- Signals should be discriminable from each other (between 1 and 2 octaves, or two to four times the frequency) and should not mask each other.
- Signals should be consistent with others already in use in the plant.
- The characteristics of the signal's sound should be attention-getting without producing traumatic sensory overload (see the section on noise in Chapter V).

b. Visual Presentation for Detection of a Signal

Warning lights are often used in conjunction with an auditory alarm to attract attention to the location of the problem. Such lights should have the following characteristics:

- Be marked with visual commands or identification of the problem.
- Be of a different luminance than the immediate background.
- Not be subject to color detection confusion.
- Be shaded or out of direct sunlight.
- Be flashing, perhaps three to ten times per second with an "on" duration of not less than 0.05 sec.
- Be located within the normal visual workplace space (see the section "Dimensions for Visual Work" in Chapter II).

3. VISUAL DISPLAYS FOR INFORMATION TRANSFER

Once an operator's attention has been called to a display, the information from that display should be readable and understandable so that the operator can take the appropriate action. There are several ways in which visual displays may be used in a production system. Table IIIB–3 includes some of these uses and suggests the display type most appropriate for each. Some examples of the displays indicated in Table IIIB–3 are shown in Figure IIIB–1.

Table IIIB–3: Types of Information Displayed and Recommended Displays for Each (Adapted from Grether and Baker, 1972.)

Information Type	Preferred Display	Comments	Examples in Industry
Quantitative Reading	Digital readout or counter	Minimum reading time Minimum error potential	Numbers of units produced on a production machine
Qualitative Reading	Moving pointer or graph	Position easy to detect, trends apparent	Temperature changes in a work area
Check Reading	Moving Pointer	Deviation from normal easily detected	Pressure gauges on a utilities console
Adjustment	Moving pointer or digital readout	Direct relation between pointer movement and motion of control, accuracy	Calibration charts on test equipment
Status Reading	Lights	Color-coded, indication of status (e.g., "on")	Consoles in production lines
Operating Instructions	Annunciator Lights	Engraved with action required, blinking for warnings	Manufacturing lines in major production systems

Six types of information to be gathered from displays are indicated in column 1. For each type the preferred display (see Figure IIIB–1 for examples) is indicated in column 2. Additional information is provided in column 3 about the reasons for choosing the preferred display or about characteristics of that display that make information transfer from display to operator more effective (e.g., color-coded lights). Some examples of industrial uses of these displays are given in column 4.

Figure IIIB–1: Examples of Visual Displays

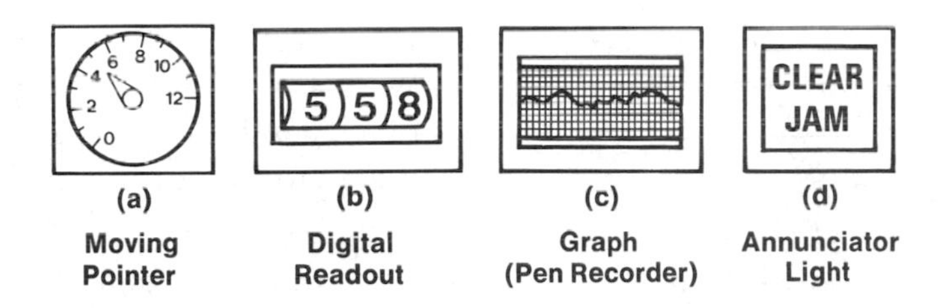

(a) Moving Pointer

(b) Digital Readout

(c) Graph (Pen Recorder)

(d) Annunciator Light

Four types of displays are shown. Part a shows a moving pointer, best for qualitative or check readings and some adjustments. Part b shows a digital readout, best for quantitative readings. Part c shows a graph (pen recording), best for detecting trends and qualitative readings. Part d shows an annunciator light, best for giving operating instructions on a control panel where many functions are monitored.

The design and installation of a visual display will affect the performance of the operator of a production system. Factors such as the distance an operator is from a display when it is read, the number of displays on a single console, the readability of the dials, and the ambient illumination should be considered when selecting and installing displays. The guidelines presented below may be used in ordering dials and other displays off the shelf and in identifying potential problems in the design of display panels in the workplace (derived in part from McCormick and Sanders, 1982).

a. Dial Design

- Avoid interpolation for quantitative readings. Choose indicators with as many gradational marks (index marks) as are needed for the degree of precision required. A maximum of nine markings between numbers is allowed.
- Make numbers progress by 1s, 2s or 5s. Definitely avoid increasing by 3s and 4s.
- Place 0 at the 9 or 12 o'clock position on a round dial with a continuous scale. If the scale does not fill the dial perimeter, locate it so that the space is at the lower part of the dial, or put the 0 at the 6 or 12 o'clock position.
- Orient numbers in the upright position, not radially.
- Choose the dial diameter (inside the scale markings) according to the number of gradations and the viewing distance (adapted from Wood-

son, 1981). A dial with 50 to 60 markings (such as a wall clock) that must be read from 3 to 6 m (10 to 20 ft) away should have an inside diameter of at least 15 cm (6 in.). Dials used for check readings of machine status (e.g., pressure gauges), which often have from 5 to 50 index markings and are read from less than 1 m (3 ft) away, can have an inside diameter of 2.5 to 3.8 cm (1 to 1.5 in.). Dials with many index markings need to be very large (up to 23 cm, or 9 in., in diameter) to be read accurately from a distance. Digital displays may be preferable in many applications.

- Select dials with target zone markings to permit more rapid reading (see Figure IIIB–2).
- Use white markings, pointers, and numbers on a black background for displays to be used in reduced ambient illumination.
- Use simple fonts and legible printing so that the displays can be easily read.

Figure IIIB–2: Target Zone Markings on Dials (Adapted from Kurke, 1956.)

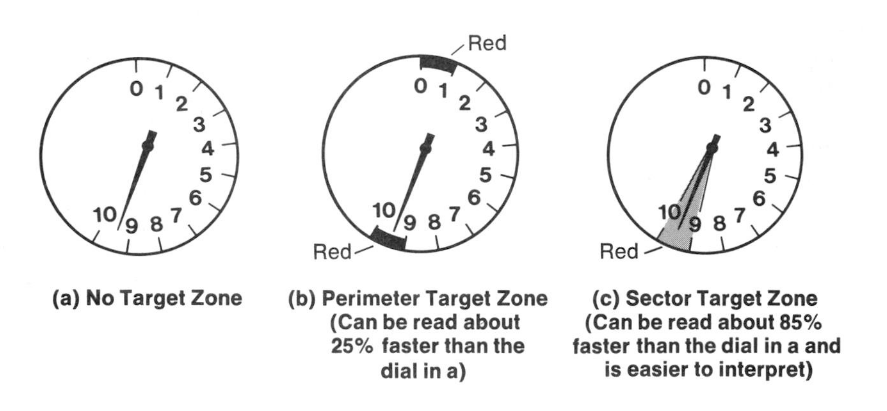

(a) No Target Zone

(b) Perimeter Target Zone (Can be read about 25% faster than the dial in a)

(c) Sector Target Zone (Can be read about 85% faster than the dial in a and is easier to interpret)

Three dials are shown, two with markings to indicate abnormal functioning or conditions to which an operator has to respond. In part a the abnormal function zone is not marked, so an operator has to be trained to recognize when a potential problem exists. In part b the two zones of concern are marked by a red rectangle at the outer edge of the dial. The pointer can be seen against the light-colored dial; its tip points to the red zone when readings indicate abnormal function. In part c the entire dial is colored red within the zone of abnormal function, making it very obvious when the pointer falls in this sector. Response time is faster for the dial in part c than for those in part a and part b because interpretation of the meaning of the pointer when it is in the red sector is immediate and demands action.

b. Viewing Distance

For quantitative (numeric) readings a digital readout is preferable to a dial; the operator does not have to consider scale markings on a digital readout, so there is less opportunity for error. The guidelines for viewing distance presented in the section on labels and signs in Chapter IV, "Information Transfer," should be used when selecting number height for a digital readout.

c. Installation of Displays: Design and Environment

Displays should be installed on a monitoring panel or on production equipment so as to minimize the potential for operator error. The following guidelines relate to factors in the display environment that should be controlled:

- Avoid shadows on the display face from adjacent protrusions or from the bezel (cover rim) of an inset indicator.
- Avoid optical distortion from the glass cover plate and glare from light sources.
- Align a group of dials uniformly when check reading is required so that all pointers are in the same position for the normal conditions (see Figure IIIB–3).
- Provide adjacent indicators on a machine control panel with the same layout of marks and numbers (see Figure IIIB–4).
- Orient indicators so that they are perpendicular to the operator's line of sight. This design should reduce parallax errors when pointers are read.
- Avoid use of color coding on the indicator if colored ambient illumination provides poor color rendition (e.g., photographic safelights or sodium vapor illumination).
- Locate frequently used indicators at standing workplaces between 107 and 157 cm (42 and 62 in.) from the floor, but as close to 152 cm (60 in.) as possible. Less frequently read indicators can be above or below this height range. For seated workplaces, locate the primary displays no higher than 50 cm (20 in.) above the work surface or 80 cm (32 in.) above seat height.
- Provide adequate levels of illumination (see the section on illumination in Chapter V, "Environment").
- Label the displays clearly. Follow the guidelines in the section on labels and signs in Chapter IV.
- Remove or cover unused displays, since they can divert attention from functioning units.

Figure IIIB–3: Alignment of Dials to Detect Normal Functioning (Adapted from Oatman, 1964.)

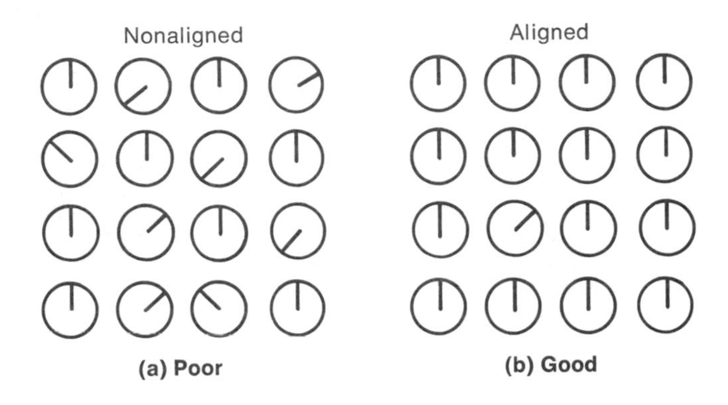

Two examples of dials located adjacent to each other on a display panel are shown. In the poor example of dial alignment in part a, the normal function ranges are located in different parts of the dials, resulting in little conformity among the displays. The operator has to check each dial to be assured that no process is out of specification. In the good example in part b, however, all of the pointers are aligned similarly for normal functioning. The operator can determine if changes need to be made by simply scanning the displays and detecting ones that do not fit the usual pattern of pointer orientation.

d. Electronic Display Design and Installation

Recent technology has provided the designer with a wide range of electronic displays, including LEDs (light-emitting diodes), liquid crystals, and characters generated in dot matrix. While these devices may save considerable space in a display panel, legibility may be a problem. Segmented numbers, such as those seen in calculator displays, are less legible than the NAMEL font and may influence the operator's accuracy in taking readings (Plath, 1970). See Figure IIIB–5.

NAMEL is an acronym for Navy Aeronautical Medical Equipment Laboratory and designates a letter and numeral font of United States Military Specification No. MIL-M-18012B (July 20, 1964).

Figure IIIB–4: Examples of Poor and Good Display Panel Dial Design (Adapted from Woodson, 1981.)

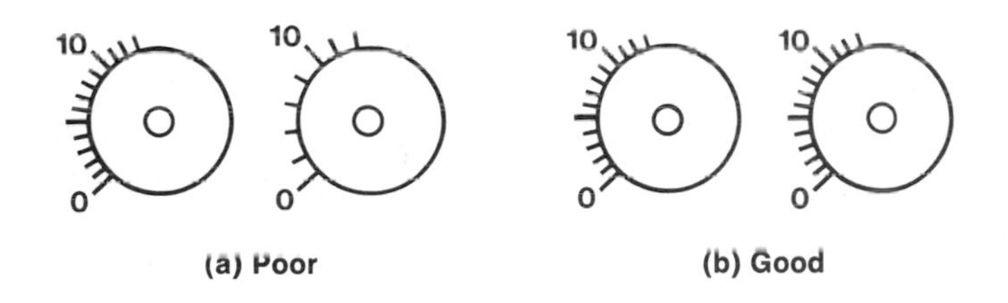

Conformity in the choice of dial markings for adjacent indicators reduces the opportunity for making errors when reading dials with moving pointers. The example in part b is good because each dial is marked with the same scale. The example in part a is poor because one scale is marked in units and the other has a mark every two units. This situation increases the possibility of misreading the dials, particularly if the processes being monitored are similar.

Figure IIIB–5: Segmented and NAMEL Font (Adapted from McCormick and Sanders, 1982; Plath, 1970.)

1 2 3 4 5 6 7 8 9 0

1 2 3 4 5 6 7 8 9 0

The upper line of numbers is in the NAMEL font, which was designed for its legibility; it should provide less opportunity for errors when operators are making readings. Segmented numbers (seen in the lower line) are less legible and may be misread in situations where fast readings are needed. For example, when numbers are changing rapidly on an electronic display, some persistence in the display phosphor may make it difficult to distinguish between the numbers 6 and 8.

The guidelines given below specify some characteristics of electronic displays and suggestions for their installation (Hinsley and Hanes, 1977; McCormick and Sanders, 1982; Snyder and Maddox, 1978).

(1) Characteristics

- Use a dot matrix character style, at least 5 × 7 and preferably 7 × 7 or 7 × 9, for the most accuracy (see Figure IIIB–6).
- Provide the following geometry for the numerals and letters displayed:
  A width-to-height ratio of about 0.6 to 0.8.
  A distance between digits of 1.1 to 1.4 times the stroke width.
  Vertical numbers rather than slanted ones.
  A dot spacing of about 0.4 to 0.6 mm (0.02 to 0.025 in.).
- Select a display that does not persist so long that an operator is unable to read current values of numbers if they are changing rapidly.
- Choose a display with lines for describing the characters that are sharp, not diffused, and have equal brightness throughout.

Figure IIIB–6: Examples of Dot Matrix Character Styles (Adapted from Gould, 1968; Snyder and Maddox, 1978; Vartabedian, 1973.)

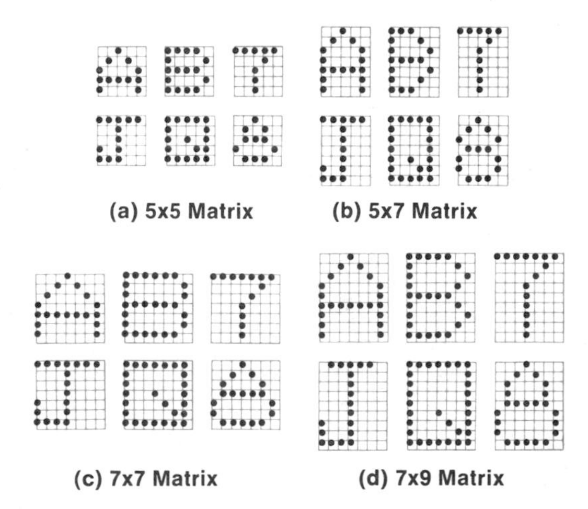

Four examples of dot matrix displays, often found in video displays are shown. Part a shows a 5 × 5 dot matrix; part b, a 5 × 7 dot matrix; part c, a 7 × 7 dot matrix; and part d, a 7 × 9 dot matrix. The larger the matrix, the more legible the numbers and letters (Gould, 1968).

(2) *Installation*

- Provide for wide-range viewing angles to ensure full visibility of all characters without any background noise. Be sure obstructions do not prevent characters from being seen from all angles.
- Minimize internal reflections, unlit images, or distractions from the background of the display unit.
- Minimize glare on the display by adjusting the direction of ambient illumination, using shields or filters or both, and locating the displays away from glare sources.

4. VIDEO DISPLAY UNIT (VDU) OR CATHODE RAY TUBE (CRT) DISPLAYS

The material in this section was developed from information in Cakir, Hart, and Stewart (1979) and W. H. Cushman and D. Crist (1979, Eastman Kodak Company).

With the increasing use of computer-controlled manufacturing systems, production equipment operators are having to learn how to interact with computer terminals and with information displayed on a screen at a central workplace. The characteristics of available computer hardware systems should be evaluated against the guidelines below.

a. Hardware Selection Characteristics

(1) *Character Size and Generation*

Modes of character generation include dot matrix, discrete stroke, monoscope (facsimile), and raster scan. Characteristics to look for when choosing one of these modes for a display follow:

- Scan lines should not be visible to the operator. Matrix dots should be large enough and spaced to fuse together rather than appear discrete.
- Optimum dot size varies from 0.8 to 1.2 mm (0.03 to 0.05 in.) for different applications. The Lincoln/Mitre font and a dot spacing/size ratio of 0.05 or less are suggested (Snyder and Maddox, 1978).
- Displays with 7 × 9 or larger dot matrix characters are significantly more legible than 5 × 7 characters (Gould, 1968).
- The width-to-height ratio for alphanumeric VDU characters should be between 0.7 and 0.8 (Gladman, 1976; Hinsley and Hanes, 1977).
- All characters and symbols should have well-defined contours and appear to be in focus. Characters should not appear to be degraded near the edges of the screen.
- The visual angle subtended by each character should be at least 15 to 20 minutes of arc. Assuming a maximum viewing distance of 71 cm (28 in.),

the minimum acceptable character height is about 4.6 mm (0.18 in.) (Stewart, Ostberg, and MacKay, 1974).

- Uppercase lettering is adequate for displays consisting of very short phrases, single words, and codes. Mixed uppercase and lowercase lettering is preferred for applications in which continuous reading from text is required (Stewart, Ostberg, and MacKay, 1974).

(2) Display Luminance, Contrast, and Flicker

- The more light in the environment around a VDU, the more difficult it is to read the display. Luminances as low as 86 candelas per square meter (cd/m^2), or 25 footlamberts (fL), may be adequate for applications in which ambient illumination is subdued, but luminances as high as 343 cd/m^2 (100 fL) may be required in environments with high ambient illumination (Gould 1968; Hayman, 1969; Stewart, Ostberg, and MacKay, 1974). See the section on illumination in Chapter V.
- Contrast affects both character visibility and legibility and is related to the measured difference in luminance of the characters and their background. A contrast of 0.88 is considered acceptable; ideally, it should be above 0.94 (Gould, 1968; Hinsley and Hanes, 1977; Howell and Kraft, 1959). Often, it is less.
- Contrast may be intensified by increasing the luminance of the characters or by decreasing the reflected ambient light. The former can be accomplished by manipulating the brightness control of the VDU; the latter by placing an antireflective filter over the screen. The filter increases contrast and legibility but decreases overall character luminance.
- Since characters written on a VDU screen begin to fade immediately, they must be refreshed frequently. For minimization of flicker the rate of regeneration for a VDU with short or medium persistence must be at least 50 hertz (Hz). A rate of 60 Hz is preferable (Gladman, 1976).
- In general, flicker may be reduced by decreasing the size or the luminance, or both, of the characters (Hinsley and Hanes, 1977) and by selecting the proper persistence for the phosphor used in the VDU.

(3) Cursor Design

The cursor on a VDU is a marker for accessing a particular piece of information. It is often a white rectangle one space wide and one space high or a moving underline one space wide. A cursor should have the following characteristics:

- It should be easily seen but should not obscure the reading of the character or symbol it marks.

- It should be easy to find at any location on the VDU screen and be easy to move from one position to another.
- It should be easily tracked as it is moved from one position to another.
- It should not be unnecessarily distracting. An override to suppress the cursor is a desirable feature.
- It should blink at about 3 Hz if it is used to attract the operator's attention on a monitoring task (Smith and Goodwin, 1971).

(4) Graphics Capability

Computer VDU terminals are available with a graphics capability that permits the presentation of information in trend curves, histograms, and similar displays. In addition, they can be formatted to display a fixed form, similar to a document at a noncomputerized station. This preformatted screen has advantages in data entry applications, eliminating the need to wait for a prompt from the computer. The graphical capability can be used effectively to show trends that may not be apparent in the myriad of data being logged at the console.

(5) VDU Terminal Alarms

The material in this section was developed from information in B. Crist, (1976, Eastman Kodak Company).

Alarms are used at the VDU interface to alert the user to information or actions that need to be taken. The following general principles should be considered when selecting or installing an alarm:

- Both auditory and visual alarms should be used.
- If the operator is likely to be away from the VDU console for any period of time, an auditory alarm is needed. The signal should be more than 200 Hz and, usually, less than 1000 Hz. A 1–8-Hz warble can be added to improve its detectability over machine noise (Hinsley and Hanes, 1977). A volume control is desirable, but it should not shut off the alarm entirely.
- If the operator's response to an alarm requires concentrated visual attention on the screen, a blinking message should not be used (unless the blinking can be overriden by the operator). Luminance coding or video reverse in conjunction with an audio signal is preferable.

b. Software Enhancement of VDU Displays

A computer display can be programmed to present information to the operator according to predetermined priorities. This software approach to the VDU display can considerably reduce the potential for operator error. Some general guidelines for improving the interface between the operator and the VDU follow (Dodson and Shields, 1979; K. Harris, 1979, unpublished report, Proctor and Gamble Company; Hinsley and Hanes, 1977; Pew, Rollins, and Williams, 1976; S. H. Rodgers, 1979, Eastman Kodak Company).

(1) Information Location and Density

(a) Location

- Locate information of a critical nature near the center of the screen.
- Display status information at the top of the screen, toward the right side. The location of this information will vary according to the type of operations performed, but it should be found in the same general part of the screen across all terminals in a manufacturing system.
- Locate error messages near the bottom of the screen; they should blink at about 3 Hz.
- Relegate code information, sampling plans, and other information that the operator may need to call up to one of the sides of the screen near the top. This material should not interfere with the information being used by the operator to troubleshoot a problem or perform a data entry task.

(b) Density

Readability of status information or other text drops off rapidly at screen-packing densities (the amount of space filled with characters) greater than 50 percent. Figure IIIB–7 illustrates the time required to locate answers to questions as a function of display densities of 30, 50, and 70 percent (shown below the curve). The differences in location time between the 70 percent density and the 30 and 50 percent densities were significant.

For the reduction of information density, the following strategies can be used:

- Develop a hierarchy of information needs; display information routinely only for the higher priorities. Relegate the other information to routines that can be called up as needed.
- Use coded information to pack more data on the screen without losing the desired spacing.
- Use graphics to display events that are changing in time (Booher, 1975; Hinsley and Hanes, 1977; Mitchell-Bishop, 1979).
- Provide a logging printer to record all status data; use symbols to identify specific status groups (e.g., all information from one machine section) on the printout. This arrangement will free the VDU interface for more interactive evaluations of manufacturing status.
- Use color displays; identify blocks or segments of information by different colors.

Figure IIIB–7: Response Time to Locate Information as a Function of Display Density (Adapted from Dodson and Shields, 1979.)

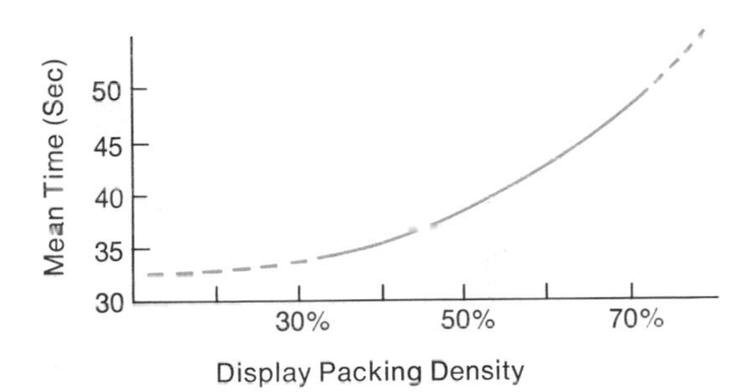

70 Percent Density Display

50 Percent Density Display

30 Percent Density Display

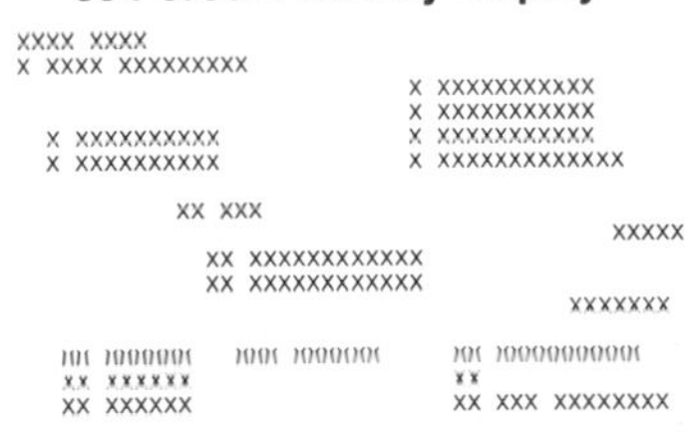

The average time, in seconds, it took video display operators to locate technical information on the screen is indicated on the vertical axis. The horizontal axis represents different screen-packing densities, also illustrated by the figures below the graph. The time to locate information increased as a function of the amount of the screen that was filled (packing density), especially once 50 percent or more of the screen was used.

(2) *Software Aids for Alarm Response*

- Errors with a structured response pattern should be handled within the computer and should not trigger alarms (Mitchell-Bishop, 1979). Only those errors that require rapid operator decision making and intervention should activate the redundant alarms. Early signs of a system going out of specification should be identified by low-level alarms since response time is not as critical.
- In some operations it may be desirable to preprogram a hierarchy of alarms that could be altered if changes were made in manufacturing specifications.
- Because the operator has to respond quickly to an alarm indicating a major failure in the manufacturing system, it is important to use the computer to assist in the problem-solving process. It should be possible to program the computer to display specific information each time one of the major system failures is detected and triggers an alarm. Building information retrieval into the alarm can halve the time taken to remedy the problematic situation, since most of the response time is spent in gathering relevant information.
- Since failure in one component of a manufacturing system may trigger failures in other components, it is important for the operator to know the chronological order of failure. Information that will get in the way of decision making, such as alarms from systems that are secondarily affected by the initial problem, should not be presented to the operator unless requested.
- Once the operator has responded to an alarm with a controlling action, some feedback should be indicated on the VDU to acknowledge that action has been taken (Galitz, 1979; Pew, Rollins, and Williams, 1976).

(3) *Error Detection and Retrieval in Data Entry*

The material in this section was developed from information in Mitchell-Bishop (1979).

- When the VDU interface is being used for data entry or controlling functions, the possibility of miskeying or entering the wrong information exists. With software aids it should be possible to detect errors of omission, commission (extra data), or confusion of numeric and alphabetic characters (see the section on coding in Chapter IV).
- Since correction of an error might require going to a different subroutine from the one being used in data entry, there should be provision for more than one error message.
- Provision of a tab, skip, or cursor control key is desirable to permit rapid

movement of the cursor to an error location on the screen.

- Provision of a temporary file location to store records with incomplete data or suspected errors should be considered. If this provision is not feasible, the records of concern should be flagged so that they are not included in summary reports generated from the central data base.

SECTION III C. CONTROLS AND KEYSETS

The design of information coming to the production system operator via labels, signs, and instructions is discussed in Chapter IV, "Information Transfer." The operator's response to this information is communicated through controls and data entry devices. A control is anything—a switch, lever, pedal, button, knob, keyboard—used by a person to put information into a system. Performance can be enhanced if the controls operate as one expects them to (they follow population stereotypes); they are dimensioned to fit the human body; and their operating characteristics are within the strengths and precision capabilities of most people.

1. POPULATION STEREOTYPES

People expect things to behave in certain ways when they are operating controls or when they are in certain environments. Although it is possible to educate people to operate systems that do not follow the stereotypes, their performance may deteriorate when placed in an emergency situation. Stereotypes often change because of changes in technology. For example, automobile emergency brakes used to be, primarily, hand brakes located to the left of the steering column and under the dashboard. They were pulled out to apply the brakes. By the 1960s a large number of car models were using a foot pedal emergency brake, located just above the floor, forward and to the left of the clutch or brake pedal. The release lever for the emergency brake was located in the same area as the older emergency hand brake, although it was much smaller. People who learned to drive on cars with the hand emergency brakes could be expected to experience difficulty responding to an emergency such as brake failure in a modern car; their reflex response would be to release, rather than set, the emergency brake.

As further examples of stereotypes, the following environmental behaviors are expected (Woodson, 1981; Woodson and Conover, 1964):

- Very loud sounds, or sounds repeated in rapid succession, and visual displays that blink or are very bright imply urgency and excitement.
- Speech sounds are expected to be at approximately head height and in front of a person.
- Seat heights are expected to be at least 40 cm (15.5 in.) above the floor in production workplaces and offices.
- Very large or dark objects imply heaviness. Small or light-colored objects

imply lightness. Large, heavy objects are expected to be at the bottom and small, light ones at the top.

- Red signifies "stop" or "danger," yellow indicates "caution," green indicates "go" or "on," and a flashing blue indicates an emergency control vehicle, such as a police car.
- Coolness is associated with blue-green colors; warmness is associated with yellows and reds.

Stereotypes relating to rotary controls should, generally, follow Warrick's principle: an operator usually moves a control so that the part of it nearest the display moves in the direction he or she is trying to move the display's indicator (Brebner and Sandow, 1976; Thylen, 1966; Warrick, 1947; Woodson and Conover, 1964). Some expectations of operations of these controls are as follows (also see Figure IIIC–1):

Figure IIIC–1: Movement Stereotypes for Rotary Controls (Adapted from Warrick, 1947 and McCormick and Sanders, 1982.)

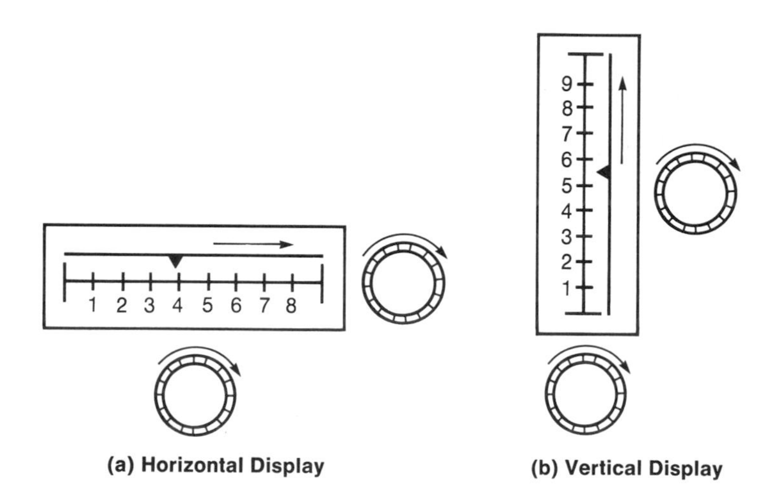

(a) Horizontal Display **(b) Vertical Display**

Four examples of the movement of a control relative to its fixed-scale display are shown. In part a the control is below or at the right end of a horizontal display; clockwise movement of either control would be expected to move the pointer to the right. In part b the control is located below or at the right of a vertical display; clockwise movement of either control would be expected to move the pointer upward.

- Knobs on electrical equipment are expected to turn clockwise for "on," to increase current, and counterclockwise for "off," to decrease current.
- Wheels or cranks to control direction of a moving vehicle are expected to use clockwise rotation to make a right turn and counterclockwise rotation to make a left turn.
- For vertical displays when the control knob is located at the bottom of the scale, the knob is expected to move in relation to the desired indicator movement. Clockwise rotation is usually associated with higher values; this rotation could be upward movement of a pointer on a fixed scale or downward movement of the scale when a pointer is fixed.

Some additional stereotypes for common controls are as follows:

- For vertical levers that move in the horizontal plane (e.g., crane controls), movement away from the body is associated with decreasing action (lowering) and movement toward the body with increasing action (raising). Movement of a lever to the left should be associated with movement of the object controlled to the left also.
- Pulling a control such as a throttle outward from a panel signifies that it has been activated (on). Pushing it in disengages it (off).
- For controls mounted overhead (e.g., on the ceiling of a control booth), pushing forward (away from the body) specifies increasing (on) activity, and pulling back specifies decreasing (off) activity.

Unlike the stereotypes for electrical equipment, the expected behaviors for valves or handles controlling liquids are counterclockwise for "open" and clockwise for "closed." Most threading on pipes follows a similar stereotype, with clockwise rotation tightening a joint and counterclockwise rotation loosening it. There may be trade- or situation-specific stereotypes (such as left-handed threads on certain gas cylinders) that should be incorporated into the design of production systems in order to minimize confusion with existing equipment.

The design and selection of controls to fit population stereotypes is particularly important for infrequently used emergency controls. Such controls should be simple to use and require minimal decision making by the operator. Since handedness could influence the expected direction of movement of a knob (Chapanis and Gropper, 1968), emergency controls should be clearly marked to indicate the proper direction of rotation.

Stereotypes may differ among countries. For example, in the United States and Germany "up" and "down" switch positions are associated with "on" and "off," respectively. In England the relationship is reversed (Murrell, 1965). If a piece of equipment is purchased that violates the local stereotype for movement, it should be very clearly labeled to indicate "on" and "off" actions. Some common United States stereotypes for up and down movement of switches, such as toggle or rocker switches, are presented in Table IIIC–1.

Table IIIC–1: United States Stereotypes for Up and Down Switch Settings (Adapted from D. Alexander, 1976, Tennessee Eastman Company.)

UP	DOWN
ON	OFF
START	STOP
HIGH	LOW
IN	OUT
FAST	SLOW
RAISE	LOWER
INCREASE	DECREASE
OPEN	CLOSE
ENGAGE	DISENGAGE
AUTOMATIC	MANUAL
FORWARD	REVERSE
ALTERNATING	DIRECT
POSITIVE	NEGATIVE

Some movement stereotypes for toggle switches used in equipment or production system control panels are given. The expected direction of activation of the switch (when mounted vertically) is given for 13 actions or conditions. These expectations are United States stereotypes and may vary in other countries.

Figure IIIC–2 illustrates some of the expected relationships between controls and their displays in a process control console. The interaction of the controls was considered when mounting them on the panel. Color coding may be used to help the operator identify groups of controls that must be operated together.

As increasingly complex control panels are assembled for centrally regulated production systems, controls for several similar machines, such as reactors, packaging machines, and boilers, may be clustered on one console.

Figure IIIC–2: Arrangements of Controlling Knobs and Their Displays on a Control Panel

The location of controls in relation to their displays is illustrated on a production machine control panel. Color or graphics can be used to indicate controls and displays that belong together. Location of the control so that it is easily identified with its display, as shown at the bottom of this picture, will reduce the possibility of the operator taking an incorrect action in an emergency situation

Operators expect each similar set of controls to operate identically. Some notably serious errors in responding to emergency situations have been attributed to incompatibility between displays and controls and to the arrangement of controls and displays on the left side of a console as mirror images of the ones on the right side (Greenberg, 1980; Seminara, Gonzalez, and Parsons, 1977). With some exceptions, as when the hands must be used simultaneously to operate controls, it is preferable to make similar sets of controls operate with the same direction of movement (i.e., as if they would be actuated with the right hand).

Labeling of controls and displays should recognize the following needs of the operator (Ely, Thomson, and Orlansky, 1963):

- The need to read the control settings while making adjustments, such as with a discrete setting on a rotary selector switch.
- The need to see the display while using the control to change a setting. Controls and displays should be close enough together to permit the operator to see both without having to take awkward postures.

- The need to identify some controls quickly in order to respond to emergencies appropriately. Color coding can be added to labeling to improve the identification of these high-priority controls.
- The need to have a consistent location for labels relative to controls and displays between work stations or machine sections. Labeling cannot overcome inconsistency problems.

For many designs the control should be mounted just below its display, permitting both to share one label denoting the function. The label is usually placed to the right of or below the control, as illustrated in Figure IIIC–2.

For concentric controls (two or three rotary knobs stacked on one another), the smallest control's display should be nearest the control and the largest knob's display should be farthest from it (Bradley, 1966). Color coding of the knobs and displays is recommended.

2. DESIGN, SELECTION, AND LOCATION OF CONTROLS

As mentioned earlier, a control is anything—a switch, lever, pedal, button, knob, keyboard—used by a person to put information into a system. Even the simplest control has several characteristics that affect the ease, speed, and accuracy of its use. The most important characteristics are the following:

- displacement, linear or angular
- operating force
- friction, inertia, or other drag
- number of positions
- direction of movement
- detents and/or stops
- appropriate identification
- compatibility with displays
- size

Displacement is the amount the control has to be moved or rotated to change a setting. All of these characteristics should be considered in relation to the specific workplace when selecting a control. Detailed suggestions for the design of controls and their relationship to the workplace will be found later in this section. First, some general guidelines for workplace layout and control selection are presented (Chapanis and Kinkade, 1972; Ely, Thomson, and Orlansky, 1963).

a. Location

Information about the appropriate heights for controls and displays can be found in Section IIA, "Layout," in Chapter II. The most frequently used controls should be within easy reach. All controls should be placed or guarded so that they will not be accidentally activated.

The following guidelines give some specific recommendations for the location of controls:

- Keep the number of controls to a minimum. The movements required to activate them should be as simple and easy to perform as is possible, except where resistance should be incorporated to prevent accidental activation.
- Arrange the controls at the workplace so that the operator can adjust posture frequently, especially if extended hours of monitoring are required.
- If one hand or foot must operate several controls in sequence, arrange the controls to allow for continuous movement through an arc (if this arrangement does not violate any of the basic rules of control location).
- Assign to the hands controls that require precision or high-speed operation. When there is only one major control that, at times, must be operated by either hand or both hands, place it in front of the operator, midway between the hands.
- Handedness is important only if a task requires skill or dexterity. If the control requires a precision movement, place it on the right, since most people (about 90 percent of the population) are right-handed (Barsley, 1970).
- Assign to the feet controls that require the application of large forces; otherwise, provide the controls with power assists.
- Distinguish between emergency controls and displays and those that are required for normal operations, using the following techniques: separation, color coding, clear labeling, or guarding. In some instances an emergency mode or special operating position can be built directly into the normal control through the use of a detent, an emergency alarm, or a spring that can only be actuated by exceeding a minimum force. Emergency controls should be easily accessible and within 30° of the operator's normal line of sight.
- If the same relative groupings for major controls and displays cannot be kept, make any exception drastic and obvious.
- To prevent accidental activation of a control, place it away from other frequently used controls, recess it, or surround it with a shield.

b. Spacing

Table IIIC–2 contains recommendations for the separation of common controls on a panel. For other combinations of controls the following factors should be considered (Ely, Thomson, and Orlansky, 1963):

- Requirements for the simultaneous or sequential use of the controls.

Table IIIC–2: Recommended Separations for Various Types of Controls (Adapted from Bradley, 1954.)

Control	Type of Use	Measurement of Separation	Recommended Separation Minimum mm	Minimum in.	Desirable mm	Desirable in.
Push Button	One Finger (Randomly)		12	½	51	2
	One Finger (Sequentially)		6	¼	25	1
	Different Fingers (Randomly or Sequentially)		12	½	12	½
Toggle Switch	One Finger (Randomly)		20	¾	51	2
	One Finger (Sequentially)		12	½	25	1
	Different Fingers (Randomly or Sequentially)		16	⅝	20	¾
Crank and Lever	One Hand (Randomly)		51	2	100	4
	Two Hands (Simultaneously)		76	3	127	5
Knob	One Hand (Randomly)		25	1	51	2
	Two Hands (Simultaneously)		76	3	127	5
Pedal	One Foot (Randomly)	D, d	d = 100 D = 203	4 8	152 254	6 10
	One Foot (Sequentially)		d = 51 D = 152	2 6	100 203	4 8

The recommended distance between two similar controls on a panel or machine is determined by the type of control (Column 1) and the way it must be activated (Column 2). The minimum and optimum separations are given in Columns 4 and 5 for each type of control. Column 3 shows how the separation distances are defined, from the centers or the sides of the controls.

- The body member being used.
- The size of the control and the amount of movement (displacement or rotation).
- Requirements for blind reaching (having to reach the control and grab it without seeing it).
- The effects on the system of inadvertently using the wrong control.
- Personal equipment, such as gloves, that might hinder control manipulation.

c. Shape Coding

Varying the shape, size, and type of controls on a complex control panel may assist the operator in identifying a specific control quickly, and can reduce the potential for error. Shape coding is desirable in areas of reduced illumination where vision is blocked, for example, by parts of production equipment or when job requirements force the operator to look elsewhere. Size coding is less satisfactory. To distinguish controls by size might make the dimensions of several of them inappropriate for the exertion of force or for the precision movements needed. With shape coding of knobs, three to five shapes can be distinguished without visual cues (Jenkins, 1947). Shape coding is discussed in Section IVA on person-to-person information transfer.

d. Control Resistance

The material in this section was developed from information in Chapanis and Kinkade (1972) and Ely, Thomson, and Orlansky (1963).

Some force must always be applied to make a control move. The resistance of the control and device to which it is coupled may be elastic (spring loading), as in a power tool trigger; static and sliding friction, as in a rheostat; viscous damping, as in a dashpot to control motion; inertial, as on a seat belt reel; or the resistance may be combinations of the above. Depending on the kind and amount of resistance, the following effects on performance can occur:

- Altered precision and speed of control operation.
- Changes in the feel of the control.
- Changes in the smoothness of control movement.
- Altered susceptibility of the control to accidental activation and the effects of shock and vibration.

The control resistance should be assessed when selecting or designing controls for specific operations. All controls should be large enough to grasp or activate without exceeding a pressure greater than 150 kilopascals (kPa), or 22 pounds per square inch (psi), on the skin (Rehnlund, 1973).

Additional guidelines for control resistance and design of the control's operation are given below (Chapanis and Kinkade, 1972; Damon, Stoudt, and McFarland, 1963):

- Design control movements to be as short as possible, consistent with the requirements of accuracy and feel. Figure IIIC–3 illustrates this principle for a bar-type knob.

Figure IIIC–3: Examples of Poor and Good Control Movement Design (Adapted from Ely, Thomson, and Orlansky, 1963.)

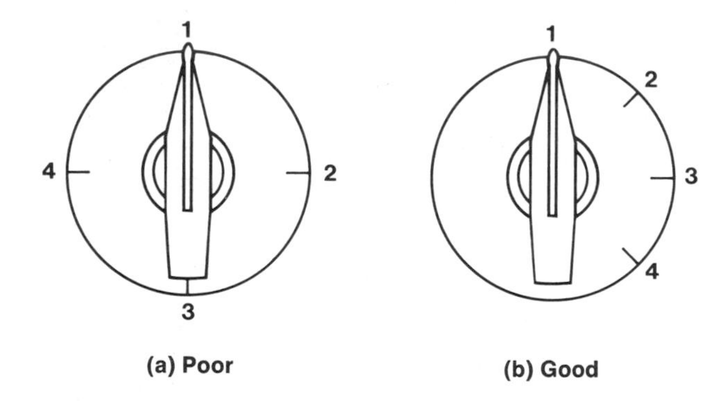

The example of "poor" control movement on the left requires the operator to make a 270-degree rotation of the dial to cover three settings. The "good" control movement alternative requires only a 120-degree rotation to cover the same range. If the differences between settings are large and an error could be critical, however, the "poor" design may be appropriate.

- Provide a positive indication of control activation so that malfunction will be obvious to the operator.
- Provide feedback to the operator from the system that the desired equipment response has taken place.
- Design control surfaces to prevent slippage by the foot, finger, or hand activating them. Knurls or indentations on knobs and roughened, rather than smooth, surfaces for foot pedals and for some buttons are desirable. The choice of knurling and indentation will be a function of the frequency of activation and the forces required.
- Provide an arm or foot support if precise, sustained positioning of controls is required. Avoid static loading of the arm or leg muscles.

- Use controls with enough resistance to reduce the possibility of inadvertent activation by the weight of a hand or foot. The force required to activate a control can be greater if it is activated infrequently or for short periods than if it must be activated for long periods continuously.
- Provide artificial resistance if power assists are used to aid the operator in activating a control.
- Provide a backrest or similar support if a seated operator must push with a force greater than 22 N (5 lbf) on a one-hand control.
- Design the workplace so that the operator can move the trunk and entire body if both hands are required to exert more than 135 N (30 lbf) through more than 38 cm (15 in.) in the fore-and-aft plane.
- Fit control design to the speed, force, and accuracy capabilities of most people, not just the most capable operators. The values given for forces in the section on types of controls (Section e, which follows) have been selected to include the less strong portion of the working population.
- Pay particular attention to the force requirements for activation of infrequently used controls, such as control valves on liquid or solvent lines. Locate these valves from 50 to 100 cm (20 to 39 in.) above the floor whenever possible, so they are accessible and maximum strengths can be applied to them.

Although the information in this section will aid in the selection and location of controls, each application will have its own requirements. An excellent approach to determining the best design is to simulate it and run an experiment to test its suitability, using psychometric measures (see the section on psychophysical scaling methods in Appendix B, Chapter VI).

e. Types of Controls

The following factors will determine which control is most suitable for a given application:

- The speed and the accuracy of response needed.
- The space available.
- The ease of use.
- The readability in an array of similar controls.
- The demands of other tasks performed simultaneously with control operation.

Table IIIC–3 rates some of the more common controls for several of these factors. Further details of control characteristics and uses follow for toggle switches, push buttons, rotary selection switches, knobs, cranks, levers, valves, handwheels, foot pedals, and data entry keyboards.

Table IIIC–3: Characteristics of Common Controls (Adapted from Chapanis and Kinkade, 1972; Damon, Stoudt, and McFarland, 1963; Murrell, 1965.)

Control	Suitability Where Speed of Operation Is Required	Suitability Where Accuracy of Operation Is Required	Space Required to Mount Control	Ease of Operation in Array of Like Controls	Ease of Check Reading in Array of Like Controls
Toggle Switch (On-Off)	Good	Good	Small	Good	Good
Rocker Switch	Good	Good	Small	Good	Fair[1]
Push Button	Good	Unsuitable	Small	Good	Poor[1]
Legend Switch	Good	Good	Small	Good	Good
Rotary Selector Switch (discrete steps)	Good	Good	Medium	Poor	Good
Knob	Unsuitable	Fair	Small–Medium	Poor	Good
Crank	Fair	Poor	Medium–Large	Poor	Poor[2]
Handwheel	Poor	Good	Large	Poor	Poor
Lever	Good	Poor (H) Fair (V)	Medium–Large	Good	Good
Foot Pedal	Good	Poor	Large	Poor	Poor

[1] Except where control lights up for "on."
[2] Assumes control makes more than one revolution.
H = Horizontal
V = Vertical

A summary of the suitability of ten different controls (Column 1) for five different conditions or job requirements (across the top) is given. The ratings are based on typical examples of each control type, not on the extremes of performance in each range. The types of controls that should be considered when the design or work situation has certain requirements, such as speed, accuracy, ease of operation, ease of reading, or limited space on a control panel, can be determined from this summary.

(1) Toggle Switches

Toggle switches are most commonly used when an operation has only two options (on or off) and when control panel space is limited. Three-position toggle switches are available (e.g., off, low, and high) but cannot be operated with as much speed as those with only two positions. Figure IIIC–4 shows the recommended dimensions and forces for toggle switches.

Figure IIIC–4: Toggle Switch Characteristics (Adapted from Department of Defense, 1974.)

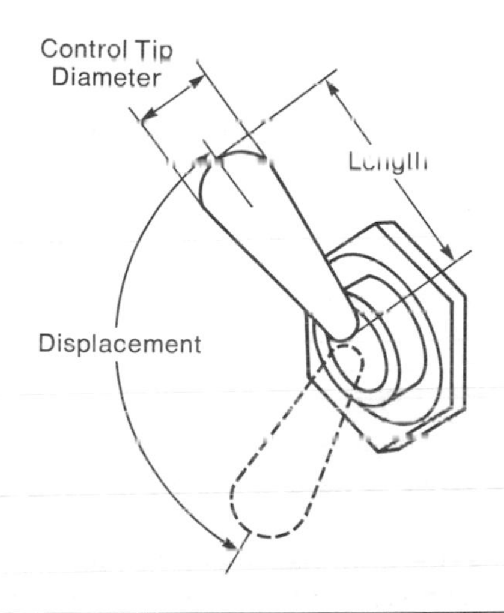

Parameter	Recommended Design Values (Minimum–Maximum)	
Control Tip Diameter	3–25 mm	0.12–1.00 in.
Length		
normal	12–50 mm	0.5–2.0 in.
if operator wears gloves	38–50 mm	1.5–2.0 in.
Displacement		
2-position switch	30°–120°	
3-position switch	18°–60°	
Resistance		
normal	3–11 N	10–40 oz
if control tip is small	3–5 N	10–16 oz

The recommended range of dimensions and resistances for two- and three-way toggle switches is given. The minimum resistance is specified to reduce the potential for accidental activation of the switch. The minimum and maximum displacements of the control are also specified.

(2) Push Buttons

Push buttons are frequently used to enter information into a piece of equipment where each button represents a separate response, as in selecting a beverage from a vending machine. Push buttons must be accompanied by a display so that the operator can observe the results of control activation. Legend switches are often used because they light up when turned on. Membrane-switch push buttons may also be used (see the discussion on keyboards for data entry later in this section).

In many applications push buttons may be operated sequentially by alternate fingers. In this type of task, guidelines for keyboard design (see later) should be used to determine spacing and force requirements.

Large push buttons are available that can be operated by the heel of the hand or the hip in assembly and packing operations. Since these push buttons can result in repeated trauma to soft tissue, they are not recommended if a foot pedal or finger-operated control can be used. Automatic counting devices employing photocells or weight checks are preferable to hand- or foot-operated controls in many of these operations.

Figure IIIC–5 contains further information about push-button characteristics.

(3) Rotary Selector Switches

Rotary selector switches are useful for applications where from 3 to 24 values must be selected and where accuracy is needed. Because there is a preset detent for each value, the selections can be made accurately and quickly. These switches require more space for operation than toggle switches do since room must be made for the fingers. The selector may be either a bar or a round knob, the former being preferred on panel boards with a large number of similar controls so that the values are easily seen. Figure IIIC–6 illustrates some rotary selector switches.

The following two guidelines will help to reduce the potential for error in using rotary selector switches:

- Avoid selections that are 180° apart. Use only as much of the control's 360° rotation as is needed to accommodate the number of values required.
- Fit stops at the beginning and end of the range of values. The stops allow the operator to count off the appropriate number of detents if visual control of the selection is not possible.

Figure IIIC–5: Push-Button Characteristics (Adapted from Department of Defense, 1974; Moore, 1975; Murrell, 1965.)

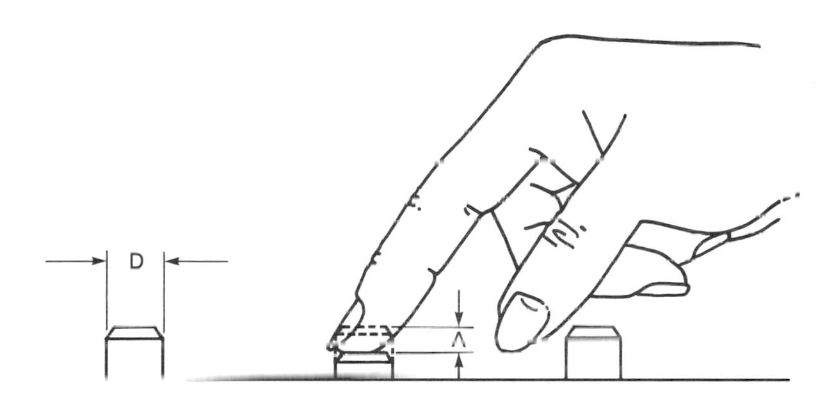

Parameter	Recommended Design Values (Minimum–Maximum)	
Diameter (D)		
Fingertip activation	10–19 mm	0.38–0.75 in.
Palm or thumb activation	19–NA mm	0.75–NA in.
Emergency push buttons	not less than 25 mm	not less than 1.0 in.
Displacement (A)		
Finger activation	3–6 mm	0.12–0.25 in.
Palm or thumb activation	3–38 mm	0.12–1.50 in.
Resistance		
Finger activation	2.8–11 N	10–40 oz
Thumb activation	2.8–22.7 N	10–80 oz

Note: NA indicates data are not available.

Recommended diameter (D), displacement (A), and resistance ranges for a standard push button are shown. Distinctions are drawn between push buttons operated with the index or middle finger and those activated by the thumb or palm. Maximum diameter is not indicated for the latter condition because it varies with the location of the push button in the workplace.

Figure IIIC–6: Examples of Rotary Selector Switches

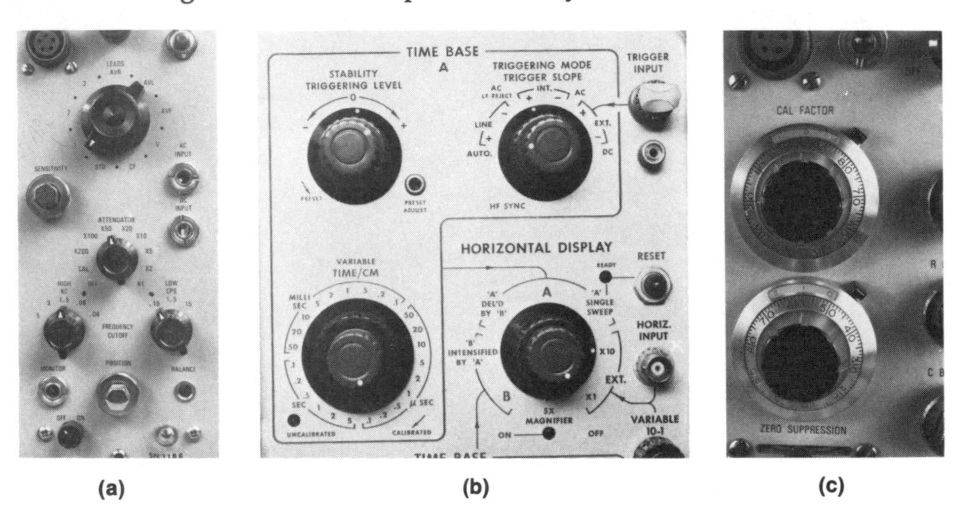

(a) (b) (c)

Several rotary selector switches are shown. The switches on the left have specific stops for each setting; the switches on the right have continuous settings. The switches in the middle of the picture are double controls: each one has an upper and lower component. The lower switch has specific stops and is a gross-setting control; the upper control is for fine adjustment and provides continuous settings.

Further specifications for bar-type rotary selectors are given in Figure IIIC–7.

(4) Knobs

Knobs extend the range of rotary selector switches since they can be rotated through more than 360° and can be moved through a continuous, rather than discrete, series of settings. They should be designed so that the fingers do not obscure the scale, and they should be mounted on the control panel with adequate clearance to allow proper grasping. Adequate clearance is particularly necessary for knobs where forces to activate them are near the maximum values (see Table IIIC–4). Figure IIIC–8 provides information on the recommended design of knobs.

The torques that can be applied to knobs of different diameters and depths are presented in Table IIIC–4. This table shows that a 5-cm (2-in.) diameter knob is preferable to a smaller, or larger, one for fingertip control, and that setting the knob out 2.5 cm (1 in.) from the panel surface improves the ability to exert force on it. The values given in Table IIIC–4 are not the recommended values but maximum torques that can be applied by most people. For frequent operation of a control, the tabulated values should be cut in half for the appropriate design limits.

Figure IIIC–7: Recommended Design Criteria for Rotary Selector Switches (Adapted from Department of Defense, 1974.)

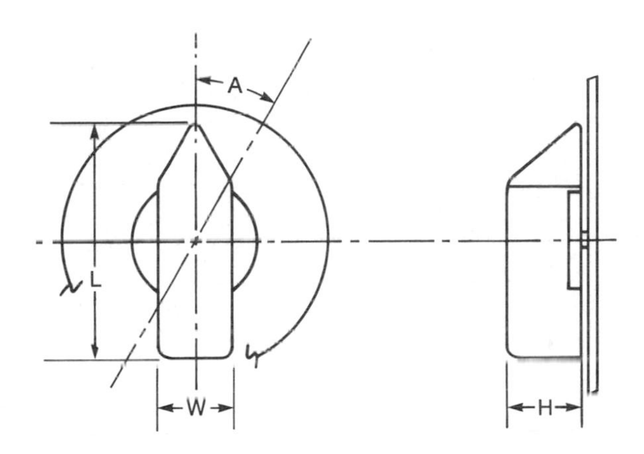

Parameter	Recommended Design Values (Minimum–Maximum)	
Dimensions		
Length (L)	25–100 mm	1.0–4.0 in.
Width (W)	NA–25 mm	NA–1.0 in.
Depth (H)	16–75 mm	0.6–3.0 in.
Displacement (A)		
Closely grouped controls	15°–40°	
Widely separated controls	30°–90°	
Resistance:	0.110–0.675 N · m	1–6 lbf · in.

Note: NA indicates data are not available.

Recommended dimensions (L, W, and H), displacement (A), and resistance ranges for bar-type rotary selector switches are given. Minimum widths (W) are not given since this value will vary with the characteristics of the material used to fabricate the switch. The marks at either end of the displacement path represent stops.

Figure IIIC–8: Knob Design Recommendations (Adapted from Chapanis and Kinkade, 1972; Kellermann, van Wely, and Willems, 1963.)

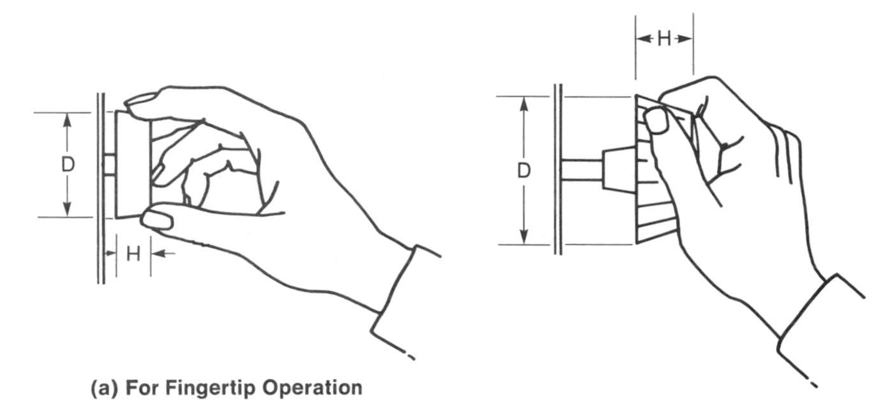

(a) For Fingertip Operation

(b) For Palm Grasp (Star Pattern or Knurled Knob)

Fingertip Operation

Parameter	Recommended Design Values (Minimum–Maximum)	
Diameter (D)	10–100 mm	0.4–4.0 in.
Diameter minimum, for very low torque	6 mm	0.2 in.
Depth (H)	12–25 mm	0.5–1.0 in.

Palm Grasp

Parameter	Recommended Design Values (Minimum–Maximum)	
Diameter (D)	35–75 mm	1.5–3.0 in.
Depth (H), minimum	15 mm	0.6 in.

The recommended diameter (D) and depth (H) for knobs operated either by fingertips (part a) or with full palmar grasp (part b) are shown. The palmar grip design is appropriate for controls operating valves or other devices where fairly large forces (or torques) have to be developed. Information about maximum forces for each type of knob operation is given in Table IIIC–4.

Table IIIC–4: Maximum Torques That Can Be Applied to a Round Knob as a Function of Knob Diameter and Depth (Developed from information in Woodson, 1981, adjusted for sex and age.)

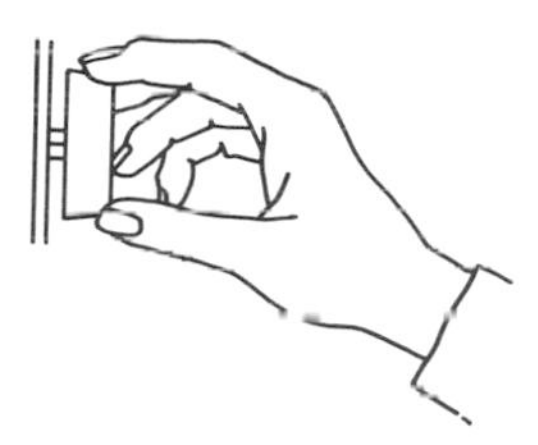

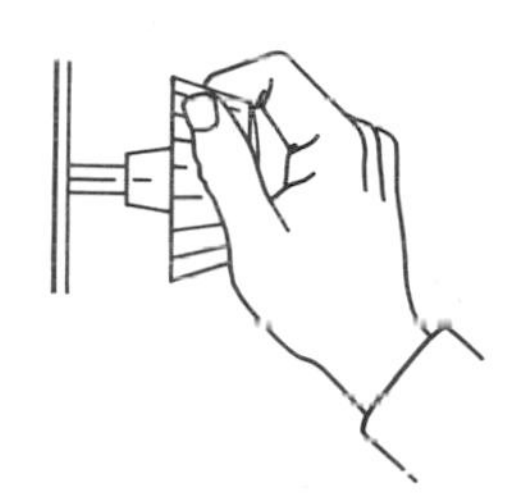

Knob Diameter	Knob Depth, Precision Grip		Knob Depth, Power Grip	
	12 mm (0.5 in.)	25 mm (1.0 in.)	12 mm (5.0 in.)	25 mm (1.0 in.)
	Maximum Torques			
12 mm (0.50 in.)	0.43 N · m (3.8 lbf · in.)	0.51 N · m (4.5 lbf · in.)	0.90 N · m (8 lbf · in.)	1.36 N · m (12 lbf · in.)
19 mm (0.75 in.)	0.51 N · m (4.5 lbf · in.)	0.68 N · m (6.0 lbf · in.)	1.70 N · m (15 lbf · in.)	2.49 N · m (22 lbf · in.)
25 mm (1.00 in.)	0.68 N · m (6.0 lbf · in.)	0.85 N · m (7.5 lbf · in.)	5.08 N · m (45 lbf · in.)	6.10 N · m (54 lbf · in.)
38 mm (1.50 in.)	1.11 N · m (9.8 lbf · in.)	1.27 N · m (11.2 lbf · in.)	7.12 N · m (63 lbf · in.)	10.17 N · m (90 lbf · in.)
51 mm (2.00 in.)	1.70 N · m (15.0 lbf · in.)	2.03 N · m (18.0 lbf · in.)	11.10 N · m (99 lbf · in.)	13.22 N · m (117 lbf · in.)
76 mm (3.00 in.)	0.51 N · m (4.5 lbf · in.)	0.51 N · m (4.5 lbf · in.)	14.24 N · m (126 lbf · in.)	16.27 N · m (144 lbf · in.)

The maximum forces (torques), in newton-meters (N · m) and pound-force inches (lbf · in.), that can be generated by most people in turning round knobs of different diameters (column 1) and depths (across the top) are presented. Both precision (columns 2 and 3) and power (columns 4 and 5) grips are shown. For precision control both very small (<25 mm, or 1 in.) and large (76 mm, or 3 in.) knob diameters put the hand at a biomechanical disadvantage, so less force can be developed than at intermediate values. For power grip control the larger knob results in more palmar support and more force development. Too little depth, however, can limit grip stability and reduce the amount of force that can be developed or maintained.

(5) Valves

Detailed information about the design of valve handles for controlling liquid flow is not available in the literature. The guidelines in this section have been developed from field observations and data on hand anthropometrics and strengths. Some typical valve types are shown in Figure IIIC–9, which also

Figure IIIC–9: Criteria for Valve Design (Adapted from Woodson, 1981.)

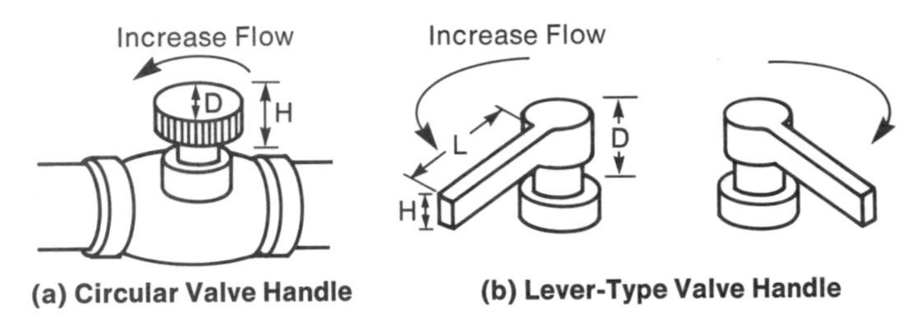

(a) Circular Valve Handle **(b) Lever-Type Valve Handle**

Circular Valve Handle

Parameter	Recommended Design Value	
Diameter (D) minimum	35 mm	1.5 in.
Depth (H) of handle, minimum	25 mm	1.0 in.
Activation force, maximum, for 35-mm (1.5-in.) handle diameter or length	7.12 N·m	63 lbf·in.

Lever-Type Valve Handle

Parameter	Recommended Design Value	
Length (L) minimum	35 mm	1.5 in.
Height (H) of lever, minimum	16 mm	0.6 in.
Depth (D) of handle, minimum	25 mm	1.0 in.
Activation force, maximum, for 35-mm (1.5-in.) handle diameter or length	7.12 N·m	63 lbf·in.

Two common types of valves are illustrated, typical of those found in plumbing and other fluid control lines. The recommended minimum dimensions for circular (part a) and lever-type (part b) valve handles are shown. In part a, the circular handle, D represents the knob diameter and H represents the distance from the top of the knob to the valve body. In part b, the lever type, L represents the lever length, H represents the handle-to-valve body depth, and D represents the vertical height of the valve handle. The maximum force for activating a valve handle is based on the 35-mm (1.5-in.) minimum dimension given.

gives recommended dimensions and maximum torques. Most valves have torques in the range of 0.56 to 4.52 N · m (5 to 40 lbf · in.) when first installed (S. H. Rodgers and R. H. Jones, 1981, Eastman Kodak Company).

Additional guidelines for the selection and installation of valves follow:

- Consistency in the direction of operation of two- and three-way valves is very important in order to minimize human error. Specification of rotation direction should be given whenever ordering valves.
- Labeling of the valve ports on multiple-way valves is recommended, but chemicals may obscure the labels if there is frequent use of the valves. Thus consistency in the assignment of valve ports to specific functions is desirable across a chemical production system.
- The way a valve is mounted (e.g., stem handle up, down, or to one side) will influence how an operator grasps it for turning. Method of grasping, in turn, will determine how much force can be applied to it to break open a corroded, or frozen, valve. Maximum muscle strength for valve activation is available at 51 to 114 cm (20 to 45 in.) above the floor and within 38 cm (15 in.) of the front of the operator's abdomen. When locating valves on equipment, the designer should specify clearances so that the operator can easily access these and other controls.

(6) Cranks

Cranks take up a large amount of space, but they have the advantage of providing either fine or coarse adjustment over a wide range. For fine adjustment the crank grip should not rotate. But for grosser adjustments the grip should rotate so that the wrist and hand can be kept in optimal alignment throughout the rotation.

The recommendations given in Figure IIIC–10 are for cranks with radii between 3.8 and 19 cm (1.5 and 7.5 in.). For larger cranks (12-to-20-cm, or 5-to-8-in., radii), the maximum torques will increase, the values being most affected by the height of the crank above the floor, the speed of operation, and the forward reach required at its furthest travel from the operator. If rapid turning is required, peripheral forces should be kept below 45 N (10 lbf), so that most people's strength capacities will not be exceeded (Chapanis and Kinkade, 1972).

(7) Levers

The material in this section was developed from information in Kellermann, van Wely, and Willems (1963).

Levers are useful in providing accurate adjustment over a small range, and they are useful in situations where simultaneous operation of controls is needed. Knobs or locking devices can be mounted on a lever to permit an operator to manipulate two controls with one hand.

Figure IIIC–10: Recommendations for Crank Design (Adapted from Ely, Thomson, and Orlansky, 1963; Murrell, 1965.)

Torque Range		Minimum Crank Radius Orientation of Handles		
		Horizontal or Vertical on Side,	Vertical and Facing the Operator	
N · m	lbf · in.	91 cm (36 in.) above floor	91 cm (36 in.)	122–142 cm (48–56 in.)
0–2.3	0–20	3.8 cm (1.5 in.)	3.8 cm (1.5 in.)	6.4 cm (2.5 in.)
>2.3–4.5	>20–40	11.4 cm (4.5 in.)	6.4 cm (2.5 in.)	6.4 cm (2.5 in.)
>4.5–10.2	>40–90	19.0 cm (7.5 in.)	11.4 cm (4.5 in.)	11.4 cm (4.5 in.)
>10.2	>90	19.0 cm (7.5 in.)	11.4 cm (4.5 in.)	19.0 cm (7.5 in.)

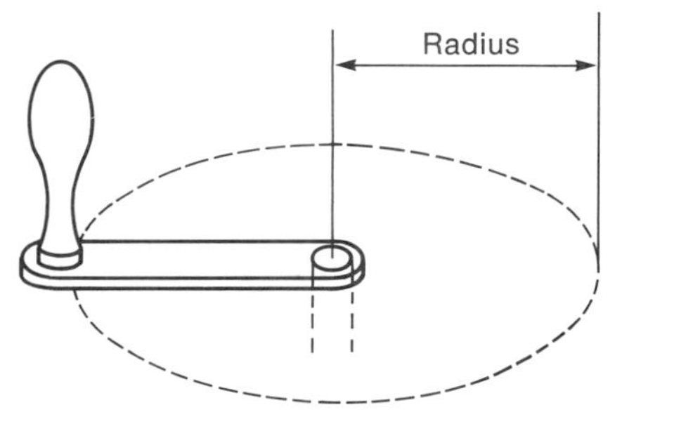

The minimum crank radius, in centimeters and inches, needed to exert force (torque) in four ranges (columns 1 and 2) and in different orientations (columns 3–5) is shown. Torque is expressed in newton meters and pound-force inches. Three different crank handle orientations or locations are shown. In column 3 the orientation is either horizontal or vertical on the side of the surface instead of in front of the operator, at 91 cm (36 in.) above the floor. In column 4 the crank is in a vertical position facing the operator, at 91 cm (36 in.) above the floor. In column 5 it is in the same position but 122–142 cm (48–56 in.) above the floor. Crank radii should not be less than the values given for each torque range, but larger cranks may be used. At radii greater than 25 mm (10 in.), vertical cranks mounted on the side of a piece of equipment may require excessive reaches for operation.

The selection of lever length depends on the task to be done. Long levers require relatively less force in operation than short ones and permit more linear arm motion. For small lever displacements (less than 30°), a straight stick is suitable. For larger displacements, a ball grip or T-handle should be used.

Levers necessitating considerable force should be activated at shoulder level for standing work, at elbow level for seated work, and preferably somewhat to one side, not directly in front, of the operator. The lever should move toward the axis of the body so that the body is subjected to as little torsion as possible. Location of the lever at all positions should be within the arm reach envelopes given in Section IIA, "Layout," in Chapter II.

Recommendations for lever design are given in Figure IIIC–11. A minimum force of 10 N (2.2 lbf) is recommended to reduce the opportunity for accidental activation of a lever that is activated with a palmar grasp.

(8) Handwheels

The material in this section was developed from information in Kellermann, van Wely, and Willems (1963) and Murrell (1965).

Handwheels are used when considerable forces have to be exerted and two hands are available to exert them; in all other cases knobs or cranks are preferable. Handwheels are slow to activate through multiple revolutions unless equipped with a flywheel. Check readings are not possible because rotations greater than 360° are used. Accuracy can be achieved, but grosser adjustments are usually made.

Recesses in the rim of a handwheel may permit more force to be applied to it than would otherwise be the case by permitting a better grip. However, these recesses should not force a small or large hand to take an abnormal position (see Section III D on hand tool selection and design). Larger wheels should be able to be grasped with the whole hand and should offer a means of support. If there is any risk of uncontrolled movement of the wheel, it should be cast in one mold without multiple spokes.

Vertically displayed handwheels should be placed between 95 and 120 cm (37 and 47 in.) above the floor for standing workplaces. Horizontally displayed handwheels should be located from 125 to 140 cm (49 to 55 in.) above the floor. Access to the handwheels should not be blocked by equipment or other structures in the production system.

Figure IIIC–12 provides criteria for the design of handwheels.

Figure IIIC–11: Criteria for Lever Design (Adapted from Chapanis and Kinkade, 1972; Kellermann, van Wely, and Willems, 1963; Murrell, 1965.)

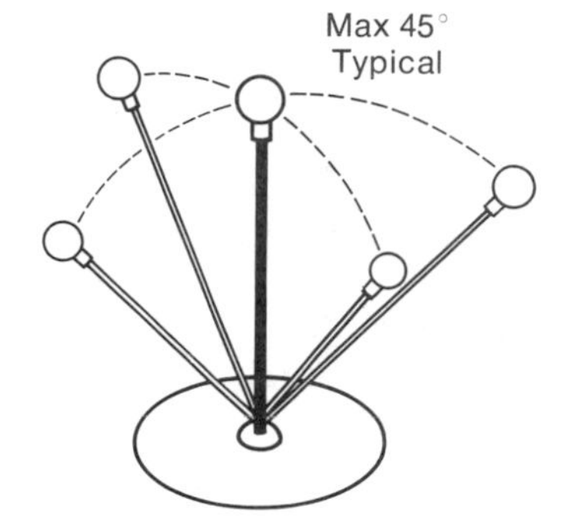

Maximum displacements and operating angles for a floor-mounted lever, such as a gearshift, are indicated, as well as the range of handle diameters for precision (finger) or power (palmar) grasp. The height of the lever handle above the floor (in neutral) is also given for seated and standing operations and the distance from the lever (in neutral) to the front of the body is specified. Maximum forces, in newtons and pounds-force, that can be developed by operators in one-handed operations from the seated position are shown for precision and power grasps.

Parameter	Recommended Design Value	
Operating angle, maximum, in each direction from neutral	45°	
Displacement, maximum		
From front to back	35 cm	14 in.
From side to side	95 mm	37 in.
Diameter of handle, minimum to maximum		
For finger grasp	12–75 mm	0.5–3.0 in.
For palm grasp	38–75 mm	1.5–3.0 in.
Height of lever handle above floor		
Seated operation	75 cm	30 in.
Standing operation	125 cm	49 in.
Distance range in front of body, lever in neutral (assumes optimal height for lever and that person can lean forward if necessary)	50–65 cm	20–26 in.
Recommended maximum forces (one-handed operation)		
Front to back, palm grasp	130 N	29 lbf
Front to back, finger grasp	9 N	2 lbf
Side to side, palm grasp	90 N	20 lbf
Side to side, finger grasp	3 N	0.8 lbf

Figure IIIC–12: Criteria for Handwheel Design (Adapted from Department of Defense, 1974; Ely, Thomson, and Orlansky, 1963; Kellermann, van Wely, and Willems, 1963.)

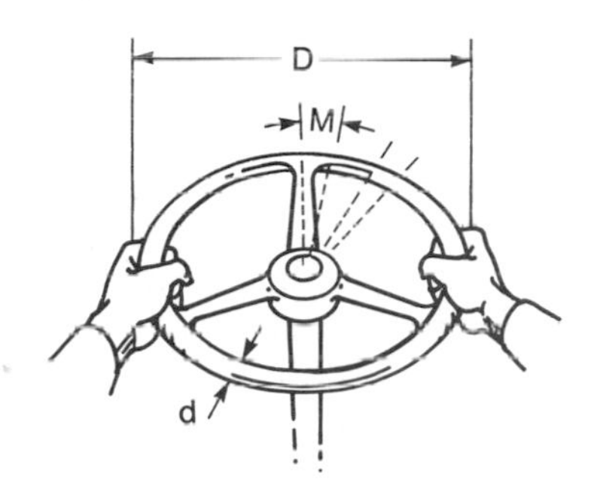

Parameter	Recommended Design Value (Minimum–Maximum)	
Handwheel diameter (D)	18–53 cm	7–21 in.
Rim diameter (d)	20–50 mm	0.8–2.0 in.
Displacement (M), from neutral	60°	
Resistance at rim (tangential force)		
One-hand operation	20–130 N	4–29 lbf
Two-hand operation	20–220 N	4–49 lbf

The recommended range of dimensions (handwheel diameter, D, and rim diameter, d) and displacement (M) for handwheels designed for operation by two hands are given. Minimum-to-maximum tangential forces needed to operate them, with one or both hands, are also shown.

(9) Foot Pedals

Foot-operated pedals leave the hands free to do other work. They are frequently used to keep count in assembly operations (switching pedals) or to operate equipment during packing or assembly tasks when both hands are occupied (operating pedals). In a switching pedal the pedal stroke is usually accomplished with the front of the foot, and small forces and strokes are used. In an operating pedal the force is applied by the whole foot, and the amount that can be exerted is dependent on the holding time and frequency of operation. Very precise force development is best relegated to the hands, not the feet.

Foot pedals are not recommended for standing work, except for very infrequent use. It is unwise to use more than two foot pedals in a seated workplace. Operations that require movement of the feet between controls, such as in playing an organ, are very tiring.

The recommended dimensions for foot pedals and some guidelines for the operating forces, or counterpressures, for switching and operating pedals are given in Figure IIIC–13.

Figure IIIC–13: Foot Pedal Dimensions and Counterpressures (Adapted from Department of Defense, 1974; Ely, Thomson, and Orlansky, 1963; Kellermann, van Wely, and Willems, 1963; Murrell, 1965.)

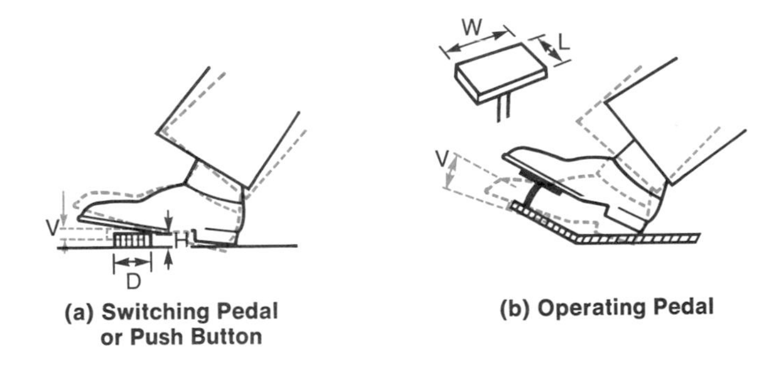

The recommended minimum diameter (D) for a foot-operated push button is given in part a. Part b gives the minimum length (L) and width (W) of a pedal, which vary with the frequency of use of the pedal. In addition, the range of pedal displacement distances (V) for pedals activated by either ankle or whole leg movement is shown, as well as the maximum height (H) above the heel rest for a push button, and the recommended range of ankle motion from the neutral position for any foot-operated control. The minimum and maximum forces, or counterpressures, needed to activate foot controls are given. Minimum forces will increase to 40 N (9.8 lbf) if the foot rests on the pedal. Maximum forces depend on which muscles can be used to activate the pedal; the large muscles of the leg are able to deliver more force than the smaller muscles controlling ankle motion.

Switching Pedal or Push Button (a) Parameter	Recommended Design Value	
Diameter (D)		
Minimum	12 mm	0.5 in.
Preferred	50–80 mm	2–3 in.
Displacement range (V)		
For ankle flexion	12–65 mm	0.5–2.5 in.
For whole leg movement	25–180 mm	1–7 in.
Maximum height of pedal above heel rest (H), lower leg vertical	8 cm	3 in.
Angle of ankle from neutral position, recommended minimum–maximum range, operator seated	20° up, 30° down	
Counterpressures, recommended minimum–maximum	15–75 N	3.3–16.5 lbf

Operating Pedal (b) Parameter	Recommended Design Value	
Minimum Length (L)		
Occasional use	8 cm	3 in.
Constant use	25 cm	10 in.
Minimum width (W)	9 cm	3.5 in.
Displacement Range (V)		
For ankle flexion	12–65 mm	0.5–2.5 in.
For whole leg movement	25–180 mm	1–7 in.
Angle of ankle from neutral position, recommended minimum–maximum range, operator seated	20° up, 30° down	
Counterpressures, recommended minimum–maximum	15–90 N	3.3–19.8 lbf

Pedals that are in constant use should be provided with an adjustable return spring to allow for differences in operator strength and variations in the nature of the work. Pedals that result in overstretching of the ankle joint (more than 25° around the resting position of the foot) are not recommended. The more frequently a foot pedal is operated, the nearer it should be to its minimum force limit.

If the operation of a foot pedal requires very high counterpressures, the pedal should be placed to allow the leg muscles, not just the ankle, to exert the force. Counterpressures greater than 400 N (90 lbf) should not be required on a frequent basis even when the leg is involved, as in operating a brake (Mortimer, 1974).

3. KEYBOARDS FOR DATA ENTRY

Keyboards are rapidly becoming incorporated into the production as well as the office workplace. Both switch-activated and pressure-sensitive keys are in wide use. The following list gives some common examples of keyboards:

- typewriters
- word-processing equipment
- computer data entry/process control equipment
  - card punching
  - computer terminals
  - magnetic tape/disk-encoding equipment
- calculators
- printers and copiers
- phototypesetting equipment

As microprocessors are built into business and consumer products, keyboards will become the major interface for operators in communicating with the equipment. Thus they will become a primary part of the transfer of information from person to equipment. Selection of a keyboard for data entry tasks is influenced by many factors, as shown below (J. A. Stevens, 1975, Eastman Kodak Company):

- nature of the task
- operator proficiency
- keyboard layout
- location of the keyboard in the workplace
- keyboard parameters
  - spacing
  - key size
  - displacement/force
  - interlocks
  - feedback
  - response time

a. The Nature of the Task

The nature of the task can be defined by the frequency of activation of the keys, the duration of use, and the relative importance of accuracy and speed. If not given specific instructions, operators will try for accuracy rather than speed (Howell and Kreidler, 1963).

If the task is highly repetitive, and certain strings of letters and/or numbers recur frequently (as in entering code data into a computer terminal), a keyboard with programmable keys is useful since it permits the string to be entered with a single keystroke. When fewer than 5000 keystrokes are made per shift, a keying operation can be considered occasional. A frequency of 10,000 keystrokes per hour has been recorded for an experienced operator (Klemmer and Lockhead, 1962). Pressure-sensitive keys, such as in touch pads and poke boards, are suitable for occasional keying tasks; they are not recommended for more repetitive operations (Klemmer, 1971).

b. Operator Proficiency

In selecting a keyboard, the designer should keep in mind the experience and proficiencies of the probable users. Experienced numeric data entry operators are three times as fast and make a tenth as many errors as inexperienced operators. About 70 percent of all errors are self-detected (Klemmer, 1971). Full-time operators take at least a year to develop maximum speed and will continue to improve their accuracy at a diminishing rate after that (Klemmer, 1971; Fitts and Posner, 1968). Provision of visual or auditory aids for the occasional or inexperienced operator can improve keying performance and productivity (see the section on video display units in Section III B.) In a laboratory study of a keying task, ten experienced numeric keyset operators produced an average raw error rate (detected plus undetected errors) of 2.2 percent, with a range of from 1 to 4 percent (J. A. Stevens and G. Evan, Eastman Kodak Company). During the test it was observed that the faster operators tended to have less eye movement and suspended their forearms in midair rather than resting them on the table.

c. Keyboard Layout

The layout of the numeric or alphanumeric keyset can influence keying performance. Location of the "0" button on a numeric keyboard should be within the thumb's normal range of motion when the fingers are poised over the keys. The "reset" (or "clear") button should be near enough to be reached by the thumb without lifting the fingers from the keys. Some reorientation of the wrists should be required so that the button will not be activated accidentally. Because keyboard use may be only one of the manual tasks done in a work station, the "0" and "reset" buttons should be accessible to both left and right thumbs.

Figures IIIC–14 and IIIC–15 (Lutz and Chapanis, 1955) illustrate the standard numeric keysets for the touch telephone and the adding machine. The touch telephone layout is associated with slightly higher speeds and lower error rates (Alden, Daniels, and Kanarick, 1972; Lutz and Chapanis, 1955). If a workplace already incorporates an adding machine layout keyset (e.g., a calculator), however, the data entry keyset of additional keyboards should be in the same format in order to be consistent.

Of alphabetic or alphanumeric keysets, the Scholes, or QWERTY, keyboard (see Figure IIIC–16) is the most familiar. With a totally inexperienced set of keyboard users (without previous typing experience), one might consider using the Dvorak keyboard (see Figure IIIC–17). This keyboard was developed to make typing more efficient; work is better balanced between the left and right hands, and more work is assigned to the home keys (the keys with double circles in Figures IIIC–16 and IIIC–17). In tests with inexperienced or specially trained operators, productivity was up to 74 percent higher on the Dvorak keyboard (Lekberg, 1972). The all-encompassing use of the Scholes keyboard has made implementation of the Dvorak keyboard difficult, however. It takes approximately 28 days for Scholes-trained typists to reach the same keying rates on the Dvorak keyboard (Strong, 1956).

For data entry keyboards on computer terminals, the availability of programmable and special-function keys should be considered if appropriate to the task (Pew, Rollins, and Williams, 1976). Redundancy of the numeric keys (across the top of the keyboard and also as a separate numeric set next to the typewriter keyset) can be helpful if a terminal may be used as a calculator as well as an alphanumeric data entry keyboard. Chord keys (Seibel, 1964), which tie together two or more keys to simplify data entry, may be appropriate for specific applications, although the programmable key is more flexible.

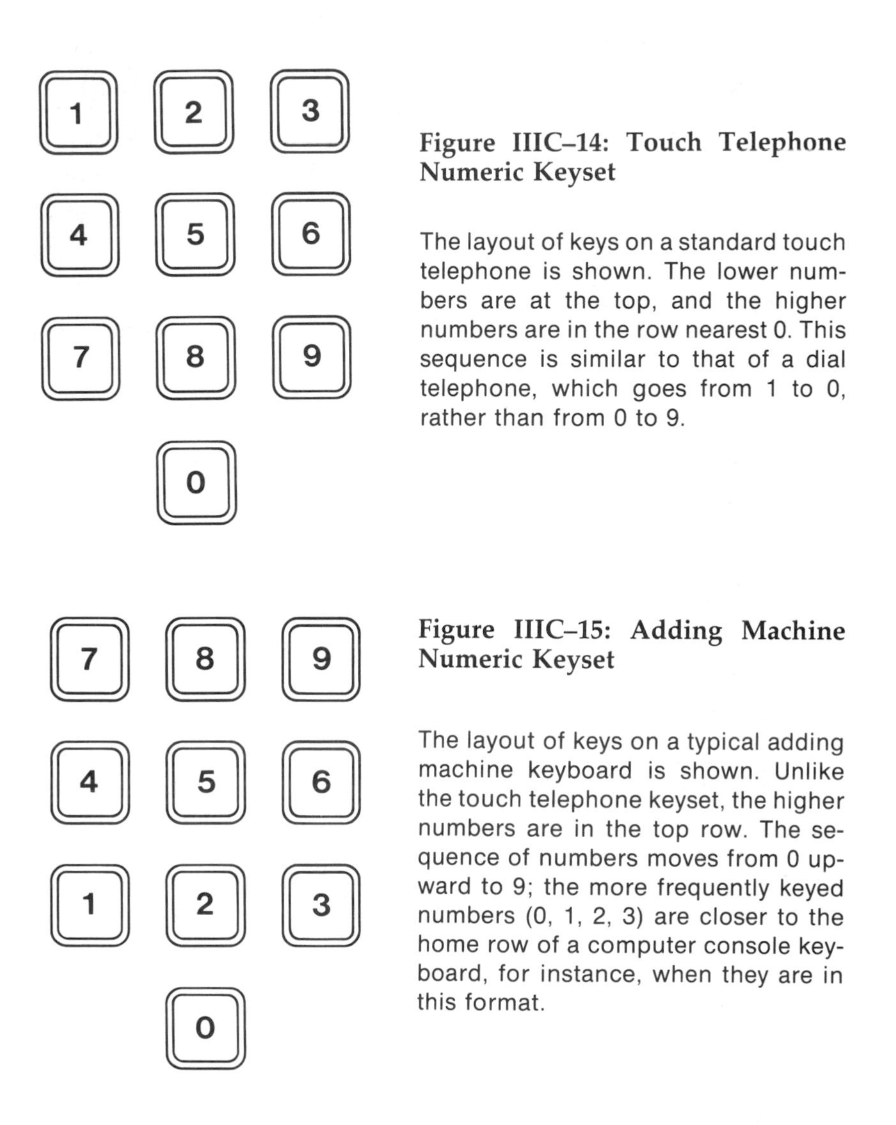

Figure IIIC–14: Touch Telephone Numeric Keyset

The layout of keys on a standard touch telephone is shown. The lower numbers are at the top, and the higher numbers are in the row nearest 0. This sequence is similar to that of a dial telephone, which goes from 1 to 0, rather than from 0 to 9.

Figure IIIC–15: Adding Machine Numeric Keyset

The layout of keys on a typical adding machine keyboard is shown. Unlike the touch telephone keyset, the higher numbers are in the top row. The sequence of numbers moves from 0 upward to 9; the more frequently keyed numbers (0, 1, 2, 3) are closer to the home row of a computer console keyboard, for instance, when they are in this format.

d. Keyboard Parameters

(1) Key Spacing

Key centers should be spaced 19 mm (¾ in.) apart. If the spacing is less than this amount, keying performance is reduced (Alden, Daniels, and Kanarick, 1972; Deininger, 1960).

Figure IIIC–16: Scholes Keyboard (Adapted from Klemmer, 1971.)

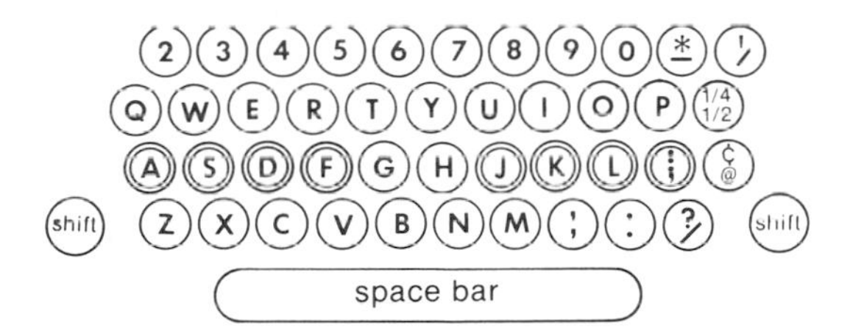

The layout of keys in the common typewriter format (QWERTY) is shown. The format was developed to physically separate keys whose letters are often found adjacent to each other in writing, such as *t* and *h* or *q* and *u*. Separating these keys reduced the possibility of the mechanical typewriter's type fonts jamming. The resting location of the fingers is indicated by double circles on the keys of the second row above the space bar.

Figure IIIC–17: Dvorak Keyboard (Adapted from Dvorak, Merrick, Dealey, and Ford, 1936.)

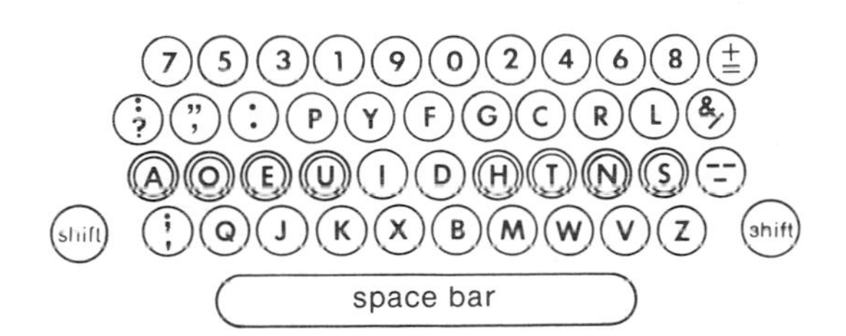

The typewriter keyboard shown here was designed to be more efficient and to permit more rapid typing. The keys are placed according to their frequency of use, with the more frequently used keys being assigned to the stronger fingers. The keys with double circles represent the locations where the fingers rest between key activations.

(2) Key Size

The key tops should be 12 mm (0.5 in.) square. Miniaturized keysets are associated with decreased speed and increased errors (Hufford and Coburn, 1961).

(3) Key Displacement/Force

Figure IIIC–18 illustrates part of a keyset and the dimensions indicated above. The displacement of the key in the vertical plane may range from 1.3 to 6.4 mm (0.05 to 0.25 in.) for a standard key (Kinkead and Gonzalez, 1969) and even less with a poke board or touch pad. It is important that displacement variability between keys be kept to a minimum.

Figure IIIC–18: Recommended Key Design

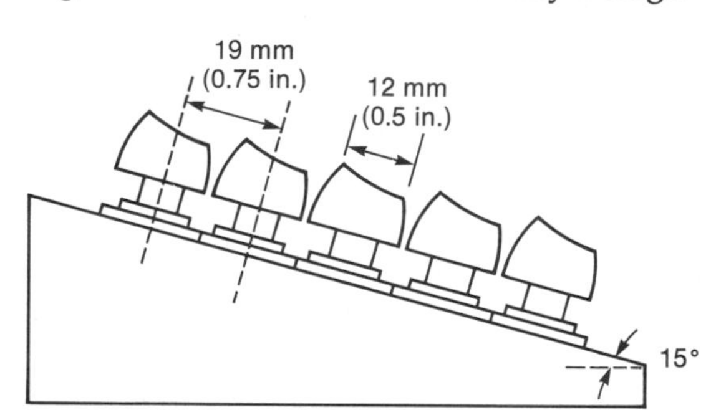

The minimum size of the square key surfaces and the recommended separation (center to center) of keys are indicated. The keyboard angle relative to the horizontal surface it sits on is also specified.

The force needed to activate a key should range from 0.3 to 0.75 N (1 to 2.5 ozf) (Alden, Daniels, and Kanarick, 1972; Kinkead and Gonzalez, 1969; Van Cott and Kinkade, 1972) for repetitive keying tasks. Finger strengths for keying and some performance capabilities have been measured. The range of acceptable forces for key activation is based on the probable frequency of activation and the percent of strength of the weakest digit.

The following information about finger attributes should be remembered in designing keysets:

- The thumb is the strongest, and the little finger the weakest, of the fingers (Haaland, Wingert, and Olson, 1963).
- The thumb and the middle finger are associated with fewer errors than the other fingers (Seibel, 1972).
- The index and middle fingers are most suited to high-frequency activation, such as tapping (Kiemer, 1929).

Poke boards or touch pads are flat keyboards with little key travel; they are often used on equipment where only occasional entries must be made. There must be some minimum activation force required in order to reduce the possibility of activating a key by lightly grazing it in passing. In addition, bounce (multiple switching when single switching was intended) is more likely to occur at lighter activation forces (J. A. Stevens, 1977, Eastman Kodak Company). Poke board forces of about 0.6 N (2 ozf) have been found to be acceptable to experienced operators (J. A. Stevens, 1975, Eastman Kodak Company). Where keying volume is high, poke boards should not be used.

(4) Keyset Interlocks

Keyset interlocks can be built into keyboards to prevent simultaneous activation of two keys at the same time. They keep the key from being triggered until 75 percent of its downward displacement has occurred. Their use has been associated with a marked increase in productivity as well as a considerable decrease in raw errors (Alden, Daniels, and Kanarick, 1972).

(5) Feedback

Feedback to the operator can be tactile, auditory, or visual; all are useful indications that a key has been activated. The tactile feedback, often used in conjunction with an auditory signal, can be a click or snap of the key. Poke boards do not give much tactile feedback and are often equipped with an auditory signal or a light that goes on when they are activated.

The appropriateness of a given keyset and the type of feedback to be used will be determined by the tasks to be performed. For instance, auditory signals indicating key depression would not be appropriate in continuous, high-frequency keying tasks. Touch pads or poke boards may be more appropriate for occasional, low-frequency uses, such as controls on some business machines (J. A. Stevens, 1977, Eastman Kodak Company). Recent advances in flat-panel or membrane keyboards feature raised keys with full travel upon activation (Bishop, 1980).

(6) Response Time

The response time of the key and its associated equipment should not be so long as to reduce the keying speed of a good operator, but it should be sufficiently long to avoid bounce (more than one activation when only one was intended). On the basis of studies with poke boards and standard keysets, the response time (or computer polling time) should be about 0.08 sec for occasional use (J. A. Stevens, 1977, Eastman Kodak Company). This result is a compromise between human poking time (0.06 sec) and the need to avoid bounce (0.10 sec). System response time, such as the time needed for the computer to initiate another prompt, should not exceed 2 sec and preferably be much less (Hinsley and Hanes, 1977).

e. Locating the Keyboard in the Workplace

Glare from the key tops, the angle of the keyboard and its height above the floor, and the issue of whether the keyset can be used by both the right and left hands are all important factors in the selection and use of keyboards. Further information on these topics can be found in Section IIA, "Layout," and in the discussion of illumination in Section VC.

SECTION III D. HAND TOOL SELECTION AND DESIGN

Hand tools are used throughout the manufacturing system to operate, assemble, or repair equipment. Their design can affect the productivity and health of an operator if they do not fit the person or task. In most instances the tools are purchased from an outside vendor. The guidelines given in this section can be used in specifying which tools should be purchased, so less acceptable ones will be avoided.

1. FACTORS OF CONCERN IN HAND TOOL DESIGN

There are five major factors that can affect the health and performance of hand tool users: static loading of arm and shoulder muscles resulting in fatigue and soreness; awkward hand position; pressure on the palm and fingers; vibration and noise exposure with power tool use; and pinch points with double-handled tools. Each factor will be considered in the discussion that follows.

a. Static Muscle Loading: Fatigue and Soreness

When tools are used in situations where the arms have to be elevated or the tools have to be held for extended periods, such as during grinding operations, muscles of the shoulder, arm, and hand may be loaded statically. This loading can result in fatigue and reduced capacity to continue the work, and it may produce soreness in the muscles within a day. The most severe manifestations of these complaints may be tendonitis, tenosynovitis, bursitis, epicondylitis, carpal tunnel syndrome, or deQuervain's syndrome. The causes of these problems are not fully understood. What is known is that forceful, highly repetitive exertions at awkward hand, wrist, or arm postures are associated with increased complaints in susceptible people (Tichauer, 1966). Examples of these situations are given below for the muscle groups most affected.

(1) Shoulder

Abduction of the shoulder (elevating the elbow) will occur if forces have to be applied, or work done, with a straight tool on a horizontal workplace (see Figure IIID–1). An angled tool reduces the need to raise the arm (see Figure IIID–2).

Figure IIID–1: Shoulder Abduction (Elbow Elevation) with Soldering Iron Use (Adapted from Greenberg and Chaffin, 1977.)

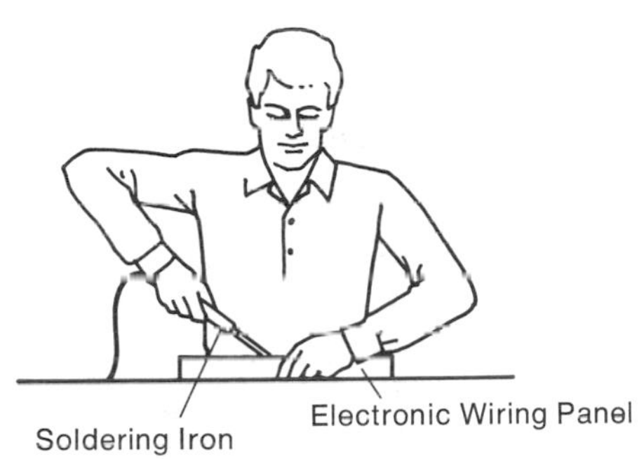

The straight tip of the soldering iron has to be held at an angle to the piece being soldered. Consequently, the operator's elbow is lifted by moving the arm out from the body (shoulder abduction). This action puts a load on the shoulder muscles that can result in fatigue if soldering is done repeatedly during the shift.

Figure IIID–2: Soldering Iron with Tip Designed to Reduce Shoulder Abduction (Adapted from Chaffin, 1973; Tichauer, 1966.)

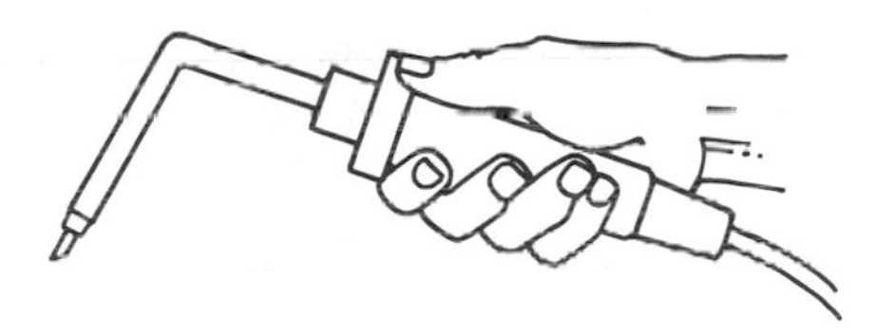

The tip of the soldering iron is bent to a 90° angle. This arrangement permits the operator to approach the piece to be soldered from an angle without having to lift the elbow. Such a tool is most suitable for soldering a piece that is vertically oriented to the work surface.

(2) Forearm

Repetitive work with the arms extended can produce soreness in the forearm when assembly tasks with force are done. For reduction of this problem the workplace can be arranged so that the elbows can be kept at about 90° during the work cycle.

Heavy gripping can be required to keep the hand from slipping off the tool during its use. For improvement of this situation the handle should be designed to minimize the need to grip. For example, the design might use flanges or a knurled surface on the handle. The grip diameter should fit the smaller hand. A handle for two-handed use will help steady a tool or counteract unequal distribution of its weight (see Figure IIID–3).

Figure IIID–3: Electric Drill with Stabilizing Handle (Adapted from Greenberg and Chaffin, 1977.)

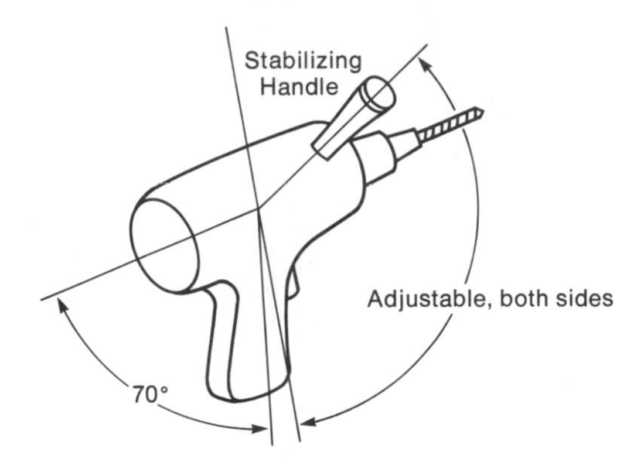

The small power drill shown has been provided with an adjustable, locking handle that is perpendicular to the drill body and at least 10 cm (4 in.) long. This design provides the operator with a handle that can be gripped with the hand that is not activating the trigger and gives increased stability during drilling operations. The handle should be adjustable to accommodate right- or left-hand operation of the trigger and to allow the operator to assume the most stable position. The angle of the primary handle (with the activating trigger) to the drill body is at its optimum position when placed at 70°.

(3) Fingers and Hand

Continuous holding or application of forces results in fatigue and loss of finger flexibility. For a solution to this problem, tool activation forces should be kept low to reduce heavy loading on the fingers. A power grip switch or a bar, instead of a single-finger trigger, is preferable (Konz, 1979a). In addition, a handle should minimize the need for gripping to prevent slippage. A spring-loaded return of a two-handled tool saves fingers from having to return the tool to its starting position.

b. Awkward Hand Positions

Strength is lost as the wrist is moved from its neutral position (see Figure IIID–4). Awkward hand positions may result in wrist soreness and difficulty in sustaining a grip on a tool. To reduce this problem, the workplace designer

Figure IIID–4: Wrist Angles During Tool Use

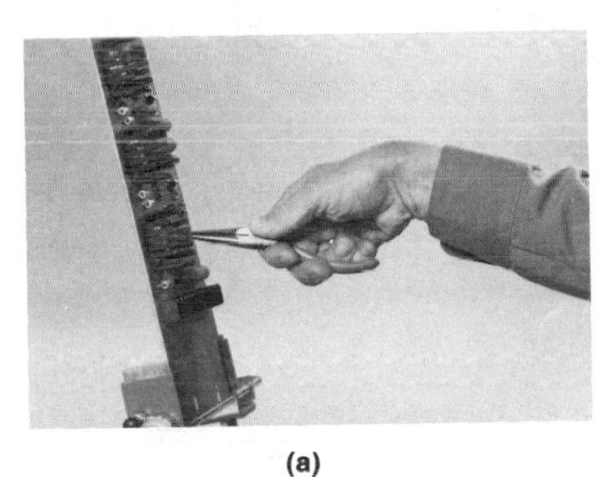

(a)

(b)

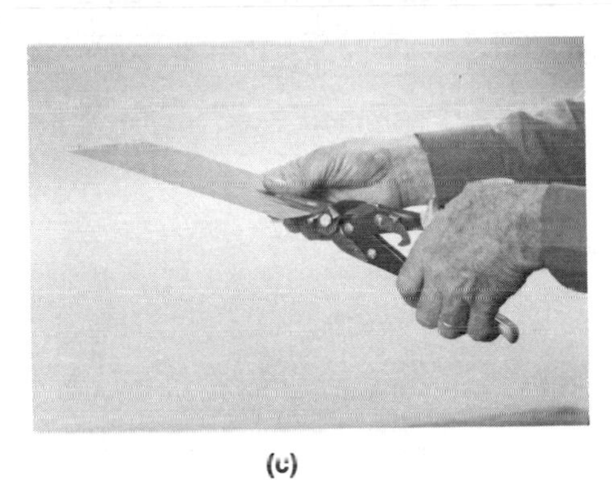

(c)

Assembly, repair, and fabrication activities may require the pliers (parts a and b) or tin snip (part c) user to cock the wrist while flexing the fingers. This posture puts a strong stretch on the tendons passing through the wrist. If the orientation of the workplace is changed, it is often possible to alter the wrist deviation and reduce the stress. A neutral wrist position is shown in Figure IIID–5.

should allow the workpiece to be positioned so as to minimize extreme wrist deviations during assembly or maintenance operations (Terrell and Purswell, 1976; Tichauer, 1966). See Chapter II, "Workplace Design," for more information.

c. Pressure on Tissues or Joints

Pressure can be transmitted to both the palm and the fingers during hand tool use, especially when large forces must be exerted.

(1) Palm

The tool can press into the palm at the base of the thumb, where blood vessels and nerves pass through the hand. This situation may result in some pain and swelling of the hand (see Figure IIID–5). For reduction of this potential for injury, handles for tools should be long enough so that they do not end in the palm, especially for tools used in high-force applications such as pliers and riveters (Greenberg and Chaffin, 1977).

(2) Finger

Force exerted by the fingers (e.g., to hold a trigger, activate a slide switch, or steady a heavy tool) may put high pressure on the skin or joint (>150 kPa, or 22 psi) (Rehnlund, 1973). The most appropriate way to solve this problem is to keep the forces low. The area over which the force is applied should be wide so that the force per unit of area is reduced.

Figure IIID–5: Pressure Point in Palm

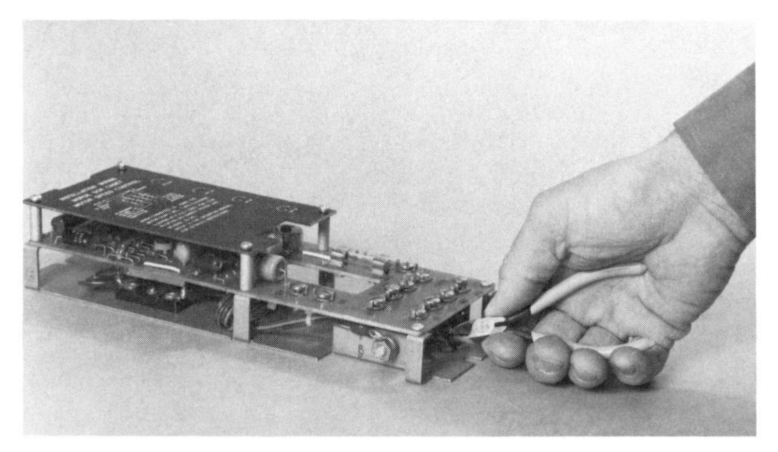

A short pair of wire clippers is used to perform a repair task. Although the pliers are small enough to get into restricted spaces, they do not extend beyond the center of the palm. As force is applied, it is transmitted to the base of the thumb, where nerves and blood vessels pass through the palm. Repeated trauma to this part of the hand may produce soreness or damage in susceptible people.

d. Vibration and Noise

Power tools, such as pneumatic chipping hammers or pavement breakers, are noisy (>75 dB) and vibrate at 60–90 Hz (Ayoub and McDaniel, 1975). At these frequencies sufficient amounts of vibratory energy may be correlated with circulatory problems in the hands (so-called vibration white fingers). The weight of the tool often contributes to vibration injury since the tool has to be gripped tightly to support it. Newer designs reduce noise amplitude by 50 percent and vibration amplitude by 90 percent and are less heavy. Pavement breakers mounted on a power truck platform are available that eliminate the need to handle the tools at all.

e. Pinch Points

The fingers or hand can be caught between the parts of a double-handled tool. This situation is of particular concern where large forces are being exerted. Stops and adequate clearance between the handles are desirable in double-handled tools (Greenberg and Chaffin, 1977); see Figure IIID–6.

f. Other Factors

In addition to considering the five factors above, one should design or select hand tools that can be used with equal ease by right- and left-handed people.

Figure IIID–6: Stops on Two-Handled Tools to Reduce Pinches

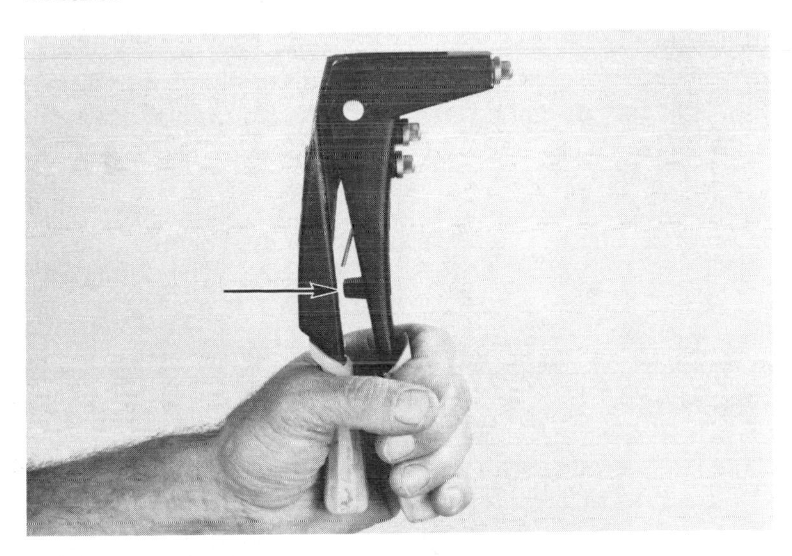

The flat projection (indicated by arrow) on the handle of this pop riveter is designed to keep the handles from meeting along their length, where the hand or fingers could get pinched. Such stops are recommended when tools are used to develop high forces.

One way to reduce the handedness of a hand tool (Laveson and Meyer, 1976) is to locate switches or controls near the center, rather than to one side of the tool. The following questions should be asked when purchasing a hand tool:

- How is the force applied?
- Can either hand hold and operate the tool?
- How is the force used?
- Can the tool be moved in the direction of force equally well by either hand?
- Is the workpiece equally visible when operating with the right and left hands?
- How is the force controlled?
- Can both the right and left hands exert the controlling force with equal ease?

2. DESIGN AND SELECTION RECOMMENDATIONS

The primary concern when selecting or designing a hand tool is to make the tool suitable for the operations done at the workplace. Each workplace may require something different, so compromises often must be made. The recommendations given below are meant as general guidelines only.

a. Handle Length

For tools that involve application of appreciable forces, such as pliers, scissors, or screwdrivers, the following guidelines are recommended (Greenberg and Chaffin, 1977):

- Length of 13 cm (5 in.).
- Minimum length of 10 cm (4 in.) for most operations.
- When tools are used with gloves, add 13 mm (0.5 in.).

b. Handle Diameter

Since grip strength diminishes as the fingers are spread apart, the handle diameter can influence a person's ability to do tasks where a power grip is needed, such as in putting screws into metal sheets, cutting plastic sheeting, or tightening a nut. For circular handles the following dimensions are recommended (R. M. Little, 1977, Eastman Kodak Company):

- Power grips: recommended diameter of 4 cm (1.5 in.); acceptable range of from 3 to 5 cm (1.25 to 2 in.).
- Precision operations: recommended diameter of 12 mm (0.45 in.); acceptable range of from 8 to 16 mm (0.3 to 0.6 in.).

For a cutout handle, such as on a handsaw or a portable electric saw, the width of the handle should follow the dimensions shown in Figure IIID–7.

Figure IIID–7: Handle Dimensions for a Cutout Handle (Adapted from R. M. Little, 1977, Eastman Kodak Company.)

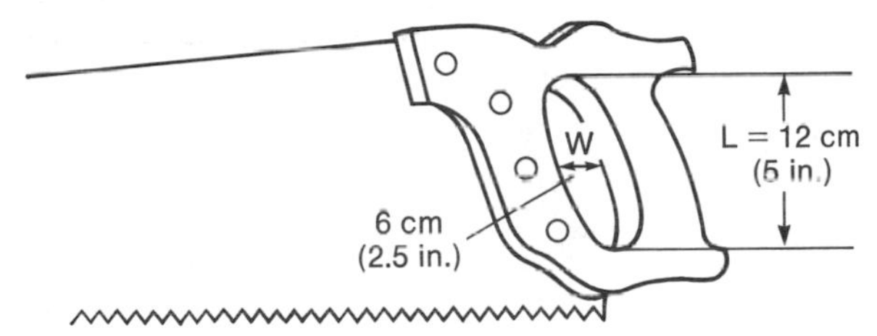

The recommended minimum dimensions (length L and width W) for a handle cut out from a solid surface, such as in a carpenter's saw, are shown. The opening may be angled up to 15° from the vertical to keep the wrist in the neutral position during tool use in the horizontal plane. The handle should allow at least 12 cm (5 in.) of spread (L) for the hand and 6 cm (2.5 in.) of clearance (W) for the gloved hand.

c. Handle Span

For tools with two handles, such as pliers, scissors, clippers, or pop riveters, the recommended distance between the handles at the point of application of the greatest force is from 6.5 to 9.0 cm (2.5 to 3.5 in.). Figure IIID–8 illustrates maximum grip forces as a function of grip span, using the average curve for industrial women. The 6.5-to-9.0-cm (2.5-to-3.5-in.) range is one at which most people should be able to develop maximum grip strengths.

The outside distance between the handles should not exceed 9.0 cm (3.5 in.) for a grip between the third finger and palm at the base of the thumb. The curve of the handles should not be greater than 13 mm (0.5 in.) over their length; note the handles of the pliers and the wire cutters in Figures IIID–4 and IIID–5 (Greenberg and Chaffin, 1977).

d. Other Features of Handle Design

The material in this section was developed from information in Greenberg and Chaffin, 1977 and Konz, 1979b.

The following handle characteristics are desirable:

- Handles should provide good electrical and heat insulation. Compressible rubber and plastic are good heat, and sometimes good electrical, insulators. Wood has a lower heat conductivity but is a poorer electrical insulator than those two substances. Handle temperature should not exceed 35°C (95°F) where prolonged gripping or body contact occurs.

Figure IIID–8: Handle Span for Force Grips (Adapted from Fitzhugh, 1973.)

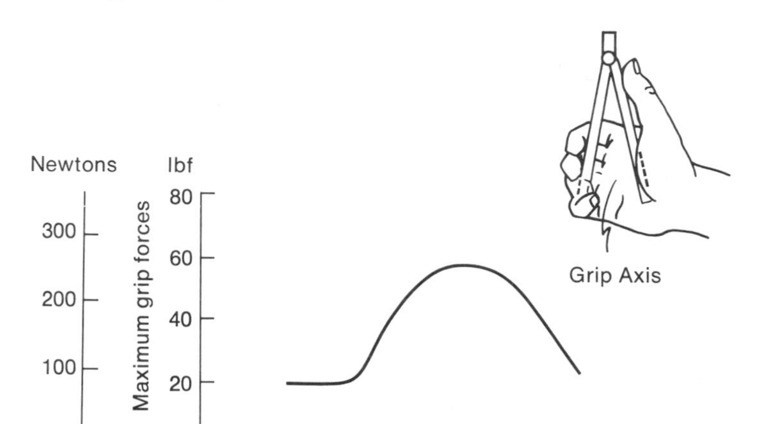

The average maximum grip forces, in newtons and pounds-force, are shown for industrial women at several different grip spans, in millimeters and inches. The force transducer used was a heavy leaf metal spring that fit comfortably between the fingers and the base of the thumb at the intermediate spans (see inset at the right). Less force can be developed voluntarily when the span is small or large because the hand is at a biomechanical disadvantage.

- Handles should not have protruding sharp edges or corners. For a good grip a compressible gripping surface is best.
- Handles should be hard enough to resist embedding of work particles or dirt in the gripping surface.
- Handles should be impervious to absorbing oil, solvents, and other chemicals.
- Handles should not have a polished or highly smooth surface. For pushing or pulling along the tool axis, a slight ripple texture aids in avoiding slippage. For twisting and rotations, shallow longitudinal grooves are best. (See Figure IIID–9.)

Handles should be designed to provide as great a force-bearing area as possible. Avoid formfitting handles with finger recesses. They can force the hand into one position and concentrate pressure between the fingers (see Figure IIID–10).

Figure IIID–9: Handle Design: Grooves and Ridges (Adapted from Greenberg and Chaffin, 1977.)

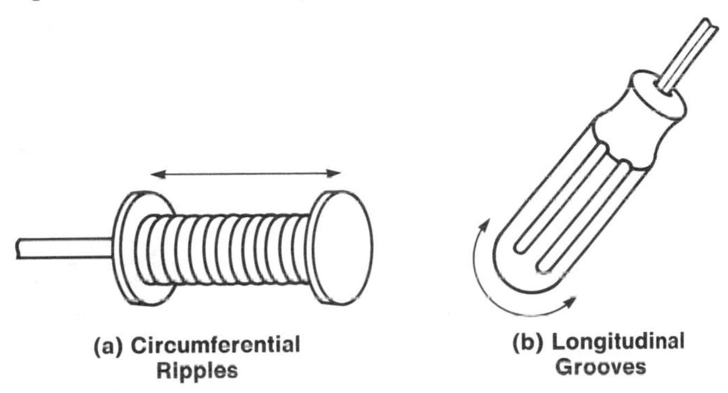

The screwdriver handle in part a has circumferential ripples that increase frictional resistance but do not dig into the fingers. The screwdriver in part b has a handle with shallow, smoothed longitudinal grooves. These grooves reduce slippage by providing additional surface area against which to fix the fingers. If the grooves are not smoothed, these edges become pressure points, making it more difficult to develop forces against them.

Figure IIID–10: Handle Design: Finger Recesses (Adapted from Tichauer, 1966.)

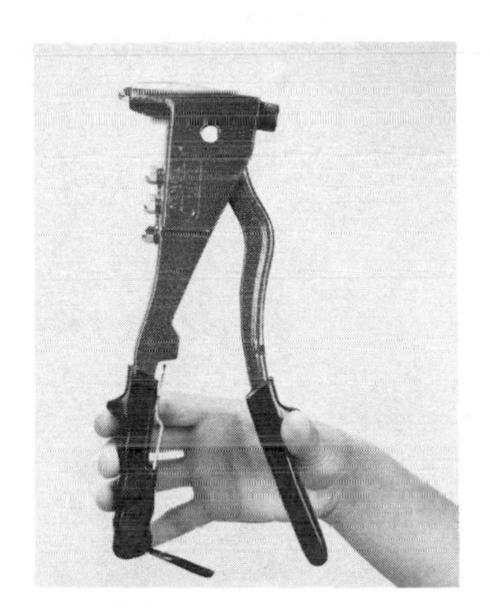

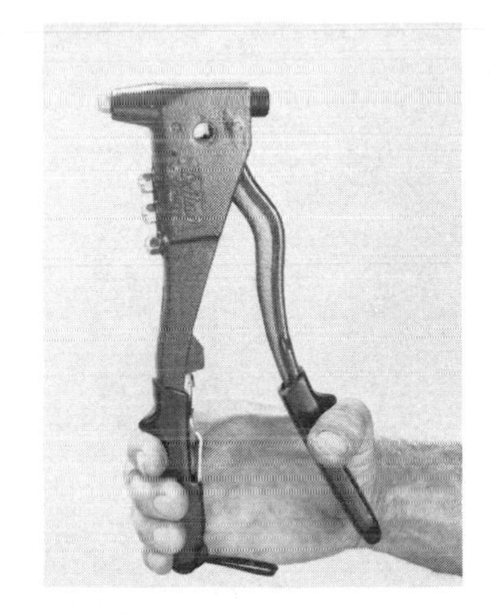

Molded finger recesses on a pair of pliers are shown on the left. The impact of these recesses on the way a person with large hands holds the pliers is illustrated on the right. The fingers may overlap the recesses, thereby pinching the soft tissue and making force exertion painful. A small hand's fingers, as shown on the left, are spread too far apart to be able to develop maximum forces.

e. Switches and Stops

Power tools with on-off switches may have a safety interlock that requires the operator to hold them in the "on" position when using the tool, as in using a portable circular saw. Location of a switch on a tool should take into consideration the need for stabilizing the tool during use. If a thumb switch is used, the tool's weight should be supportable with a power or hook grasp (see Chapter VI, Appendix A).

In the design of the location of power tool triggers, attention should be paid to the center of gravity of the tool relative to the gripping and triggering requirements. Power drills are easier to use if they have an additional handle (see Figure IIID–3) through which the hand not operating the trigger can stabilize the tool (Greenberg and Chaffin, 1977). Table IIID–1 summarizes the weights and trigger-activating forces of some common air and electric hand tools. Lighter tools and lower activation pressures are preferable, so that tool use is easier for most people.

A stop or guard on a hand tool used to exert heavy forces, such as a knife or screwdriver, can take the effort out of trying to stabilize or control it by heavy gripping (Konz, 1979b). Figure IIID–11 illustrates a thumb stop on a pair of pliers.

f. Tool Weight and Suspension

The weight of a hand tool will determine how long it can be held or used and how precisely it can be manipulated. A tool such as a grinder or polisher that

Figure IIID–11: Thumb Stop on a Tool

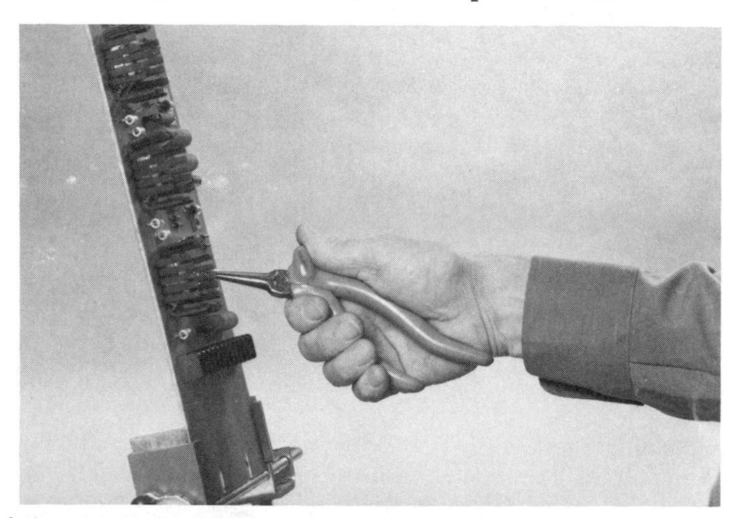

A thumb stop is shown on a pair of needle-nosed pliers designed for use on a vertically displayed assembly task. The stop increases the stability of the hand on the tool, especially when exerting forces tangentially.

Table IIID–1: Power Tool Weights and Trigger Activation Forces (Adapted from R. M. Little, 1981, Eastman Kodak Company.)

Tool Type	Weight		Trigger Type	Grip	Average Force to Activate Trigger	
	kg	lb			N	ozf
$\frac{1}{4}$-in. electric hand drill	2.3	5	Index finger trigger	Pistol	17–22	62–80
$\frac{3}{8}$-in. electric hand drill	4.3	9.5	Index finger trigger	Pistol	30	108
$\frac{1}{2}$-in. electric hand drill	4.5	10	Index finger trigger	Pistol	52	189
7-in. Sander Grinder	7	16	Thumb bar	Straight	33–36	120–130
Air Hammer	7	16	Thumb bar	Straight	10	37
Chopper Hammer	8	17	Thumb bar	Straight	32	115
Air Saw	3	7	Thumb bar	Straight	16	56
Air Saw	3	7	Index finger trigger	Pistol	10	37
Air Angle Drill	3	7	Thumb bar	Straight	9	34

The weights (Columns 2 and 3, in kilograms and pounds) and trigger activation forces (Columns 6 and 7, in newtons and ounces) of several common electric and air tools (Column 1) are given. Trigger types (Column 4) and grip types (Column 5) are also specified. The tools weighing more than 3 kg (7 lb) and requiring activation forces in excess of 10 N (38 ozf) are difficult for many people to use.

may have to be held away from the body in certain operations (e.g., metal-finishing tasks) should be counterbalanced, if possible, to reduce shoulder and arm fatigue.

In general, any tool weighing more than 2.3 kg (5 lb) that has to be operated while supported by the arms and has to be held out from the body in awkward postures is likely to fatigue the small muscles of the forearm and shoulders. For precision operations tool weights greater than 0.4 kg (1.0 lb) are not recommended unless a counterbalancing system is available that does not restrict tool movement by the operator. See Section II B, "Adjustable Design Approaches," in Chapter II for further information. For some jobs, such as drilling concrete, heavier tools may be necessary to help absorb impact vibrations; these tools can often be mounted on a truck to reduce effort for the operator.

g. Special-Purpose Tools

For operations where highly repetitive actions are necessary, where existing tools are not optimal, and where the workplace cannot be adjusted, provision of a specially designed tool for the job should be considered. For example, Figure IIID–2 illustrated an angled soldering iron head developed to reduce awkward wrist positions in some electronic assembly operations. Another special-purpose tool is a toolholder, which permits a person to hold a tool, such as a chisel, away from the fingers (see Figure IIID–12).

Special tools, such as O-ring positioners and terminal-wrapping aids, are often designed to help in assembly operations where very high dexterity would otherwise be needed. The following situations are *not* appropriate ones in which to use a special-purpose tool:

- If the tool is only appropriate for a short part of a multitask cycle and thus becomes an extra tool to pick up and set down frequently.
- If it takes much longer to use the tool than to go ahead with another, less suitable one already in use.
- If the operation is not a continuous one and occurs only occasionally in the work shift.
- If the workspace around the operator is limited and there is no place to set tools between uses.

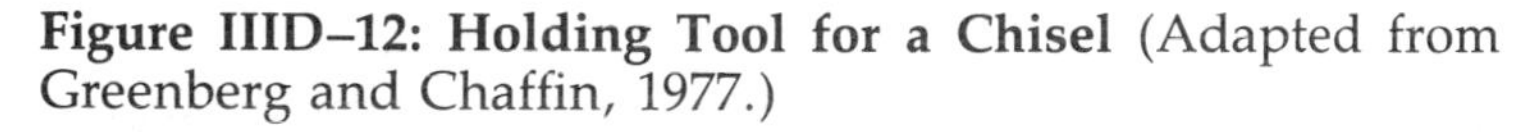

Figure IIID–12: Holding Tool for a Chisel (Adapted from Greenberg and Chaffin, 1977.)

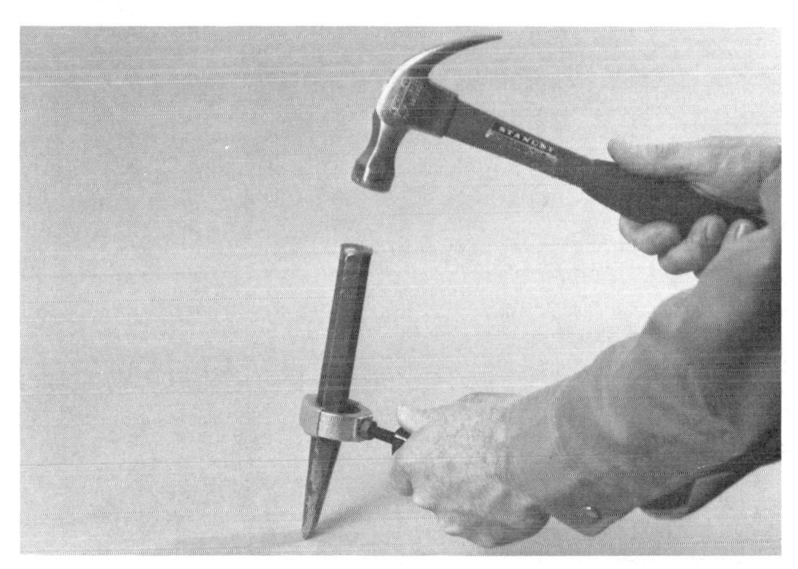

The holding tool is an adjustable clamp. It provides stability for the chisel by allowing the operator to use a power grip to hold the clamp rather than a pinch grip to hold the chisel itself. The hand is kept well away from the hammer's action in driving the chisel into the workpiece or surface.

REFERENCES FOR CHAPTER III

Alden, D. G., R. W. Daniels, and A. T. Kanarick. 1972. "Keyboard Design and Operation: A Review of the Major Issues." *Human Factors, 14 (4):* pp. 275–293.

Ayoub, M. M. 1974. *Hand Held Tools.* Lubbock, Tex.: Texas Tech University, Department of Industrial Engineering, 32 pages.

Ayoub, M. M., and J. W. McDaniel. 1975. *Criteria and Standards for Non-Electric Hand Tools; A Literature Review and Recommendations.* Lubbock, Tex.: NIOSH, DHEW/PHS, Texas Tech University, 41 pages.

Barsley, M. 1970. *Left-Handed Man in a Right-Handed World,* London: Pitman.

Bishop, A. 1980. "Membrane Keyboards Adopt Raised Keys." *Electronics, 44* (August 14, 1980).

Booher, H. R. 1975. "Relative Comprehensibility of Pictorial Information and Printed Words in Proceduralized Instructions." *Human Factors, 17 (3):* pp. 266–277.

Bradley, J. V. 1954. *Control-Display Association Preferences for Ganged Controls.* WADC-TR-54-379, Wright-Patterson AFB, Ohio: Aerospace Medical Research Labs.

Bradley, J. V. 1966. "Control-Display Association Preferences for Concentric Controls." *Human Factors, 8 (6):* pp. 539–543 (condensation of Bradley, 1954).

Brebner, J., and B. Sandow. 1976a. "The Effect of Scale Side on Population Stereotype." *Ergonomics, 19 (5):* pp. 571–580.

Brebner, J., and B. Sandow. 1976b. "Direction-of-Turn Stereotypes—Conflict and Concord." *Applied Ergonomics, 7 (1):* pp. 34–36.

Cakir, A., D. J. Hart, and T. F. M. Stewart. 1979. *The VDT Manual.* Darmstadt: IFRA, 251 pages (also New York: Wiley).

Chaffin, D. B. 1973. "Localized Muscle Fatigue—Definition and Assessment." *Journal of Occupational Medicine 15 (4):* pp. 346–354.

Chapanis, A. 1951. "Studies of Manual Rotary Positioning Movements." *Journal of Psychology, 31:* pp. 51–65.

Chapanis, A., J. S. Cook III, J. D. Folley, and J. W. Altman. 1963. "Design for Ease of Maintenance." Chapter 9 in Morgan et al. (1963), pp. 367–410.

Chapanis, A., and B. A. Gropper. 1968. "The Effect of the Operator's Handedness on Some Directional Stereotypes in Control-Display Relationships." *Human Factors, 10:* pp. 303–319.

Chapanis, A., and R. Kinkade. 1972. "Design of Controls." Chapter 8 in Van Cott and Kinkade (1972), pp. 345–379.

Crawford, B. M., and J. W. Altman. 1972. "Designing for Maintainability." Chapter 12 in Van Cott and Kinkade (1972), pp. 585–631.

Damon, A., H. W. Stoudt, and R. A. McFarland. 1963. "Design Recommendations for Hand and Foot Controls." Section 6.3 in Morgan et al. (1963), pp. 262–275.

Deatherage, B. H. 1972. "Auditory and Other Sensory Forms of Information Presentation." In Van Cott and Kinkade (1972), pp. 123–160.

Deininger, R. L. 1960. "Human Factors Engineering Studies of the Design and Use of Push Button Telephone Sets." *Bell Systems Technical Journal, 39* (July 1960): pp. 995–1012.

Department of Defense. 1974. *Human Engineering Design Criteria for Military Systems, Equipment and Facilities.* MIL-STD 1472B, May 15, 1970, 239 pages.

Dodson, D. W., and N. E. Shields, Jr. 1979. "Development of Display Design and Command Usage Guidelines for Spacelab Experiment Computer Application." In *Proceedings of the 1979 Human Factors Society Meeting,* Boston, Mass., Oct. 29–Nov. 1, 1979, pp. 70–74. Copyright by the Human Factors Society, Inc., Santa Monica, Calif., and reproduced by permission.

Dvorak, A., N. L. Merrick, W. L. Dealey, and G. C. Ford. 1936. *Typewriting Behavior.* New York: American Book, 521 pages.

EEOC, Civil Service Commission, Department of Justice, and Department of Labor. 1978. "Uniform Guidelines on Employee Selection Procedures." No. 6570-06, Part 1607. *Federal Register, 43 (166):* pp. 38290–38345.

Ely, J. H., R. M. Thomson, and J. Orlansky. 1963. "Design of Controls." Chapter 6 in Morgan et al. (1963), pp. 247–280. "Layout of Workplaces." Chapter 7, in Morgan et al. (1963), pp. 281–320.

Fitts, P. M., and M. L. Posner. 1968. *Human Performance.* Monterey, Calif.: Brooks/Cole Publ., pp. 17–19.

Fitzhugh, F. E. 1973. *Grip Strength Performance in Dynamic Gripping Tasks.* Occupational Health and Safety Engineering Rpt. Ann Arbor: Univ. of Michigan, Department of Industrial and Operations Engineering. Cited in Greenberg and Chaffin (1977).

Galitz, W. O. 1979. "Debut II—The CNA Data Entry Utility." In *Proceedings of the 1979 Human Factors Society Meeting,* Boston, Mass., Oct. 29–Nov. 1, 1979, pp. 50–54.

Galitz, W. O. 1980. *Human Factors in Office Automation.* Atlanta: Life Office Management Assoc., Inc., 236 pages.

Gladman, R. 1976. "Human Factors in the Design of Visual Display Units." Chapter 6, pp. 137–152. In *Visual Display Units and Their Application,* edited by D. Groves. Guilford Surrey, England: IPC Science and Technology Press.

Gould, J. D. 1968. "Visual Factor in the Design of Computer-Controlled CRT Displays." *Human Factors, 10 (4):* pp. 359–376.

Greenberg, J. 1980. "Human Error: The Stakes Are Raised." *Science News, 117* (February 23, 1980): pp. 122–125.

Greenberg, L., and D. B. Chaffin. 1977. *Workers and Their Tools, A Guide to the Ergonomic Design of Hand Tools and Small Presses.* Midland, Mich.: Pendell Publishing Co., 143 pages.

Grether, W. F., and C. A. Baker. 1972. "Visual Presentation of Information." Chapter 3 in Van Cott and Kinkade (1972), pp. 41–121.

Haaland, J., J. Wingert, and B. Olson. 1963. "Forces Required to Activate Switches, Maximum Finger Pushing Force, and Coefficient of Friction of Mercury Gloves." Honeywell Memorandum, February 23, 1963. Cited in Alden, Daniels, and Kanarick (1972).

Hayman, E. 1969. "Design Criteria for CRT Alphanumeric Displays." In *Proceedings of the International Symposium on Man-Machine Systems, IEEE,* Cambridge, England, 16 pages. Harlow Essex: Cossou Electronics Ltd. September 8–12, 1969.

Hinsley, D. A., and L. F. Hanes. 1977. *Human Factors Design Considerations for Graphic Displays.* Westinghouse R & D Center Rpt. No. 77-IC57-GRAFC-R, 72 pages.

Howell, W. C., and C. L. Kraft. 1959. *Size, Blur, and Contrast as Variables Affecting the Legibility of Alphanumeric Symbols on Radar-Type Displays.* WADC Tech. Rpt. 59-536. WADC Wright-Patterson AFB, Ohio:

Howell, W. C., and D. L. Kreidler. 1963. "Informational Processing Under Contradictory Instructional Sets." *Journal of Experimental Psychology, 65:* pp. 39–46.

Hufford, L. E., and R. Coburn. 1961. *Operator Performance on Miniaturized Decolmantery Keysets.* NEL Report No. 1083. San Diego: U.S. Naval Electronics Laboratory, Dec. 1961.

Jenkins, W. O. 1947. "The Tactual Discrimination of Shapes for Coding Aircraft-Type Controls." In *Psychological Research on Equipment Design,* edited by P. Fitts. Research Rpt. No. 19. Army Air Force, Aviation Psychology Program. Cited in McCormick and Sanders (1982).

Kellermann, F. T., P. A. van Wely, and P. J. Willems. 1963. *Vademecum—Ergonomics in Industry.* Eindhoven, Netherlands: Phillips Technical Library, 102 pages.

Kiemer, E. 1929. "A Revised Keyboard for the Typewriter." Master's thesis, New York University. Cited in Dvorak et al. (1936).

Kinkead, R. D., and B. K. Gonzalez. 1969. *Human Factors Design Recommendations for Touch-Operated Keyboards—Final Report.* Document 12091-fr. Minneapolis: Honeywell.

Klemmer, E. T. 1971. "Keyboard Entry." *Applied Ergonomics, 2:* pp. 2–6.

Klemmer, E. T., and G. R. Lockhead. 1962. "Productivity and Errors in Two Keying Tasks: A Field Study." *Journal of Applied Psychology, 46:* pp. 401–408.

Konz, S. 1979a. *Work Design.* Columbus, Ohio: Grid Press, 592 pages.

Konz, S. 1979b. "Design of Handtools." In *Proceedings of the Human Factors Society Meeting,* Boston, Mass. Oct. 29–Nov. 1, 1979, pp. 293–300.

Kroemer, K. H. E. 1971. "Foot Operation of Controls." *Ergonomics, 14 (3):* pp. 333–361.

Kurke, M. I. 1956. "Evaluation of a Display Incorporating Quantitative and Check-Reading Characteristics." *Journal of Applied Psychology, 40:* pp. 233–236.

Laveson, J., and R. Meyer. 1976. "Left-out 'Lefties' in Design." London: Taylor and Francis, In *Proceedings of the 6th Congress of the International Ergonomics Association,* University of Maryland, July 11–16, 1976, pp. 122–125.

Lekberg, C. 1972. "The Tyranny of QWERTY." *Saturday Review,* September 30,1972, pp. 37–40.

Lutz, M. C., and A. Chapanis. 1955. "Expected Location of Digits and Letters on Ten-Button Keysets." *Journal of Applied Psychology, 39 (5):* pp. 314–317.

McCormick, E. J., and M. S. Sanders. 1982. *Human Factors in Engineering and Design.* 5th ed. New York: McGraw-Hill, 512 pages. Copyright by the Human Factors Society, Inc., Santa Monica, Calif., and reproduced by permission.

Mitchell-Bishop, V. 1979. "Human Operator Performance in Computer Based Message Switching Systems: A Case Study." In *Proceedings of the 1979 Human Factors Society Meeting,* Boston, Mass., Oct. 29–Nov. 1, 1979, pp. 45–49.

Moore, T. G. 1975. "Industrial Pushbuttons." *Applied Ergonomics, 6 (1):* pp. 33–38.

Morgan, C. T., J. S. Cook III, A. Chapanis, and M. W. Lund. 1963. *Human Engineering Guide to Equipment Design.* New York: McGraw-Hill, 609 pages.

Mortimer, R. G. 1974. "Foot Brake Pedal Force Capability of Drivers." *Ergonomics, 17 (4):* pp. 509–513.

Murrell, K. F. H. 1965. *Human Performance in Industry.* New York: Reinhold Publishing Company, 478 pages.

Oatman, L. C. 1964. "Check Reading Accuracy Using an Extended Pointer Dial Display." *Journal of Engineering Psychology, 3:* pp. 123–131.

Pew, R. W., A. W. Rollins, and G. A. Williams. 1976. "Generic Man-Computer Dialogue Specification: An Alternative to Dialogue Specialists." London: Taylor and Francis, In *Proceedings of the International Ergonomics Association 1976,* pp. 251–254.

Plath, D. W. 1970. "The Readability of Segmented and Conventional Numerals." *Human Factors, 12 (5):* pp. 493–497.

Rehnlund, S. 1973. *Ergonomi.* A. B. Volvo Bildungskoncern, 87 pages. Translated from the Swedish by C. Soderstrom.

Seibel, R. 1964. "Data Entry Through Chord, Parallel Entry Devices." *Human Factors, 6:* pp. 189–192.

Seibel, R. 1972. "Data Entry Devices and Procedures." Chapter 7 in Van Cott and Kinkade (1972), pp. 311–344.

Seminara, J., W. Gonzalez, and S. Parsons. 1977. *Human Factors Review of Nuclear Power Plant Control Room Design.* Electric Power Research Institute Rpt. EPRI NP 309, Palo Alto, Calif.

Singleton, W. T. 1962. "The Industrial Use of Ergonomics." In *Ergonomics for Industry: 1.* London: Department of Scientific and Industrial Research, 16 pages.

Smith, S. L., and N. C. Goodwin. 1971. "Blink Coding for Information Display." *Human Factors, 13 (3):* pp. 283–290.

Snyder, H. L., and M. E. Maddox. 1978. *Information Transfer from Computer-Generated Dot-Matrix Displays.* Rpts. DAFC04-74-G-0200 and DAAG 29-77-G-0067. Research Triangle Park, N.C.: U.S. Army Research, 276 pages.

Stewart, T. F. M., O. Ostberg, and C. J. MacKay. 1974. *Computer Terminal Ergonomics, A Review of Recent Human Factors Literature.* 2nd ed. Rpt. Dnr 170/72-5. Stockholm: Statskontoret.

Strong, E. P. 1956. *A Comparative Experiment in Simplified Keyboard Retraining and Standard Keyboard Supplementary Training.* Washington, D.C.: Civil Services Administration, 41 pages.

Terrell, R., and J. Purswell. 1976. "The Influence of Forearm and Wrist Orientation on Static Grip Strength as Design Criteria for Hand Tools." London: Taylor and Francis, In *Proceedings of the International Ergonomics Association,* July 11–16, 1976, pp. 28–32.

Thylen, J. O. 1966. "The Effects of Initial Pointer Position Relative to the Control on Directional Relationships in the Presence of Two Conflicting Stereotypes. *Ergonomics, 9:* pp. 469–474.

Tichauer, E. R. 1966. "Some Aspects of Stress on the Forearm and Hand in Industry." *Journal of Occupational Medicine, 8 (2):* pp. 63–71.

U.S. Air Force. 1955. *Handbook of Instructions for Aerospace Personnel Subsystem Design.* AFSCM 80-3. Washington, D.C. USAF.

U.S. Army. 1975. *Human Factors Engineering Design for Army Material.* MIL-HDBK-759. Washington, D.C. Dept. of Defense.

U.S. Army. 1978. *Human Engineering Equipment Design.* MIL-STD-1472B. Huntsville, Ala. USATDC

Van Cott, H. P., and R. G. Kinkade, eds. 1972. *Human Engineering Guide to Equipment Design.* Rev. ed. Washington, D.C.: American Institutes for Research, 752 pages.

Van Cott, H. P., and M. J. Warrick. 1972. "Man as a System Component." Chapter 2 in Van Cott and Kinkade (1972), pp. 17–40.

Van Nes, F. L., and H. Bouma. 1980. "On the Legibility of Segmented Numbers." *Human Factors, 22 (4):* pp. 463–474.

Vartabedian, A. G. 1970. "Human Factors Evaluation of Several Cursor Forms for Use on Alphanumeric CRT Displays." *IEEE Transaction on Man-Machine Systems, 11 (2):* pp. 132–137.

Vartabedian, A. G. 1973. "Developing a Graphic Set for Cathode Ray Tube Display Using a 7 × 9 Dot Pattern." *Applied Ergonomics, 4 (1):* pp. 11–16.

Warrick, M. J. 1947. "Direction of Movement in the Use of Control Knobs to Position Visual Indicators." In *Psychological Research on Equipment Design,* edited by P. M. Fitts. Research Rpt. No. 19. Army Air Force, Aviation Psychology Program. Washington, D.C. Government Printing Office.

Woodson, W. E. 1981. *Human Factors Design Handbook.* New York: McGraw-Hill, 1072 pages.

Woodson, W. E., and D. W. Conover. 1964. *Human Engineering Guide for Equipment Designers,* 2nd ed. Berkeley: Univ. of California Press. Copyright © 1964 by the Regents of the University of California and reprinted by permission of the University of California Press.

Chapter IV INFORMATION TRANSFER

Contributing Authors

Stanley H. Caplan,
M.S., Industrial Engineering

Richard L. Lucas,
Ph.D., Psychology

Thomas J. Murphy,
Ph.D., Psychology

Chapter Outline

Section IV A. Person-to-Person Transfer

1. Instructions
2. Forms
   a. Sequence
   b. Readability and Comprehensibility
   c. Space and Content
3. Questionnaires
   a. Design of the Questions
   b. Design of the Questionnaire Format
   c. Use of the Questionnaire Data
4. Labels and Signs
   a. Comprehensibility
   b. Legibility
   c. Readability
5. Coding
   a. Alphanumeric Coding
   b. Shape Coding
   c. Color Coding

Section IV B. Product-to-Person Transfer: Visual Inspection

1. Factors Influencing Inspection Performance
   a. Measures of Performance
   b. Individual Factors
   c. Physical and Environmental Factors
   d. Task Factors
   e. Organizational Factors
2. Guidelines to Improve Inspection Performance

Like every other component in a system, the human element is not perfectly reliable; all parts of the human-machine system are likely to malfunction in time. Using analytical techniques, one can frequently calculate the performance reliability of a system by multiplying the probabilities for error of each of the components (Meister, 1971; Swain and Guttmann, 1980). Human error is affected by a number of variables both within and between operators, and it is more difficult to quantify than a system error.

Information can be transferred from person to person, from equipment to person (displays), from product to person (inspection), or from person to equipment (controls). Displays and controls were discussed in Chapter III, "Equipment Design." Thus this chapter focuses on design guidelines for improving the transfer of information between people and in inspection tasks and for reducing the potential for human error.

Qualitative examples of error opportunities for operators are shown in Table IV–1. By identifying these opportunities in each step of the manufacturing process, the designer can usually identify effective approaches to reduce human error.

SECTION IV A. PERSON-TO-PERSON TRANSFER

The transfer of information between people via written instructions, forms, questionnaires, codes, labels, and signs can be subject to error unless the sender and receiver are each interpreting it in the same way. As a person gains more experience in a work situation, misinterpretation of information from instructions or signs will be less likely to occur. However, communications should be designed for the novice or casual visitor to a work area. In addition, since emergency situations often result in reflex reactions rather than analytical troubleshooting, well-designed written information is needed for people with experience in the workplace as well.

1. INSTRUCTIONS

The material in this section was developed from information in S. H. Caplan (1978, Eastman Kodak Company).

There are many profitable areas for human factors input in the design of instruction sets, especially in the choice of words and the use of graphics. Instruction sets should be simple, concise, and clear while giving enough information to allow the operator to make judgments about specific problems that may not be addressed. In the development of instructions for complex systems, the following questions should be considered:

- Who are the users?
- What is the relative importance of each instruction?
- What is the proper location for the instructions?
- How much information redundancy is appropriate?

- What is the most effective method and format of presentation (diagrams, photographs, prose)?
- How intelligible are the instructions? Is the appropriate nomenclature used?

Table IV–1: Examples of Error Opportunities (Adapted from S. H. Caplan, 1975, Eastman Kodak Company.)

Task	Error	Opportunity for Error
Labeling	Mislabel	Illegible product ID* Inadequate label storage or retrieval system
Packaging	Mix Product	Simultaneous handling of similar product Switched control cards Product change: improper clearing of line Shift change: communications failure
Order Picking	Mix Order	Similar items adjacent to each other Similar product IDs*
Sorting	Mix Kinds	Incorrectly identified Poor handwriting Distraction during the task
Transcribing, Keying	Substitute, Transpose, Omit	Poor handwriting Memory overload Incompatibility of format Look-alike characters
Following Instructions	Misinterpret	Poor comprehensibility Poor legibility Poor readability
Filling Out Forms	Enter Information at Wrong Place, Enter Wrong Information	Too much information on form Poor layout of form Lack of instructions
Looking Up Tables	Mistrack Across Columns	Excessive spacing between columns Lack of tracking aids (spaces or lines)
Monitoring Control Panels	Misread Dial, Misjudge Trend	Parallax problem (difficult to line up) Look-alike dials with different scales Information overload: inadequate sampling Inadequate knowledge of system

* ID = identification number or code.

The first two columns specify the task and the error that is most likely to occur. The third column (Opportunity for Error) gives characteristics of the information display or work layout that contribute to errors in common industrial tasks.

Where instructions are to be used by a large part of the general population, it is very important to use words that are easily understood. The *Word Frequency Book* (Carroll, Davies, and Richman, 1971) can be consulted to find out how often certain words are used.

Some general guidelines for the design of instructions are as follows:

- The sequence of the instructions should follow the sequence of actions required.
- Information should be located in specific places in order to avoid misuse or unsafe operations.
- Instructions should be integrated into the equipment or the production work sheet rather than set aside on a separate sheet (Szlichcinski, 1979).
- Short sentences, flow diagrams, algorithms, lists, and tables are superior to prose (Miller, 1975).
- It is usually best to use the active tense and be affirmative than to use the passive tense and be negative. The main topic of the instruction should appear at the beginning of the sentence (Broadbent, 1977).
- The choice between formats is situation-dependent.
- All instructions should be tested on naive users before being finalized.

Table IVA–1 gives an example of the two-column format, which is an appropriate design for many-step instructions. The second column (Further Information) is especially useful during an operator's learning phase on the equipment.

In a list of specifications for service or supply, more than a part number should be given. For instance, the specification should be supplemented with a short description (#23417—2-oz ink remover), which relieves the operator from memorizing the five-digit code when following the instructions. If there are any hazards to warn the operator about, highlight the warnings and describe the potential consequences of not following the instructions.

2. FORMS

The proliferation of paperwork in industry is responsible, in part, for the large number of poorly designed and improperly maintained forms, which often last beyond their useful life cycle. For instance, production-system-monitoring data are logged on special forms to show output; service schedules; temperatures of water, steam, and products; and other information required to control the process. With computer-controlled systems has come the generation of a preformatted screen that makes data entry simulate the act of filling in a form. Before a form is developed for information transfer in the workplace, there are questions one should ask about the need for the form. The Office of Records Management of the U.S. government has issued a forms

Table IVA–1: Use of a Two-Column Format in an Instruction Set (Adapted from S. H. Caplan, 1981, Eastman Kodak Company.)

ACTION STEP	FURTHER INFORMATION
1. Press two-sided button	Check display on panel to see that two-sided light is on
2. Copy first page	Copies will automatically exit into top hopper unless the side hopper light is on
3. Reload copies	Supply drawer unlocks automatically; do not change copy orientation
4. Copy second page	Copies will be delivered to selected exit
5. Repeat steps 2–4	Do for each pair of remaining originals

The first column gives each step in a concise and clear verb-noun format. The second column gives additional information pertinent to each instruction, which may not be needed by an experienced operator.

analysis handbook (Anon. 1960), which gives guidelines for form generation and evaluation. Some of the questions one should ask when developing a form are as follows:

- Is the form needed? Does a similar form already exist?
- Is every item on the form needed? Is it available elsewhere?
- Is every copy (duplicate, triplicate) of the form needed?
- Can the form be combined with another one?
- Is there periodic analysis of the office or shop procedures in which the forms are used?
- Is there a functional file of forms to aid in the review of a new form?
- Is operator involvement solicited in the design of new forms?

Once the need for a new form has been established, the design guidelines presented in the following sections should be used to optimize the information collection and flow (S. H. Caplan, 1978, Eastman Kodak Company).

a. Sequence

- Sequence the items on the form in a logical and easy-to-follow way. Follow standardized item locations where they exist.
- Make the sequence of items on the form follow the sequence of the source document or the production process.
- Consider the clerical routines when determining the order of items on the forms.

b. Readability and Comprehensibility

- Provide clear instructions for filling out the form.
- Make sure all captions are easy to understand and are legible under all conditions of use.
- Use color coding or other highlighting techniques to facilitate handling, checking, routing, or dispatching of the forms.
- Make the margins and filing data correspond to the characteristics of the filing equipment or binders that the forms are stored in. For computer terminals, try to keep the forms to one page (full screen).

c. Space and Content

- Keep the amount of writing to a minimum.
- Provide sufficient space for each answer.
- Provide for overflow or continuation pages if the designated space may be inadequate.
- Design the answer space to take advantage of typewriter or printer characteristics.

3. QUESTIONNAIRES

The material in this section was developed from information in C. Amoroso (1980, Eastman Kodak Company).

Questionnaires may be used in industry to gather data on people's attitudes toward job situations or environmental stresses, such as shift work schedules or thermal comfort. The design of these questionnaires can determine how accurate the data are and how widely the results can be applied. As in the development of psychometric scales (see Chapter VI, Appendix B), any questionnaire should be tested in a pilot study before being used in a large-scale study. This pilot test gives the data collector an opportunity to identify potentially unclear wording or interpretation problems with questions. In addition, any problems of data analysis can be identified early enough to make changes for the final study.

The guidelines presented in the following sections are generally applicable to the design of questionnaires for attitude surveys or information collection. (Guidelines for the physical layout of questionnaires were covered in the section on forms above.) These guidelines are grouped into three categories: design of the questions, design of the questionnaire format to ensure completion and efficient analysis, and use of the data collected.

a. Design of the Questions

- Use short, active, affirmative sentences. Lead with a verb whenever possible (Wright and Barnard, 1975).
- Use indefinite, rather than definite, articles to avoid biasing the response (Loftus and Zanni, 1975). For example, use "Did you see *a* broken window?" instead of "Did you see *the* broken window?"
- Use words that are familiar to the questionnaire respondent. Define terms and abbreviations. Avoid ambiguous terms like "significant" (Freed, 1964).
- Express one thought per question. If there is a following thought, make it a new question (Wright and Barnard, 1975).
- For multiple-choice questions:

  Use independent categories with no overlap.

  For precise information collection, provide seven or nine options. For attitude surveys, three or five options are sufficient.

  Provide equal numbers of options on either side of the neutral point (Douglas and Anderson, 1974).

  Provide an option of "Don't know," "Don't remember," or "No opinion" (Selltiz, Jahoda, Deutsch, and Cook, 1959).

b. Design of the Questionnaire Format

The material in this section was developed from information in Douglas and Anderson (1974) and Goode and Hatt (1952).

- Provide brief and clear instructions in boldface type on the first page of the questionnaire.
- Group items coherently and logically. Move from simple questions to more complex ones.
- Design the questionnaire for easy data analysis. The answers should line up, preferably along the right margin; only one way of responding (circling, checking, or underlining) should be used throughout.
- Design the questionnaire so that it takes no more than 30 minutes to complete; preferably, one should be able to complete it within 15 minutes.

c. Use of the Questionnaire Data

- Identify whether the population surveyed has the necessary information to respond to the questionnaire.
- Provide space on the first page of the questionnaire for information about the respondent (job classification, years of experience on the job, sex, age, height, weight, etc.). This information is needed to analyze the group's responses and to characterize the sample.
- Include a description of the sample that filled out the questionnaire with the report of the responses. Do not extrapolate beyond the population sampled.
- Report the percentage of the original sample that did not respond to the questionnaire, and, if possible, determine why they did not.
- Use analysis techniques that recognize multiple variables affecting a result (multivariate). Multivariate analyses are often superior to univariate (frequency counts) analyses of data; attitude is rarely determined by a single factor.

4. LABELS AND SIGNS

Labels and signs are short messages used to transfer information about policies or equipment use between people. Communication can be enhanced by proper attention to the components of the written communication process (see Table IVA–2). The objective of a communication is to have the receivers understand what the sender meant. The sender is the message designer and must consider the discriminative, interpretative, and recall skills of the receivers and the environmental conditions under which they will receive the message.

There are three factors in message design for labels and signs that enhance communication: comprehensibility, legibility, and readability. Guidelines for improving each of these factors follow.

a. Comprehensibility

Comprehensibility is a measure of how reliably the receiver interprets a message. Among other things, it depends on the person's prior knowledge of a situation and his or her language skills. A message designer can improve comprehensibility by following these guidelines:

- Have in mind the purpose and intended meaning of the message.
- Gear the language to the least knowledgeable user in the probable user population; keep it simple, in any case (Payne, 1951).
- Make the message brief and concise.
- Avoid ambiguous words.

- Use examples to clarify meaning, especially for novel situations.
- Use standardized signs or labels wherever possible.
- Reread messages with a critical eye to be sure that a person seeing them for the first time can get the necessary information on the first reading.
- Avoid using symbols out of context or for inexperienced viewers. Validate the use of symbols experimentally whenever possible (Cahill, 1975).
- When constructing a sentence, use one clause, active (rather than passive) voice, and affirmative (rather than negative) words.

An example of an ambiguous instruction is illustrated in the following incident, which occurred in a paint shop (Chapanis, 1965). A piece of equip-

Table IVA–2: Factors Affecting the Written Communication Process (Adapted from S. H. Caplan, 1975, Eastman Kodak Company.)

Design of Message by Sender →	Factors Affecting Message Transmission →	Elements Influencing Receipt of the Message
Comprehensibility	Environment	Discrimination
• Purpose	• Viewing Distance	• Visual Abilities
• User Knowledge	• Viewing Angle	Interpretation
• Brevity	• Illumination	• Language Skills
• Accuracy	• Deterioration	• Situation Knowledge
• Clarity	• Competing Displays	Recall
Legibility	• Timing Pressure	• Time Delay
• Font Style		• Interference
• Font Size		
• Colors		
Readability		
• Borders		
• Layout		
• Abbreviations		
• Spacing		
• Case		

The sender of a message must design written material so that it is comprehensible, legible, and readable (Column 1). Factors that influence each of these design objectives are listed. As the message is transmitted through the environment (Column 2), other factors may affect how well it can be picked up by the receiver. The receiver's characteristics (Column 3) also influence the accuracy with which information is communicated. Attention to the factors listed in each part of the process should reduce the potential for errors in written communication.

ment arrived with these instructions: "Finish 198 all over, but may have 684B on areas designated. . . ." Paint shop personnel could have interpreted these instructions in four ways:

1. Finish 198 or 684B on the designated areas, and finish 198 everywhere else.
2. Finish 198 all over, even if 684B were applied first on the designated areas.
3. Finish 198 all over first. The 684B is optional afterward on the designated areas.
4. Finish 684B first on the designated areas, then 198 all over.

A listing of what was expected, in the proper order, would have eliminated this ambiguity.

b. Legibility

The material in this section was developed from information in Berger (1944), Cornog and Rose (1967), and McCormick and Sanders (1982).

Legibility affects the user's ability to discriminate among or recognize letters or numbers. It is affected by the characters' shape, size, contrast, color, and quality of reproduction. Use of the following guidelines should improve the legibility of messages on labels and signs as well as in other communication forms, such as printed forms.

1. Keep fonts simple; avoid curlicues and flourishes. See Figure IVA–1.
2. Use the guidelines below for letter and number size to improve legibility in normal lighting conditions. Figure IVA–2 illustrates the width, height, and stroke width of a character.
   - Stroke width should be $\frac{1}{6}$ of the height for black letters or numbers on a white background.
   - Letter width should be $\frac{3}{5}$ of the letter height, except for I, which should be one stroke width, and M and W, which should be $\frac{4}{5}$ of the height.
   - Number width should also be $\frac{3}{5}$ of the number height, with the exception of 1, which should be one stroke width.
   - Letter or number height will depend on the viewing distances and the criticalness of the information. For situations where illumination is adequate, say greater than 108 lux (lx), or 10 footcandles (fc), on the label, Table IVA–3 can be used to determine the appropriate letter or number height.
   - Characteristic openings or breaks in a letter or number should be readily apparent.
3. In unusual lighting conditions, modify the previous guidelines as follows:
   - In darkrooms or other reduced-light locations, white letters on a black

Figure IVA–1: Examples of Fonts (Adapted from T. W. Faulkner, 1968, Eastman Kodak Company.)

When a printed label or message must be read quickly and easily, it is important to choose a plain and simple design of type font. There are some slightly more complex designs that can be easily read because they are familiar from wide use. USE OF ALL UPPER CASE LETTERS REDUCES LEGIBILITY. **LESS FAMILIAR DESIGNS MAY RESULT IN ERRORS ESPECIALLY IF THEY ARE READ IN HASTE.** FONTS DESIGNED PRIMARILY FOR AESTHETIC REASONS ARE VERY POOR CHOICES. OBVIOUSLY EXTREMES LIKE OLD ENGLISH SHOULD NEVER BE USED. **AVOID COMPLEX FONTS** **Keep It Simple.**

Several styles of typefaces are shown to demonstrate that simple, unadorned fonts in both uppercase and lowercase are easier and faster to read than more complex fonts. This is true for most types of communication as well as for labels and signs.

Figure IVA–2: Definitions of Font Characteristics

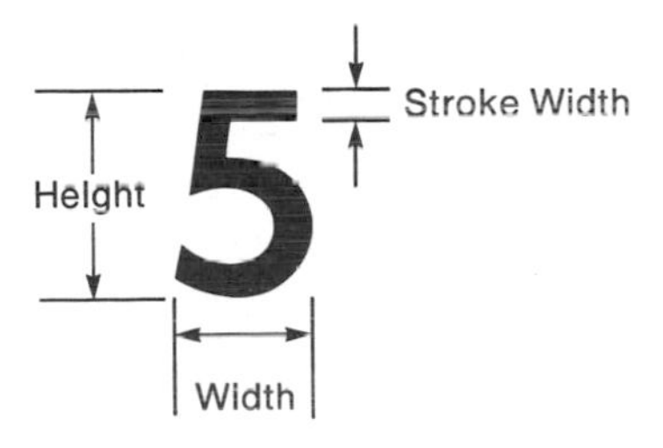

Height is measured from the top to the bottom of the character, and width across its widest part. Stroke width is the thickness of the line used to generate the letter or number.

Table IVA–3: Letter or Number Height Versus Viewing Distance For Labels (Adapted from Peters and Adams, 1959; Smith, 1979; Woodson and Conover, 1964.)

Viewing Distance	Critical Labels	Routine Labels
0.7 m (28 in.)	2 to 5 mm (0.1–0.2 in.)	1 to 4 mm (0.04–0.2 in)
0.9 m (3 ft.)	3 to 7 mm (0.1–0.3 in.)	2 to 5 mm (0.1–0.2 in.)
1.8 m (6 ft.)	7 to 13 mm (0.3–0.5 in.)	3 to 10 mm (0.1–0.4 in.)
6.1 m (20 ft.)	22 to 43 mm (0.9–1.7 in.)	11 to 33 mm (0.4–1.3 in.)

The distance from the operator to the display when it is read (Column 1) will determine how high the letters or numbers should be for legibility (Columns 2 and 3). Critical labels refer to key control or component identifiers and to position markers on such controls (Column 2). Routine labels refer to overall instrument identifiers or any markings required only for initial familiarization (Column 3).

background tend to be more visible. In this case the stroke width should be $\frac{1}{8}$ of the height. The characters should be about 50 percent larger than the values shown in Table IVA–3.

- If the sign or label is more than 200 cm (79 in.) above the floor, the character dimensions should be altered for better legibility. For instance, if a label or sign will be placed substantially above head height and must be read by people working at ground level, character height should be increased in relation to width.

4. Avoid the use of colored print. But if colored letters or numbers must be used in order to take advantage of color coding, note that legibility may be reduced. Table IVA–4 illustrates different combinations of colors and their legibility in normal lighting conditions. Use of colors in reduced-light areas is less satisfactory. If colored light is used, color combinations should be tested in that condition to assess their legibility.
5. Tailor the materials and methods used for constructing labels and signs to the environmental conditions. For instance, engraved labels should not be used in an area where dirt is likely to fill in the indentations. Paper labels should be given protective coatings if used in areas where corrosive chemicals are present.

Table IVA–4: Legibility of Color Combinations in White Light (Adapted from Woodson and Conover, 1964.)

Legibility	Color Combination
Very Good	Black characters on a white background Black on yellow
Good	Yellow on black White on black Dark blue on white Green on white
Fair	Red on white Red on yellow
Poor	Green on red Red on green Orange on black Orange on white
Very poor	Black on blue Yellow on white

The ease with which written information can be distinguished from its background is indicated in the first column for different color combinations (Column 2).

6. Locate labels and signs on the equipment or in the workplace so that glare, reflections, and shading do not make them difficult to read. If a sign or a label is placed out-of-doors, pay attention to the changing direction of the sun when locating it in order to improve its visibility. Matte surface paints may also be used to reduce reflections.
7. Size and locate labels or signs that are placed on curved surfaces (such as piping or drums) so that the lettering remains readable from one viewing location.

c. Readability

Readability refers to the ease of reading words or numbers, assuming that the individual characters are legible. It is affected by the use of uppercase or lowercase, spacing, borders, and layout. The following guidelines give ways to improve readability of labels and signs:

- Use capital letters for headings or messages of a few words only. Use lowercase letters for longer messages. Do not use italics except when they are needed to add emphasis to specific words or short phrases. Underlining is an alternative method for adding emphasis.

- Avoid abbreviations. Use standard ones if they must be used. If no standard abbreviation exists, test the newly developed one on inexperienced subjects in order to determine its appropriateness.
- Leave a minimum of one stroke width between characters.
- Use a border to improve readability of a single block of numbers or letters (see Figure IVA–3).

Figure IVA–3: Spacing and Borders that Improve Readability

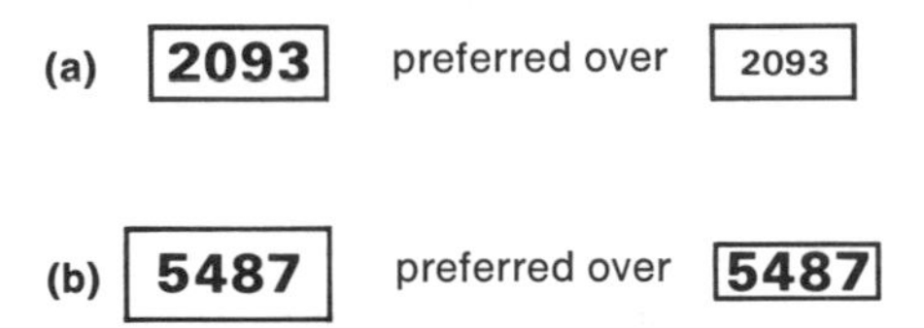

If space is limited and the character size is critical (as in part a), it is preferable to fill most of the space within the border. If space is not critical (see part b), a larger surrounding border contributes to even better readability.

- If several labels or messages are clustered in the same area, put distinctive borders around the critical ones only. Keep the embellishments to a minimum, since each one reduces the effectiveness of display of the others.
- Install labels and signs in locations where they will not be damaged by painting or routine maintenance procedures.
- Make the signs and labels accessible and easy to change if new procedures or equipment are likely to be added to the system. Permanently attached fixtures into which current labels and signs can be inserted are preferable to labels attached directly to the equipment or surrounding workplace.

5. CODING

The material in this section was developed from information in Caplan (1975).

In most production systems operators use coded information, such as part numbers, operation sequences, and lot numbers, to communicate about the process. There are many ways in which this information may be misinterpreted resulting in errors of varying consequence to the operation. The design of the codes may often be a major contributor to these errors. Code systems should be designed to minimize errors and to make sure that those that do occur are quickly detectable.

a. Alphanumeric Coding

The basic types of errors are omission, addition, substitution, and transposition of characters. The guidelines listed below are general and affect more than one type of error:

- An all-digit code should be used where possible and should not exceed four to five digits in length.
- Where longer codes are necessary, the digits should be grouped in threes and fours and separated by a space or a hyphen.
- If a numerical code system contains several digit sequences that occur very frequently, they should comprise the first or last section of the code.
- In tabular listings when a digit sequence occurs repeatedly at the start of a many-digit number, only the last digits for subsequent entries should be printed. For example:

7580170		7580170
7581010	should be	1010
7582030		2030
7591000		7591000

- Alphanumeric codes should have the letters grouped together rather than interspersed throughout the code.
- The letters B, D, I, O, Q, and Z and the numbers 0, 1, and 8 should be avoided in alphanumeric codes (McArthur, 1965).
- For long alphanumeric codes digits should be used in the last few positions.
- Simple fonts with clearly distinguishable characters should be used for the codes.
- Bold printing and high contrast should be used for all codes on labels or displays. Faded characters on a card or sheet should be avoided, especially if they need to be read under low-light conditions. Use color combinations that make codes easy to read (see Table IVA–4).
- Digits or letters should not be obliterated by keypunch holes.

Codes that combat each error type are shown in Table IVA–5.

Table IVA–5: Error Opportunities in Coding (Adapted from Caplan, 1975.)

Error Type	Preventive Feature in Code Design	Reason
Omission or addition of characters to code	Use uniform length and composition	Omission or addition will result in code immediately recognized as nonexistent
Substitution between numbers and letters	Use consistent location for numbers and letters	Substitution will result in code immediately recognized as nonexistent
Transposition of letters	Use a familiar acronym or pronounceable word (instead of random letters) that is visually and audibly distinct	It will be remembered as one element rather than individual elements, as random letters are remembered
Transposition of numbers	Introduce a rule for the relationship between adjacent numbers in the string	Transposition will yield a code with a pair of digits out of order
Illegibility	Use consistent number and letter locations Control handwriting by providing individual box for each character	Poor handwriting more easily deciphered

Common types of errors in written information are shown in column 1. Techniques to reduce the probability of making each error are given in column 2; an explanation of how they prevent errors is given in column 3.

Because it is often very difficult to change a coding system once the code has become part of the production process, it is important to implement these guidelines at the time the new production system is developed. Furthermore, even well-designed codes may be difficult for some people to use. The use of automatic data entry and retrieval systems in production may relieve some of the demands on the operators for retaining codes in selected operations. However, the space requirements for display of uncoded information on cathode ray tubes (CRTs) or video display units (VDUs) may be prohibitive, thereby creating a need for additional coding of information relating to production control.

b. Shape Coding

The material in this section was developed from information in Bradley (1969), Hunt (1953), and Jenkins (1947).

In addition to alphanumeric coding, the designer of production equipment systems may also use shape to transmit information efficiently. Shape coding is a useful technique to employ when visual control is limited. It has also been used effectively on controls used under low-light conditions (see Figure IVA–4) and on signs for traffic control. The shapes chosen should follow accepted standards, where they exist, or should bear some resemblance to the function or component they mark.

Figure IVA–4: Shape Coding of Controls (Adapted from Hunt, 1953.)

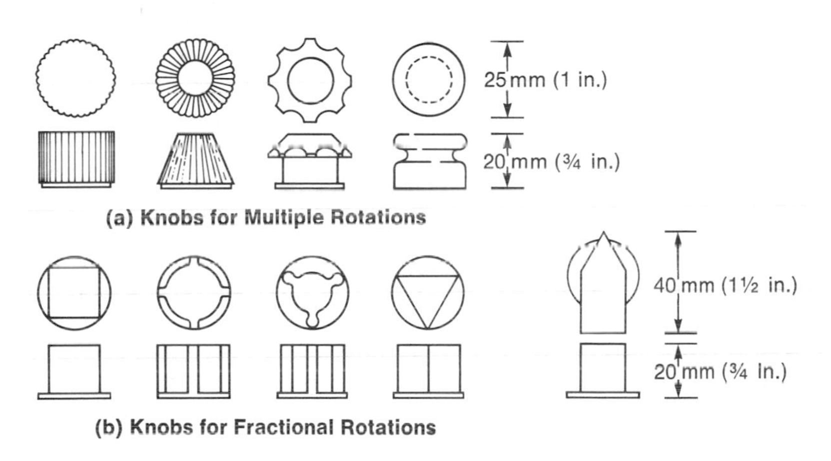

The nine knobs shown are shapes that will be distinguishable by touch alone even when an operator is wearing gloves. The knobs in part a are used when multiple rotations may be required. The first four knobs in part b are used in fractional rotation applications. The last knob in part b is a detent-positioning knob.

It is not advisable to use more than one size of a shape code in order to increase the number of items coded (usually five), particularly if the shape code has to be viewed from a distance or if discrimination between the two components similarly shape-coded is critical.

If two components may occur either separately or together, the shapes should be compatible so that they can be displayed together without losing the distinguishing shape of either. For instance, a circle and a triangle can be displayed together if the circle is open in the center and the triangle is small enough to fit in the open area. In some chemical work areas an open circle has been used to designate the location of a safety shower and a triangle used to indicate where an eye bath is. For cleansing stations where both shower and eye bath are present, the triangle and circle symbols can be superimposed.

Guidelines for choosing shapes for visual displays are developed from studies of people's ability to discriminate shapes or symbols (Sleight, 1952). Some examples follow:

- Under normal viewing conditions, such as white light and daylight, people use area and jaggedness most frequently to describe shapes (Mavrides, 1973).
- Under poor viewing conditions, such as subdued light, glare, or fading, area and the largest dimension are used to distinguish different shapes (Casperson, 1950).
- As measured by the time required to search out and sort one shape out of 21, the 11 most discriminable symbols can be ranked as follows (Sleight, 1952):

1. swastika
2. circle
3. crescent
4. airplane
5. cross
6. star
7. ellipse
8. rectangle
9. diamond
10. triangle
11. square

On the basis of these findings, and excluding shapes having additional meanings, such as the swastika, cross, crescent, and airplane, a discriminable set of shapes is the following: circle, star, ellipse, square.

c. Color Coding

The material in this section was developed from information in Brown and Hull (1971), Christ (1975), and Smith and Thomas (1964).

Color coding is more often used, and permits the designer a wider range of applications, than shape coding. There are from 8 to 15 colors that can be absolutely discriminated at least 90 percent of the time (Jones, 1962). It is possible to increase the number of options by varying a color's brightness and saturation (Feallock, Southard, Kobayoshi, and Howell, 1966). The following guidelines should be used in the design of color codes:

- Two levels of luminance (brightness) are probably the maximum that can be identified absolutely if error-free performance is required.
- Luminance is not a reliable cue in color perception because dark hues are difficult to recognize; use it sparingly.
- The accepted colors should be used for detection by people with color-defective vision (Anon, 1959).
- When there is a choice, colored lights are preferred over paints.

- When color codes are used for sizes, red should be used for the largest size. There is a strong association between color and size. However, a rainbow order from red to white could be used as an easily learned code for large to small, respectively (Poulton, 1975).

SECTION IV B. PRODUCT-TO-PERSON TRANSFER: VISUAL INSPECTION

The material in this section was developed from information in T. J. Murphy (1975, Eastman Kodak Company).

In industry, relaying information from product to person usually involves the product's quality level. This task is commonly known as inspection. In a continuous manufacturing system several inspections may be made as the product is assembled in order to reduce inspection complexity and minimize waste. In some staged assembly routines, inspection is a subsidiary task; each person checks the previous work on the product while performing his or her assembly task on the production line. Some production workers, however, spend a large majority of their shift doing inspection tasks, primarily in quality control functions. People with the following job responsibilities fit the latter classification:

- product acceptance
- process control
- batch release

The methods used for inspection may include direct visual scanning, manual measurement, or automatic measurement. As more complex computer controlled manufacturing systems are employed, stereotyped inspection routines will be done automatically, not by hand; varied tests and inspection of low-volume product will probably continue to be done manually. Guidelines in Chapter II, "Workplace Design," and Chapter V, "Environment," can be used to assist in the design of inspection workplaces where manual inspection is required.

For a large number of industrial manufacturing systems, visual scanning for defects is the primary inspection method. The product to be inspected (the intermediary stages as well as the final product) is often moving on conveyors or is automatically fed to an inspection workplace for static inspection. The process control inspector is under time pressure to identify and record defects so that quality problems can be rectified quickly. The more rapidly the product is manufactured, the more critical is the response time of the inspector. The product acceptance inspector is less paced by production equipment than by production goals. He or she still needs to respond in a timely manner, however, if defects in appearance or function of a product are found at the end of a manufacturing cycle. The batch release inspector assimilates information from many parts of the manufacturing and quality control stations and, from this information, decides whether a production run is ready to be passed on to the next operator. Computerization of such information will be of benefit to this inspector, but a visual check of the product will still be necessary to judge its quality.

In this section factors that affect inspection performance are discussed, with emphasis on visual rather than automatic or manual inspection methods.

1. FACTORS INFLUENCING INSPECTION PERFORMANCE

Table IVB–1 lists the individual, environmental, task, and organizational factors that may influence the performance of an inspector. These data are based on both industrial and laboratory studies. Although validation of the importance of each of these factors in the workplace has not always been scientifically done, all factors have been shown to influence inspection performance in the field.

a. Measures of Performance

The material in this section was developed from information in T. J. Murphy (1975, Eastman Kodak Company).

For determination of how much these factors affect inspection performance, it is necessary to define the methods for assessing that performance. The following three approaches are most appropriate to the industrial situation:

- Review of customer complaints.
- Audits of previously inspected product.
- Job sample tests (off-line inspection of known samples to determine individual inspector accuracy).

Customer complaints often trigger the audits and job sample tests. The latter are more likely to be done when a person is being trained for a new job or when a new product is coming on-line with a new set of defects to be detected.

Information should be collected on both the quantity and quality of performance. There are two errors of concern: accepting bad product and rejecting good product. The following measures should be made to evaluate defect inspection performance:

- percentage detected
- percentage correctly rated
- percentage of false alarms
- waste
- percentage correctly named
- time per inspection
- units inspected per time period (hour, day, week)

Table IVB–1: Factors that may Influence Inspection Performance (Adapted from Megaw, 1979.)

Individual Factors	Physical and Environmental Factors	Task Factors	Organizational Factors
Visual acuity	Lighting	Inspection time	Number of inspectors*
Static*	General*	Stationary*	Briefing/instructions
Dynamic	Surround luminance	Conveyor-paced*	Feedback*
Peripheral	Lighting for color	Paced versus unpaced	Feed forward*
Color vision*	Specialized*	Direction of movement	Training*
Eye movement scanning	Aids	Viewing area	Selection*
strategies*	Magnification*	Shape of viewing area	Standards*
Age*	Overlays*	Density of items*	Time on task*
Experience*	Viewing screen*	Spatial distribution of	Rest pauses
Personality	Closed-circuit TV	items	Shift*
Sex	Partitioning of display	Defect probability*	Sleep deprivation
Intelligence	Automatic scanner	Defect mix	Social factors
Subjective probability of	Background noise	Defect conspicuity*	General*
defect occurrence	Music-while-you-work*	Product physical factors*	Isolation of inspectors*
	Workplace design	Complexity	Working in pairs
		2- or 3-dimensional	Effects on sampling scheme*
		Specularity	Motivation*
		Hue	Incentives*
		Size	Product price information
		Defect physical factors	Job rotation*
		Shape	
		Size	
		Specularity	
		Contrast	

The ability of a person to perform an inspection task may be influenced by individual (column 1), environmental (column 2), task (column 3), or organizational (column 4) factors. Those factors that have been identified in industrial experiments are indicated by an asterisk (*); the other factors have been shown to influence inspection performance in laboratory tasks or military studies.

* Identified in industrial experiments.

Once these measures have been collected on a specific inspection task, it is easier to identify which of the four major performance-influencing categories (individual, task, environmental, or organizational) needs attention. For instance, by improving the detectability of a defect, one can reduce the error potential for misses or false alarms. By changing the rejection criterion level, one can alter the impact of errors; the impact depends on the relative cost of each type of error. If one were inspecting atom bombs, for instance, one might judiciously choose a rejection criterion that resulted in many acceptable bombs being rejected in order to ensure that a defective one is not released.

b. Individual Factors

The visual acuity, both for stationary (static) and moving (dynamic) objects, and peripheral vision capabilities of a person can influence his or her performance on an inspection task. Before the person can recognize a defect, he or she has to detect it. Visual losses of a severe nature, such as cataracts, tunnel vision, uncorrectable loss of acuity, or blurry vision, will make detection less probable. If an inspection task requires fine color discrimination, a person with color-deficient vision or with eye changes that alter color perception (yellowing of the lens or clouding of the eye humors, for example) will be at a considerable disadvantage. An individual who must wear bifocals or trifocals for an inspection task may experience neck muscle fatigue if the inspection field is not in a narrow band in the front of the workplace (such as when large-sized sheets or objects are inspected at a seated workplace).

Although some tests of visual function can be made to assess why an individual has difficulty on a specific inspection task, performance is not solely based on detecting defects. A study of inspectors in two different jobs indicated that 85 to 89 percent of the defects presented to them were detected, whereas only 43 to 58 percent were reported. Table IVB–2 summarizes how the additional 27 to 46 percent of the defects were classified. The data indicate a need for repeated training of inspectors to improve the reporting of defects.

An individual's ability to detect some defects has been linked to performance on tests designed to measure spatial visual perception. These tests use either embedded figures or slightly altered figures in a densely displayed series of drawings (Harris, 1964; Thurstone and Jeffrey, 1956). Because the problems of validating such a test to job requirements are so prohibitive (EEOC, 1978), it is often not feasible to choose inspectors according to this ability. However, the tests may be helpful in determining why an individual has trouble with a specific inspection task.

If several defects have to be detected simultaneously, particularly in a process control operation where the product is moving by the inspection station at a fixed pace, inspection performance may be affected by the individual's ability to keep all the defects in mind and to search effectively for each. This ability may be complexly tied to personality and intelligence variables, and there is no evidence that one can test for these variables to fit each inspection situation. Since most people will only be able to search effectively

Table IVB–2: Reasons for Not Reporting Defects by Two Groups of Inspectors (Adapted from T. J. Murphy, 1975, Eastman Kodak Company.)

Category	Group A	Group B
% Detected and reported	43	58
% Misidentified as nondefects	6	7
% Intended to be considered part of another reported defect	8	13
% Decided were unreportable (by invalid rules)	26	7
% Not seen at all	11	15
% Made a mistake; should have called, no reason given	6	0
Total	100%	100%

Two groups of inspectors in different jobs (groups A and B) on high-volume production lines were interviewed after their performance was monitored for several hours. Their reasons for not reporting defective product are tabulated here. Column 1 gives the categories of response, and columns 2 and 3 give the percentage of responses in each category for each group.

for a few defects at a time in a paced inspection task, environmental or organizational aids should be considered to assist in multiple-defect identification. Experience on the job, while helping the inspector assess the probability of occurrence and the criticality of certain defects, does not always result in improved accuracy (T. J. Murphy, 1975, Eastman Kodak Company).

With products that are manufactured a few times yearly, instead of daily or weekly, it is advisable to provide reminders of defect types. These aids can reduce the warm-up time for that product for experienced inspectors as well as for novices.

c. Physical and Environmental Factors

The lighting available for visual inspection tasks can influence performance and productivity significantly. Inadequate illumination, both qualitatively (shadows, glare) and quantitatively (too little or too much), can make the discrimination of a defect difficult. A detailed discussion of special-purpose lighting for use in inspection tasks is given in the section on illumination in Chapter V, "Environment."

Workplace design can also influence inspection performance through its effect on body posture. An inspection job that requires constant elevation of the head or a rigid posture, such as having to detect surface irregularities, can produce static muscle fatigue. This muscle fatigue will result in both distraction and a need to assume less optimal postures, which might further reduce defect detectability. Guidelines for workplaces where visual work predominates are presented in Chapter II, "Workplace Design."

Frequently, the lighting and the workplace design interact to provide a less-than-optimal inspection environment. For example, it can be difficult to provide adequate light for an inspection task where moving product is located at 75 cm (30 in.) above the floor and the inspector must remain standing to perform machine control operations. The inspector's shadow may fall on the product, making some defects more difficult to detect. If critical defects are subtle, the visual distances may be too great for some people, and those persons may have to crouch or bend to get closer to the product. When task lighting is provided and the product line is raised, the inspection task can be made considerably easier to perform.

Depending on the type of defects to be detected, competition from other environmental factors can affect inspection performance. For instance:

- If an auditory signal is an important part of the inspection process, background noise may make it difficult to detect.
- If a part is to be picked out from a large number of similar parts, the density and complexity of the background should be kept to a minimum. Simply increasing the contrast of the product with the background is not always appropriate, particularly if the background then tends to draw the eye away from the piece to be inspected.
- Luminescent, colorful paints on conveyor systems and line process equipment may brighten up the workplace and please the eye aesthetically, but careful attention should be paid to how these colors affect an inspector's performance if they become part of the process control inspection station.
- The mistake is sometimes made of using a background that contrasts with the inspected piece. This design practice is incorrect. The goal is to reduce background contrast so that the contrast between the defect and the rest of the inspected piece is at a maximum.

Aids that permit an inspector to compare a product sample to a standard, instead of having to make an absolute judgment, are useful, especially in product acceptance inspection operations. Such aids are particularly effective for color discrimination, including variations in hue, brightness, and saturation, where many thousands of differences can be detected by using a comparator, but only 8 to 15 colors can be accurately identified on an absolute basis (Feallock, Southard, Kobayoshi, and Howell, 1966; Halsey and Chapanis,

1954; Jones, 1962). Table IVB–3 summarizes data from the literature about the number of levels and dimensions of visual and other sensory modalities that can be discriminated on an absolute basis. Comparisons extend these capabilities manifold.

Table IVB–3: Amount of Information in Absolute Judgments of Various Stimulus Dimensions (Adapted from McCormick and Sanders, 1982).

Stimulus Dimension	Number of Levels That Can Be Discriminated on an Absolute Basis Under Optimum Conditions	Source
Color, Surfaces		
Hues	8–9	Halsey and Chapanis, 1954; Jones, 1962
Hue, saturation, and brightness	24 or more	Feallock, Southard, Kobayoshi, and Howell, 1966
Color, Lights	10 (3 preferable)	Grether and Baker, 1972
Geometric Shapes	15 or more (5 preferable)	Jenkins, 1947
Angle of Inclination (indicating direction, angle, position on dial)	24 (12 preferable)	Muller et al, 1955
Size of Forms (e.g., squares)	5–6 (3 preferable)	McCormick and Sanders, 1982
Brightness of Lights	3–4 (2 preferable)	Grether and Baker, 1972
Flash Rate of Lights	2	McCormick and Sanders, 1982
Sound		
Intensity (pure tones)	4–5	Deatherage, 1972; Garner, 1953
Frequency	4–7 (when intensity is at least 30 dB above threshold)	Pollack, 1953
Intensity and Frequency	8–9	Deatherage, 1972
Duration	2	Pollack and Ficks, 1954

The ability of people to distinguish among absolute levels of color, shape, position, size, brightness, and sound without comparisons available is given in column 2. The number of discriminable levels increases markedly if these factors are combined and if comparisons are available. The third column indicates the literature from which the absolute judgment data are drawn.

Comparators can be as simple as photographs showing defect types or as complicated as stereo microscopes that superimpose the standard and sample images. These approaches permit classification of defect severity or identification of a defect in a complex field.

Two experiments (Harris and Chaney, 1969) illustrate the benefits of comparison aids in inspection operations. In the first, people who were inspecting electronic chips were asked to assess the color of interference rings that indicated acceptability of the product for release. They were given three aids: a verbal description of the colors, a color scale on a piece of paper, and a standard set of color chips, which were located on one side of a two-stage microscope. The percentage errors (rejecting acceptable chips on the basis of the color criterion) are shown in Figure IVB–1.

Figure IVB–1: Percentage Errors in Using Three Aids to Assist in Color Inspection of Electronic Chips (Adapted from Springer and Harris, 1967.)

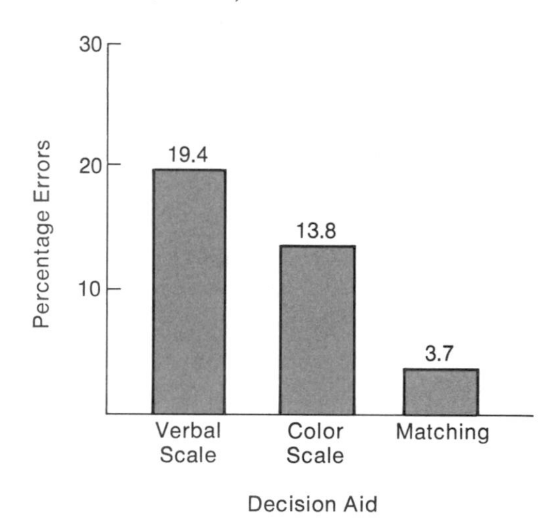

Defective electronic chips were identified by color changes. The errors made by inspectors using three different inspection aids (Decision Aids) are expressed per 100 chips inspected (percentage errors) on the vertical axis. The verbal scale was a written description of the color changes and their relationship to product quality. The color scale was a set of photographs of the color changes in relationship to common defects. The matching aid was a series of color chips indicating defects, which were placed on one side of a dual-stage microscope; this placement permitted the inspector to do a direct matching of the colors. Color matching improved the inspector's detection of defective chips, thereby decreasing errors and reducing waste, compared with the same inspector's performance with the verbal or color scales as aids.

The second experiment involved people who were inspecting solder joints in an electronics firm. When these people used photographs of eight graded samples of unacceptable to acceptable solder joints, their performance was significantly improved. The consistency of responses among inspectors was also 100 percent better. The photographs helped to establish the minimum acceptable level of the joints. Figure IVB–2 illustrates the results of this study.

Figure IVB–2: Increase in Agreement Among Inspectors Resulting from Use of Photographic Aids (Adapted from Thresh and Frerichs, 1966.)

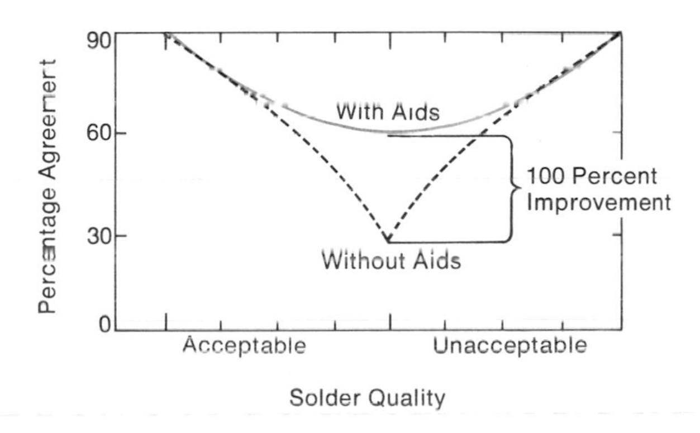

The same defective solder joints on electronic circuit boards were inspected by several inspectors. The lower curve shows the percentage of time that the inspectors agreed on solder quality when no inspection aids were provided. The least agreement is found at the junction between acceptable and unacceptable, where judgment is most needed. When photographs of acceptable joint soldering were provided at each workplace, the agreement among inspectors increased 100 percent (upper curve).

In general, aids are useful under the following conditions:

- If they obviously improve the inspector's ability to perform the task.
- If they do not take more time to use than would otherwise be needed.
- If they are readily available to each inspector.

d. Task Factors

All of the following factors affect inspection performance.

- The complexity and variety of the product.
- Its distribution in space in the inspection area.

- Whether it is moving or stationary.
- How long the inspector is given to detect and report a defect.
- How frequently defects occur.

Machine-paced inspection, usually found in process control operations, is generally more difficult than self-paced inspection, especially if multiple defects are present.

Inspection operations where defects are rare are difficult to perform since a lag in the inspector's attention can result in missed defects (Smith and Lucaccini, 1968). As the number of defective parts per hundred inspected, or the defect rate, falls below 5 percent, false reports increase rapidly (see Figure IVB–3). Defect rates below 1.5 percent result in reduced detection performance as well. Rotating people between such an inspection task and another less visually demanding task every 30 minutes can improve overall inspection performance.

Several studies have illustrated the effect of task factors on inspection performance. Table IVB–4 presents the results of some of these studies that relate to industrial problems.

Figure IVB–3: Defect Rate and Inspection Accuracy (Adapted from Harris and Chaney, 1969.)

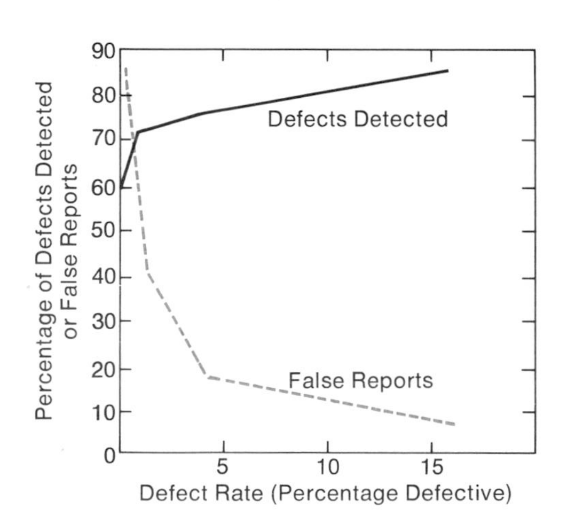

The rate at which defects occur (horizontal axis) in an inspection task will determine how many are detected (upper curve) and how many are improperly classified (lower curve, false reports). Detection increases and false reports fall with increased defect rates, generally. At very low defect rates detection is less good and false reports increase rapidly.

Table IVB–4: Effects of Task Variables on Inspection Performance (Adapted from T. J. Murphy, 1968, Eastman Kodak Company; Purswell, Greenshaw, and Oats, 1972.)

Task Variable	Explanation of Inspection Task	% Errors	Conditions	% Detection
1. Complexity	a. Students identified defects in a series of geometric shapes on a 10 × 13-cm (4 × 5-in.) target. The shapes were 10 mm ($\frac{3}{8}$ in.) high on the 5 × 5 grid and 6 mm ($\frac{1}{4}$ in.) high on the 7 × 7 grid.	4.7	5 × 5 grid	—
		11.2	7 × 7 grid	—
	b. Process control inspectors on two inspection tasks searched for 50 or 200 defects on a product.	—	50 defects	81
		—	200 defects	42
2. Velocity	Students searched for defects on targets (geometric shapes) moving at 13 or 18 cm/sec (5 or 7 in./sec).	5.1	13 cm/sec (5 in./sec)	—
		10.8	18 cm/sec (7 in./sec)	—
3. Spacing	Students searched for defective shapes at 13 cm/sec (5 in./sec) with different spacings between the targets (25, 38, and 50 cm, or 10, 15, and 20 in.).	10.1	25-cm (10-in.) spacing	—
		4.5	38 cm (15 in.)	—
		2.4	50 cm (20 in.)	—
4. Direction	Inspectors looked for defects on a moving web. Their position relative to the web's direction of movement was varied.	No significant difference	Left to right versus right to left	—
		—	Movement toward versus away from inspector	Toward better than away from

Four task variables (column 1) that affect inspection performance are presented in terms of errors (column 3) and detection rates (column 5) for some specific inspection tasks (columns 2 and 4). Error rates increased with increasing complexity (1a), or information density, and with higher speeds of product being inspected (2). The more defects being searched for, the lower was the percentage detected (1b). In a dynamic task, increased spacing (up to 50 cm or 20 in. between defects) resulted in fewer errors (3). Detection of defects was better when the product moved toward the inspector (4).

In general, the more time a person has to make an inspection, the better is his or her detection performance (Drury, 1973). As time shortens, fewer scanning options (sequential rather than simultaneous search for multiple defects, for example) are available, and there is less accommodation for individual inspector skill levels.

e. Organizational Factors

Training (including feedback on performance), work/rest cycles, shift schedules, and social factors can affect inspector performance. Initial training may be adequate to allow a person to inspect product on a high-volume line, but reinforcement is needed to ensure that defects are being detected. Routine audits of product passing or failing inspection are often used to assess individual retraining needs, but this feedback may be delayed. Rapid feedback on performance will maintain motivation, provide information, aid training, and maintain standards. Without feedback an inspector may not be aware of missing defects or may reject more product than necessary because of a desire not to make errors. In a study of a complex glass inspection where feedback was experimentally introduced, the number of defects missed was reduced by about 50 percent over the previous rates (Drury and Addison, 1973). One technique for giving rapid feedback to inspectors is to perform random on-line audits.

Social factors tend to determine rejection rates in operations where people are doing similar inspection tasks in adjacent workplaces. In one study a group of inspectors was given product to inspect that had a known defect rate of 8 percent (T. J. Murphy, 1975, Eastman Kodak Company). When working alone, the inspectors reported about 8 percent defects. They were then placed in an area near another group of inspectors who had product with a 16 percent defect rate. In a short time the 8 percent defect rate group was reporting 13.7 percent defects. The increase was attributed to their observation of the higher rejection rate of the adjacent inspectors, which influenced them to reject more of the borderline cases. If feedback were provided to inspectors, these inappropriate social infuences could be minimized.

The training for inspectors should include teaching them how to separate a defect from background noise; background noise is a characteristic of the product that looks like a defect but has no impact on product quality. Also, it may take one to two years to train an inspector to identify the defects on several products (T. J. Murphy, 1975, Eastman Kodak Company). If many products are manufactured on a line, and some of them occur very infrequently, training may not be complete for several years. Furthermore, products change, and new training is required with each item. Also, rare defects may not be recognized any better by experienced inspectors than they are by inexperienced inspectors. Each inspector usually finds his or her own system for reducing the number of defects to be detected to five to ten classes. When the defects are classified into a few groups on a chart or code sheet at the workplace, more consistency in performance between inspectors can result.

For further information on codes and their use, see the discussion on coding in Section IVA.

Because inspection is often a highly repetitive activity and may be externally paced, attention should be paid to work/rest cycle needs. In a continuous, paced web inspection task, where five to six defect types were being inspected at 76 m/min (250 ft/min), a 2 percent miss level in detection performance was noted in the first 10 minutes of the task. This rate increased to 3 percent by 30 minutes and to 5 percent by 40 minutes (T. J. Murphy, 1975, Eastman Kodak Company). Similar responses have been shown in an experiment where inspection performance during a 60-minute session was compared with that in two 30-minute sessions separated by a 5-minute rest break (Colquhoun, 1959). The subjects without a rest break missed about 5 percent of the targets by the end of 50 minutes, compared with 1 percent for the group with a rest break.

Shift work, particularly work on the midnight-to-early-morning shift, may influence inspection performance through its effect on subjective fatigue. The body is at a low point with respect to many of its 24-hour rhythms in the early morning hours, and alertness is reduced. Therefore an inspector's ability to detect rare defects could be impaired. Design of continuous, paced inspection tasks with low defect rates should incorporate aids for the inspector that reduce attention requirements. Despite these concerns, industrial audit data that we are aware of show no time-of-day effect on inspection performance.

A further social factor that may influence inspection performance is the diffusing of responsibility for quality inspection by providing several inspections for the same series of defects (T. J. Murphy, 1975, Eastman Kodak Company). An inspector earlier in the system who knows that someone else will also be inspecting the product may not be thorough; the later inspector, knowing that the defect has already been looked for, may also not be careful. Establishing definite responsibility for defect detection is a preferred alternative.

2. GUIDELINES TO IMPROVE INSPECTION PERFORMANCE

From the above information the following guidelines have been developed for the design of inspection tasks (T. J. Murphy, 1968, Eastman Kodak Company):

- Make the angle of viewing downward instead of straight ahead or upward.
- Allow moving product to move toward rather than away from the inspector (see Figure IVB–4). It does not appear to make much difference if the product moves from right to left or from left to right across the inspector's field of vision. At least 30 cm (1 ft) of viewing area should be available for every 18 m/min (60 ft/min) of product movement.

Figure IVB–4: Direction of Viewing for Inspection of Moving Product (Adapted from T. J. Murphy, 1968, Eastman Kodak Company.)

Inspection of a moving product is most accurately done when the product moves toward the operator, rather than from side to side or away from him or her.

- Evaluate the vision of inspectors who have to detect fine color differences or low-contrast defects when they are using actual job samples. Do not evaluate their vision by abstract color vision or acuity tests. Evaluate inspectors who wear bi- or trifocals under the specific task requirements with consideration of reasonable accommodations. If the viewing distance is kept under 50 cm (20 in.), most people should be able to perform the tasks with their usual vision corrections.
- Whenever possible, remove the process control inspector from direct machine pacing. Permit the operator to vary the time for inspection within a small range, if needed.
- If products are changed frequently, provide information (such as a previous defect review) to the inspector to refresh his or her memory at each product change. This review technique should decrease the warm-up period at product changes.
- If the defect rate is low, provide environmental aids to assist the inspector in detecting the defect. The aids may be special lighting or an alerting alarm. Sampling inspection may also be preferable to 100 percent inspection.

- Use different types of lighting to improve the detectability of different defects. Permit the inspector to adjust the lighting to fit his or her body size and visual requirements.
- Provide feedback to inspectors about errors or misses in detecting defects. Perform regular audits to help in this process and to identify individual training needs.
- Develop a standardized training program for inspectors; do not rely fully on one-on-one training by existing inspectors. The standardized program should include procedures for continued training and quality performance checks.
- Provide photographic or other standardized aids for comparative evaluations, especially at those inspection stations where multiple and sometimes rare defects are to be detected.
- Provide the inspector with an opportunity to break up the inspection periods with auxiliary activities (paperwork, supplies procural, etc.), so that continuous monitoring is not required for more than 30 minutes at a time.
- Make the importance of quality inspection clear to the people doing it by providing them with information about the impact of both errors and misses. Identify the accountability of each inspection operation, and eliminate unnecessary steps that might diffuse responsibility for detecting a defect.
- Minimize environmental distractors in inspection stations.

REFERENCES FOR CHAPTER IV

Anon. 1959. *Munsell Book of Color*. Baltimore: Munsell Color Book. Cited in McCormick and Sanders, (1982).

Anon. 1960. *Forms Analysis, Managing Forms, Records Management Handbook.* Federal Stock No. 7610-655-8220. Washington, D.C.: General Services Administration, National Archives and Record Service and Office of Records Management, 62 pages.

Berger, C. 1944. "Stroke Width, Form, and Horizontal Spacing of Numerals as Determinants of the Threshold of Recognition." *Journal of Applied Psychology, 28:* pp. 208–231 and 336–346. Cited in Murrell (1965), pp. 190–192.

Bradley, J. V. 1969. "Desirable Dimensions for Concentric Controls." *Human Factors, 11 (3):* pp. 213–226.

Broadbent, D. E. 1977. "Language and Ergonomics." *Applied Ergonomics, 8 (1):* pp. 15–18.

Brown, I. D., and A. J. Hull. 1971. "Testing Colour Confusability Among a New Range of Decimal Stamps." *Applied Ergonomics, 2 (2):* pp. 92–97.

Cahill, M. C. 1975. "Interpretability of Graphic Symbols as a Function of Context and Experience Factors." *Journal of Applied Psychology, 60 (3):* pp. 376–380. Cited in McCormick and Sanders (1982).

Caplan, S. H. 1975. "Guidelines for Reducing Human Errors in the Use of Coded Information." In *Proceedings of the 1975 Human Factors Society*, 1975 Human Factors Society Meetings, October 14–16, 1975, Dallas, Texas. Santa Monica: Human Factors Society, pp. 154–158. Copyright by the Human Factors Society, Inc. and reproduced by permission.

Carroll, J. B., P. Davies, and B. Richman. 1971. *Word Frequency Book.* New York: Houghton Mifflin Company (American Heritage), 856 pages.

Casperson, R. C. 1950. "The Visual Discrimination of Geometric Forms." *Journal of Experimental Psychology, 40:* pp. 668–681.

Chapanis, A. 1965. "Words, Words, Words." *Human Factors, 7 (1):* pp. 1–17. Copyright by the Human Factors Society, Inc. and reproduced by permission.

Christ, R. E. 1975. "Review and Analysis of Color Coding Research for Visual Displays." *Human Factors, 17 (6):* pp. 542–570.

Colquhoun, W. P. 1959. "The Effect of a Short Rest Pause on Inspection Efficiency." *Ergonomics, 2:* pp. 367–372.

Cornog, D. Y., and F. C. Rose, 1967. *Legibility of Alphanumeric Characters and Other Symbols. II: A Reference Handbook.* National Bureau of Standards 262-2. Washington, D.C.: U.S. Government Printing Office, 460 pages.

Deatherage, B. H. 1972. "Auditory and Other Sensory Forms of Information Presentation." Chapter 4, in Van Cott and Kinkade (1972), pp. 123–160.

Douglas, R., and J. Anderson. 1974. *Questionnaires: Design and Use.* Metuchen, N.J.: Scarecrow Press, Inc., 225 pages.

Drury, C. G. 1973. "The Effect of Speed Working on Industrial Inspection Accuracy." *Applied Ergonomics, 4:* pp. 2–7.

Drury, C. G., and J. L. Addison. 1973. "An Industrial Study of the Effects of Feedback and Fault Density in Inspection Performance." *Ergonomics, 16:* pp. 159–169.

EEOC, Civil Services Commission, Department of Justice, and Department of Labor. 1978. "Uniform Guidelines on Employee Selection Procedures." No. 6570-06, Part 1607. *Federal Register, 43 (166):* pp. 38290–38315.

Feallock, J. B., J. F. Southard, M. Kobayoshi, and W. C. Howell. 1966. "Absolute Judgements of Colors in the Federal Standards System." *Journal of Applied Psychology, 50:* pp. 266–272. Cited in McCormick and Sanders (1982).

Freed, M. N. 1964. "In Quest of Better Questionnaires." *Personnel and Guidance Journal. 43 (2):* pp. 182–188.

Garner, W. R. 1953. "An Informational Analysis of Absolute Judgement of Loudness." *Journal of Experimental Psychology, 46:* pp. 373–380.

Goode, W. G., and P. Hatt. 1952. *Methods in Social Research.* Text ed. New York: McGraw-Hill, 386 pages.

Grether, W. F., and C. A. Baker. 1972. "Visual Presentation of Information." Chapter 3, in Van Cott and Kinkade (1972), pp. 41–121.

Halsey, R. M., and A. Chapanis. 1954. "Chromaticity Confusion Contours in a Complex Viewing Situation." *Journal of the Optical Society of America, 44:* pp. 442–454. Cited in Van Cott and Kinkade (1972).

Harris, D. F., and F. B. Chaney. 1969. *Human Factors in Quality Assurance.* New York: John Wiley & Sons, 234 pages.

Harris, D. H. 1964. "Development and Validation of an Aptitude Test for Inspectors of Electronic Equipment." *Journal of Industrial Psychology, 2:* pp. 29–35.

Harris, D. H. 1969. "Effect of Defect Rate on Inspection Accuracy." *Journal of Applied Psychology, 53:* pp. 77–79. Cited in Harris and Chaney (1969).

Hunt, D. P. 1953. *The Coding of Aircraft Controls,* WADC Tech. Rpt. 53-221. Wright-Patterson AFB, Ohio: Wright Air Development Center. Cited in McCormick and Sanders (1982).

Jenkins, W. O. 1947. "The Tactual Discrimination of Shapes for Coding Aircraft-Type Controls." In *Psychological Research on Equipment Design,* edited by P. Fitts. Research Rpt. No. 19. Army Air Force, Aviation Psychology Program. Cited in McCormick and Sanders (1982).

Jones, M. R. 1962. "Color Coding." *Human Factors, 4:* pp. 355–365. Cited in McCormick and Sanders (1982).

Loftus, E. F., and G. Zanni. 1975. "Eyewitness Testimony: The Influence of the Wording of a Question." *Bulletin of the Psychonomic Society, 5 (1):* pp. 86–88.

McArthur, B. N. 1965. *Accuracy of Source Data Human Error in Hand Transcription.* Rpt. FMC-R-2234, Contract AF-33-615-1276. Santa Clara, Calif.: FMS Corporation.

McCormick, E. J., and M. S. Sanders. 1982. *Human Factors in Engineering and Design.* 5th ed. New York: McGraw-Hill, 512 pages.

Mavrides, C. M. 1973. "Codability of Polygon Patterns." *Perceptual and Motor Skills, 37:* pp. 343–347.

Megaw, E. D. 1979. "Factors Affecting Inspection Accuracy." *Applied Ergonomics, 10 (1):* pp. 27–32.

Meister, D. 1971. *Human Factors: Theory and Practice.* New York: Wiley Interscience, 415 pages.

Miller, E. E. 1975. *Designing Printed Instructional Materials: Content and Format.* Rpt. RP-WD(TX)-75-4. Washington, D.C.: U.S. Army, Human Resources Research Organization, 61 pages.

Muller, P. F., Jr., R. C. Sidorsky, A. J. Slivinski, E. A. Alluisi, and P. M. Fitts. 1955. *The Symbolic Coding of Information on Cathode Ray Tubes and Similar Displays.* Tech. Rpt. 55-375. Wright-Patterson AFB, Ohio: Wright Air Development Center, October, 1955.

Murrell, K. F. H. 1965. *Human Performance in Industry.* New York: Reinhold, 478 pages.

Payne, S. L. 1951. *The Art of Asking Questions.* Princeton, N.J.: Princeton Univ. Press, 249 pages.

Peters, G. A., and B. B. Adams. 1959. "These Three Criteria for Readable Panel Markings." *Product Engineering, 30 (21):* pp. 55–57, May 25, 1959. Cited in McCormick and Sanders (1982).

Pollack, I. 1953. "The Information of Elementary Auditory Displays." *Journal of the Acoustical Society of America, 25:* p. 745.

Pollack, I., and L. Ficks. 1954. "Information of Elementary Multidimensional Auditory Displays." *Journal of the Acoustical Society of America, 26:* pp. 155–158.

Poulton, E. C. 1975. "Colours for Sizes—A Recommended Ergonomic Colour Code." *Applied Ergonomics, 6 (4):* pp. 231–235.

Purswell, J. L., L. N. Greenshaw, and C. Oats. 1972. "An Inspection Task Experiment." In *Proceedings of the 1972 Human Factors Society,* October, 1972, Los Angeles, Santa Monica: Human Factors Society, pp. 297–300.

Selltiz, C., M. Jahoda, M. Deutsch, and S. W. Cook. 1959. *Research Methods in Social Relations.* Rev. ed. New York: Holt, Rinehart and Winston, 622 pages.

Sleight, R. B. 1952. "The Relative Discriminability of Several Geometric Forms." *Journal of Experimental Psychology, 43:* pp. 324–328.

Smith, R. L., and L. F. Lucaccini. 1968. "Vigilance Research: Its Application to Industrial Problems." In *Human Aspects of Man-Machine Systems,* edited by S. C. Brown and J. N. T. Martin. New York: Open Univ. Press, 1977.

Smith, S. L. 1979. "Letter Size and Legibility." *Human Factors, 21 (6):* pp. 661–670.

Smith, S. L., and D. W. Thomas. 1964. "Color Versus Shape Coding in Information Displays." *Journal of Applied Psychology, 48:* pp. 137–146. Cited in McCormick and Sanders (1982).

Springer, R. M., and D. H. Harris. 1967. "Human Factors in the Production of Microelectronic Devices." Paper presented at the Eighth Annual IEEE Symposium on Human Factors in Electronics, Palo Alto, Calif., May 4, 1967. Cited in Harris and Chaney (1969).

Swain, A. D., and A. E. Guttmann. 1980. *Handbook of Human Reliability Analysis with Emphasis on Nuclear Power Plant Applications.* NUREG/CR 1278. Washington, D.C.: U.S. Nuclear Regulatory Commission, 480 pages.

Szlichcinski, K. P. 1979. "Telling People How Things Work." *Applied Ergonomics, 10 (1):* pp. 2–8.

Thresh, J. L., and J. S. Frerichs. 1966. "Results Through Management Application of Human Factors." Paper presented at the American Society of Quality Control Technical Conference, New York, June 1966. Cited in Harris and Chaney (1969).

Thurstone, T. G., and T. E. Jeffrey. 1956. *Closure Flexibility (Concealed Figures).* TMNF-119. Chicago, Ill.: Industrial Relations Center.

Van Cott, H. P., and R. G. Kinkade, eds. 1972. *Human Engineering Guide to Equipment Design.* Rev. ed. Washington, D.C.: American Institutes for Research, 752 pages.

Woodson, W. E., and D. W. Conover. 1964. *Human Engineering Guide for Equipment Designers.* 2nd ed. Berkeley: Univ. of California Press.

Wright, P., and P. Barnard. 1975. "'Just Fill in This Form'—A Review for Designers." *Applied Ergonomics, 6 (4):* pp. 213–220.

Chapter V ENVIRONMENT

Contributing Authors

Kenneth G. Corl,
M.A., Psychology

Brian Crist,
M.A., Psychology

William H. Cushman,
Ph.D., Psychology

Richard M. Little,
M.S., Electrical Engineering

Richard L. Lucas,
Ph.D., Psychology

Thomas J. Murphy,
Ph.D., Psychology

Suzanne H. Rodgers,
Ph.D., Physiology

Chapter Outline

Section V A. Electric Shock

1. Types of Hazards in the Workplace
2. The Effects of Electric Shock
   a. The Body's Resistance to Shock
   b. Threshold Perception
   c. Pain
3. Prevention of Electric Shock
   a. Protecting the Worker
   b. Workplace Design Considerations

Section V B. Noise and Vibration

1. Noise
   a. Criteria for Noise Exposure
   b. Performance Effects of Noise
   c. Approaches to Reducing Noise in the Workplace
   d. Special Considerations
2. Vibration
   a. Effects on the Body
   b. Ways to Reduce Vibration Exposure

Section V C. Illumination and Color

1. Illumination
   a. Recommended Light Levels for Different Tasks
   b. Selection of Higher-Efficiency Light Sources
   c. Direct and Indirect Lighting
   d. Task Lighting
   e. Minimizing Glare
   f. Special-Purpose Lighting
2. Color

Section V D. Temperature and Humidity

1. Body Heat Balance
2. The Comfort Zone
   a. Zone Definition
   b. Factors Affecting the Feeling of Comfort Within the Zone
3. The Discomfort Zone
   a. Guidelines for Acceptable Heat Exposures in the Discomfort Zone
   b. Guidelines for Acceptable Cold Exposures in the Discomfort Zone
   c. Techniques to Reduce Discomfort in the Hot and Cold Discomfort Zones

4. The Health Risk Zone
   a. Zone Definition
   b. Cold Injury
   c. Heat/Humidity Illness and Injury
5. Working in Wet Environments
   a. Wet Clothing
   b. Hand and Forearm Water Immersion
   c. Using Safety Showers and Eye Baths
6. Surface Temperatures
   a. Hot Surfaces
   b. Cold Surfaces

The workplace environment is of concern to specialists in industrial hygiene, medicine, safety, and industrial relations as well as to those in human factors. This chapter will not attempt to cover the chemical or radiation environments; rather, it will focus on the following physical environmental factors:

- electric shock
- noise and vibration
- illumination and color
- temperature and humidity

The ways that these factors affect performance in the workplace, some guidelines for appropriate levels, and some ways to reduce their negative impact are covered in each section. The performance effects are usually seen at lower values than the levels at which health is affected. Through application of ergonomic principles to the design of environments, the negative influence of these factors on performance can often be removed.

Performance can deteriorate for a number of reasons. For example:

- excessive heat or humidity can reduce work capacity.
- glare or high noise levels can reduce the ability to detect defects or perform the task.
- specific vibration levels or cold or heat stress can reduce motor manipulation skills.
- temperature, noise, and lighting levels can produce discontent, resulting in increased time spent talking about the environmental distractor.

Table V-1 summarizes some of the health and performance effects of the environmental factors for which design guidelines are given in this chapter.

As a general rule, performance should be improved when the operator is relieved from distractions that compete for attention with the main task. A workplace that is not within the guidelines presented in this chapter may be contributing to reduced productivity. When mental or physical effort is devoted to dealing with an environmental distractor, less energy is available for productive work.

SECTION V A. ELECTRIC SHOCK

Electric shock is the term used to describe the subjective feeling or the physiological responses, or both, when electric current from an external source flows through the body. The physiological response varies from local nerve and muscle activation to tissue destruction and death. In this section information and guidelines for evaluating and eliminating electric shock hazards in maintenance operations, the workplace, and equipment and product design are presented.

Table V–1: Health and Performance Effects of Some Physical Environmental Factors

Environmental Factor	Physiological or Health Effects	Performance Effects
Electric Shock	Electrocution Electrical Burns Muscle Contraction	Distraction Interference with Manipulation Tasks
Noise	Hearing Loss Fatigue	Interference with Communication Interference with Signal Detection
Vibration	Muscle, Joint, Organ Pains Nausea Hand Circulatory Disturbances Sensory Loss in Fingers Reduced Gripping Strength	Interference with Manipulative Tasks Interference with Visual Tasks
Illumination	Headache Muscle Discomfort Fatigue Reduced Visual Acuity Eye Injury	Distraction Glare Interference Reduced Detection of Defects
Color	None Known	Influence on Appearance of Other Objects in the Area Influence on Mood
Heat/Humidity	Heat Illness: Circulatory Collapse Muscle Cramps Burns Discomfort	Distraction Interference with Manipulation Tasks (from sweaty hands, hot surface temperatures)
Cold	Hypothermia Frostbite Shivering Loss of Flexibility in the Fingers	Interference with Manipulation Tasks (through shivering, finger flexibility loss) Distraction

The physical environmental factors described in this chapter are shown in column 1. Exposure to excessive amounts of these factors may produce the physiological or health effects shown in column 2. The intensity and duration of the exposure and the susceptibility of the person involved will determine the effects on performance (column 3). The seriousness of these performance effects also depends on the nature of the exposure and individual capacities to accomplish the specific tasks.

1. TYPES OF HAZARDS IN THE WORKPLACE

Line-to-ground electrical hazards are the most common type of hazard. They are difficult to detect because the electric device may continue to operate normally. Typical examples of grounding failures are as follows (R. M. Little, 1980, Eastman Kodak Company):

- Ground wire of a power cord broken or not connected.
- Use of a two-prong adapter plug (cheater cord or plug) with equipment having a three-wire power cord.
- Ungrounded wall receptacles.
- Polarity reversed on an equipment power cord or a wall receptacle.
- Electrical insulation failure in a heating element.
- Power line short circuit to the equipment case.

So that these hazards are reduced, proper maintenance and installation of devices should be provided. For example, one should follow the standards for electric current leakage developed by the manufacturing and electric power community (see, for instance, ANSI, 1973, 1981). According to these standards a maximum value for permissible leakage current is 0.5 milliampere (mA) for two-wire appliances and three-wire, cord-connected portable appliances (ANSI, 1973).

2. THE EFFECTS OF ELECTRIC SHOCK

The outcome of a person's exposure to contact with electricity is a function of his or her body's resistance to shock and the amount of current or charge delivered.

a. The Body's Resistance to Shock

The flow of electric current through the body is determined by the amount of applied voltage and by the impedance of the body to current flow. Mathematically, this result is represented by Ohm's law, as follows:

$$I = \frac{V}{Z}$$

where I is the current, in amperes (A), V is the applied voltage, in volts (V), and Z is the impedance or resistance, in ohms (Ω).

Pathways for electric current may be confined to the limbs that contact the live circuit, or, more critically, the current may pass through the body, as in the case of hand-to-hand or hand-to-foot contact (see Figure VA–1). About 10 percent of the current from a hand-to-foot pathway flows through the heart. This pathway has the most critical effect on heart function (Kouwenhoven, 1969).

Figure VA–1: Examples of Current Flow in Contacts with Electricity

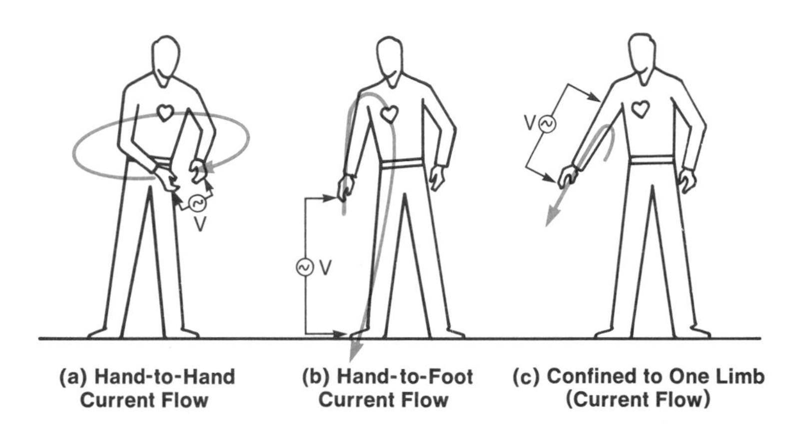

The pathways for current flow are shown for three types of contact with electricity: across the hands (part a), from hand to foot (part b), and within one limb (part c). The first two types of contact result in electric current passing through the body, making disorders of heart rhythms more possible. The voltage *(V)* applied is signified by a circle in which the symbol for current (~) is shown; the two arrows indicate the source and the sink (closing of the circuit) of the current flow.

Individual differences in body resistance to electric shock are considerable and overwhelm any differences that might be related to body size, sex, or age. Exposure to electric shock in the presence of moisture on the skin is more likely to have serious consequences because of the profound drop in skin resistance. Examples of situations in the workplace that result in wet skin include work, especially heavy work, in hot and humid environments and circumstances that contribute to anxiety and its concomitant perspiration.

The resistance of the outer layer of the skin decreases with increasing contact pressure and increasing contact area, and it differs with the part of the body contacted (Turner, 1972). Some common conditions for contact with electric current and the body skin resistances to them are presented in Table VA–1.

Within the body, muscle contraction, sensory processes, heart action, and other processes are mediated and controlled by electrical activity. These voltage potentials are measured in the millivolt (a thousandth of a volt) range. External voltage can affect these natural processes in a variety of ways and result in the following conditions:

- threshold perception
- pain
- sustained muscle contraction
- ventricular fibrillation
- cardiac arrest
- convulsions
- burns

Table VA–1: Skin Resistance for Various Contact Conditions (Adapted from Kleronomos and Cantwell, 1979.)

	Measured Skin Resistance	
Contact Condition	**Dry**	**Wet**
Finger touch	40 kΩ–1 MΩ	4–15 kΩ
Hand holding wire	15–50 kΩ	3–6 kΩ
Finger-thumb grasp	10–30 kΩ	2–5 kΩ
Hand holding pliers	5–10 kΩ	1–3 kΩ
Palm touch	3–8 kΩ	1–2 kΩ
Hand around 4-cm ($1\frac{1}{2}$-in.) pipe	1–3 kΩ	0.5–1.5 kΩ
Two hands around 4-cm ($1\frac{1}{2}$-in.) pipe	0.5–1.5 kΩ	250–750 Ω
Hand immersed	—	200–500 Ω
Foot immersed	—	100–300 Ω
Human body, internal, excluding skin	—	200–1000Ω

The resistance in ohms (Ω), kilohms (kΩ), and megohms (MΩ) of skin when dry or wet, from sweating or contact with a liquid, is given in columns 2 and 3. Column 1 specifies the contact conditions. The lower the resistance, the more current will flow for a given voltage applied to the skin. The presence of moisture on the skin reduces resistance fivefold or more in most of these conditions.

Table VA–2 summarizes the electric contact characteristics that can produce each of these responses. Examples of situations that could occur in the workplace are given along with the current thresholds and frequency characteristics of the electric contact. Short descriptions of the perception and pain thresholds are given below, because these are the types of exposure to electricity of most concern to human factors practitioners.

Table VA–2: Electric Shock Thresholds (Information assembled by R. M. Little, 1980, Eastman Kodak Company, from the following sources: Carter and Coulter, 1942; Dalziel, 1953; Dalziel and Lee, 1969; Hodgkin, Langworthy, and Kouwenhoven, 1972; Keesey and Letcher, 1970; Kouwenhoven, 1969; Plutchik and Bender, 1966; Turner, 1972; Watson and Wright, 1973.)

Effect	Threshold Current	Frequency	Examples of Situations Where Contact Could Occur
Perception			
warmth	0.2 mA/cm² dc	DC	Charging car battery
tingling or startle reaction	0.35–1.19 mA/cm² rms	60 Hz	Touching an improperly grounded appliance
Pain	<0.9 mA	DC	Contacting a charged surface with hand
	3–10 mA	60 Hz	Grasping an improperly grounded power tool
Muscle Contraction			
Let-go threshold (maximum current one can tolerate and still release an object by using the muscles being directly stimulated)	29 mA	DC	Shorting out a plating electrode with hand
	6 mA	60 Hz	Gripping a faulty tool equipped with a cheater plug
Respiratory paralysis	18–22 mA	60 Hz	Contacting a charge with a hand or foot and having it pass through the body to another limb

High voltage/high current	>5 A or 500 V	DC, AC, or impulse	Contacting a power transmission line or the high-voltage section of a television set
Cardiac Effects	54 J discharge	non-oscillating, $t < 1$ sec	Grabbing a high-voltage capacitor while standing in water
Ventricular fibrillation (uncoordinated heart activity);	54 J discharge	oscillatory, $t < 1$ sec	Discharging an electrical circuit across the body
whole-body muscular contraction	>67 mA	60 Hz	Using a badly grounded power tool in wet footing conditions
Cardiac Arrest	>5 A	60 Hz	Contacting a power transmission line
Convulsions, Head	100 mA	60 Hz	Contacting an electric power line

The effects of electric shock on the body are shown in column 1. The current at threshold for each response is shown in column 2, and its frequency is given in column 3. In the fourth column some situations where these current contacts may occur are listed. The values given are for adults and generally represent very sensitive people (0.5 percentile responses). The current is expressed in milliamperes (mA) per square centimeter (cm^2), and the value given for alternating current is the root mean square (rms) value (consult the Glossary for further information). Frequency indicates either direct current (DC) or alternating current (AC, or 60 Hz). In the discharge of a capacitor, t is the time of continuous application of the current, and the discharge is expressed in joules (J).

b. Threshold Perception

When low levels of direct current are applied to the hand, a slightly warm sensation is reported; with alternating current a tingling or prickly sensation results (Dalziel and Lee, 1969; Hodgkin, 1974). Varying thresholds for sensation exist, depending upon the body location selected and the nature of the electric contact.

In a typical situation a person may contact a live wire when moving equipment or get a shock while operating and gripping a power hand tool or appliance. As the area of skin contacted increases, higher current levels are required for perception (Turner, 1972).

The tingling or warm feeling resulting from threshold current contact is not harmful, but it may indirectly result in an accident. For instance, a person operating an electric device and sensing a mild shock may have a reflex response to draw the hand away, and he or she may drop or strike something in the workplace in the process.

c. Pain

The sensation of pain is highly subjective. Pain is usually sensed at the point of contact of an external electric source with the body. The sensation is noted as either a "tingling" or a "pin prick," or it occasionally may be described as "burning" (Turner, 1972). The current levels for pain range from 0.3 to 10 mA rms (milliamperes root mean square) at 60 Hz AC. The longer the duration of the stimulation, the less current is needed to produce pain.

The pain threshold is quite uniform over the body, but the size of the area contacted may affect the current level at which pain is perceived (Turner, 1972). In most studies the contact area is a circle with a diameter of 0.3 to 10 mm (Plutchik and Bender, 1966).

3. PREVENTION OF ELECTRIC SHOCK

The material in this section was developed from information in R. M. Little (1980, Eastman Kodak Company).

Electric shock injuries can be reduced by protecting the worker, selecting and maintaining the appropriate equipment, and designing the workplace or equipment to minimize the possibility of contact with electricity. Guidelines for reducing electric shock injuries are given below.

a. Protecting the Worker

To prevent accidental contact with live circuits while servicing them:

- Instruct workers to observe safe use of any electric interlock systems.
- Use warning labels to designate exposed high-voltage circuits.
- Insulate or guard high-voltage terminals and components that are subject to accidental contact during servicing.

- Discharge capacitor energy storage devices with bleeder resistors or electronic or mechanical short-outs, where feasible.
- Locate test points away from exposed high-voltage circuits.
- Protect electric circuits from accidental leakage or spillage of liquids.
- Warn workers to remove metallic jewelry and rings from hands and wrists when working

Where medical monitoring is done, special precautions are needed to prevent electric shock from improperly grounded devices. In this situation acceptable leakage currents are less than the values given earlier. Further information can be found in the *National Electric Safety Code* (ANSI, 1981).

b. Workplace Design Considerations

Good electric ground integrity is the first line of defense in protecting workers against electric shock in the workplace. A second defense is the use of ground fault interrupters (GFIs), which interrupt power if a current path to ground occurs (ANSI, 1981). GFIs can be useful in areas where water may be present, such as in the following situations:

- Outdoor electric receptacles where appliances or power tools are used.
- Power receptacles near sinks and tubs in laundries.
- Wet work areas where hoses or liquid chemicals are used.

In addition, they are useful at electronic service benches where inexperienced people work.

The installation and maintenance of electric equipment should be standardized, and periodic testing of current leakage should be done. Purchased and borrowed electric devices should be tested for leakage prior to use in the field.

Electric outlets should be located conveniently for use and testing in the workplace. They should be of sufficient number to prevent overloading of circuits or extensive use of extension cords. Electric power tools should be inspected regularly to identify fraying of the power cord or other potential current leakage damage.

SECTION V B. NOISE AND VIBRATION

The operation of production equipment and other machines often introduces noise and vibration into the workplace. Noise is a common environmental factor in industry, and techniques to reduce it are constantly under development. Vibration is a concern in jobs where people drive heavy equipment or in some construction tasks, such as pavement repair, but it is not encountered as frequently as noise in the workplace.

1. NOISE

Noise may affect people in the workplace in any of the following ways:

- It may contribute to hearing loss.
- It may interfere with communication.
- It may annoy or distract the people nearby.
- It may alter performance on some tasks.

Noise from other people in the work area may also be distracting, especially in offices. A summary of the types of noise complaints in one industry is presented in Table VB–1, along with a breakdown of the type of workplace involved.

Table VB–1: Noise Complaints Over a 20-Year Period (Adapted from B. Crist, 1980, Eastman Kodak Company.)

Type of Complaint	Percentage of Requests to Human Factors for Noise Evaluations
Annoyance, distraction	38
Speech difficulty	37
Hearing loss potential	10
Perceived discomfort	9
Performance concern	6
Type of Workplace	
Production areas, shops, warehouses	29
Business machines	19
Offices in office areas	17
Offices near production areas, shops	12
Quality control	7
Laboratories	7
Other	9

The table summarizes the types of complaints about noise (upper part) and the types of workplaces involved (lower part) in the requests for noise evaluations brought to the Human Factors Section over the past 20 years. The values in column 2 are the percentage of total noise-related requests fitting each category. Annoyance, distraction, and difficulty in hearing others speak were the most common complaints from both the production and office areas.

Multiple concerns are often expressed about workplace noise. About half of the concerns were expressed as, "We have trouble talking to each other, and it is hard to concentrate." Because noise from production equipment, business machines, and construction tasks is a common environmental element for many industrial workers, guidelines for exposure to it have been developed to minimize health and performance effects on people. These guidelines are discussed in the sections that follow.

a. Criteria for Noise Exposure

Levels of noise in the workplace can be defined for each of the potential effects mentioned above: hearing loss, communication interference, annoyance or distraction, and altered performance.

(1) Hearing Loss

The health effects of noise are a primary concern of industry, and active hearing protection programs were established several years ago (NIOSH, 1972a). Hearing loss, although it occurs gradually, represents irreversible damage to the inner ear. The degree to which hearing is affected depends on the intensity, frequency spectrum, and duration of noise exposure, plus individual susceptibility. Noise-induced hearing loss usually produces loss of the high-frequency components first, resulting in reduced quality, clarity, and fidelity of sounds. This hearing loss combined with normal aging processes in the hearing mechanism, may result in handicapping hearing losses.

Research has indicated that extended exposure, about 8 hours, to noise levels in excess of 85 dBA (decibels on the A scale of a sound-level meter, discussed in Chapter VI, Appendix B) may cause hearing loss (NIOSH, 1972a). Thus engineering controls or hearing protection should be implemented in work areas where noise levels routinely exceed 85 dBA. Duration of exposure will affect the acceptability of different noise intensities; however, short exposures to higher noise levels may be acceptable. The recommended noise exposure guidelines (intensity and duration) for protection against hearing loss are presented in Figure VB–1.

A sound-level meter can be used to measure sound levels to assess potential hearing loss situations, providing the individual's noise exposure is constant. If the worker is exposed to fluctuating sound levels because of changes in work location or varying machine noise in the production cycle, a noise dosimeter is preferable for evaluating individual noise exposure. A dosimeter integrates the varying sound levels and presents the results as a percentage of the daily permissible limits. Further discussion of noise measurement techniques is presented in Chapter VI, Appendix B.

(2) Annoyance and Distraction

Although noise levels below 85 dBA probably do not contribute to hearing loss problems, they may contribute to performance decrements due to distraction or annoyance. Noise from office or production equipment can reduce the effectiveness of communications and make it difficult for people to con-

Figure VB–1: Guidelines for Noise Exposure to Protect Hearing; Recommended Maximum Duration Versus Noise Level (Adapted from ACGIH, 1982; NIOSH, 1972a.)

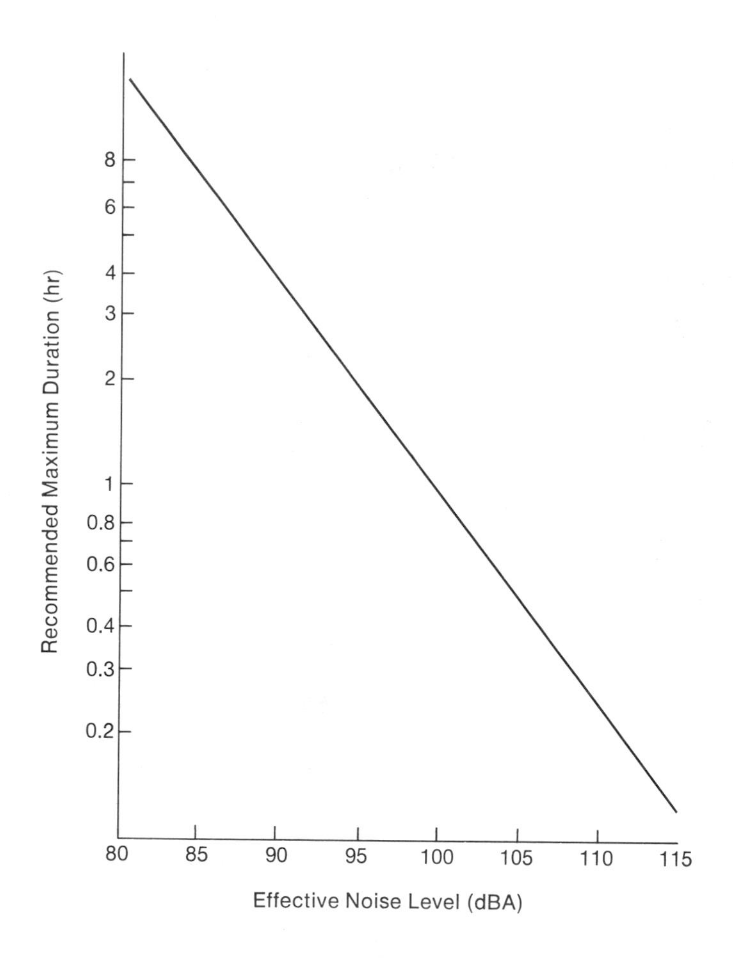

The recommended maximum duration of exposure (in hours, hr, on the vertical axis) to noise of different intensities (in decibels, dBA, on the horizontal axis) is given. The higher the noise level, the less time a person should be exposed to it in order to reduce the risk of hearing damage. Noise levels above 115 dBA should be avoided; levels below 80 dBA are not known to contribute to hearing loss over extended exposure times.

centrate in some types of tasks. In other situations lack of noise may be undesirable. This effect has been demonstrated in some landscaped offices where white noise had to be piped into the area in an attempt to improve comfort for office personnel (J. Cardoza, 1969, personal communication, Eastman Kodak Company). The white noise served to mask some of the speech sounds from neighboring areas; it also provided a steady background against which intermittent sounds were less disturbing. White noise levels of 48 dBA can be effective in masking some office sounds, but levels above 52 dB may be distracting and annoying (Nemecek and Grandjean, 1973). In areas with hard walls, such as cafeterias and break rooms, speech itself is the noise. Absorptive treatment of the room may be needed to reduce speech interference.

In setting criteria for the acceptability of noise in the workplace, one has to consider the needs for both communication and speech privacy. The preferred noise criterion (PNC) curves (see Figure VB–2) provide guidelines for ambient sound pressure levels in each of nine octave bands. They are based on the subjective ratings of office workers and on experiments involving the preferred tonal qualities of noise, speech communication requirements, and the spectra that generated complaints.

To use the PNC curves, one measures the workplace noise with an octave band analyzer and plots the results in association with the appropriate PNC curve. The selection of the appropriate curve is made from the category classifications shown in Table VB–2. These categories are for indoor activities with a steady background noise level.

An example of the use of the PNC curves to evaluate a noise complaint follows. A production foreman's office was located near a chemical mixing pump; the pump's noise made face-to-face communications and phone conversations difficult. In addition, the noise was distracting to the foreman and made concentrated performance of paperwork more difficult to do. Because the foreman was rarely in the office, the PNC–60 curve was selected for workplace comparisons. The resulting curves, with and without the pump on, are shown in Figure VB–3.

Inspection of these curves shows the following results:

- Pump noise is only above the PNC–60 curve at frequencies above 500 Hz.
- The PNC–60 curve is exceeded in three bands. If a curve is exceeded in any band, the entire curve is considered to be exceeded.
- Reduction of the pump noise in the 1000-, 2000-, and 4000-Hz bands would bring the office environment within the recommended guidelines. This reduction was achieved by enclosing and insulating the pump motor.

Engineering approaches to reduce noise can be evaluated by using the same measures and can be tailored to the specific noise spectra that produce

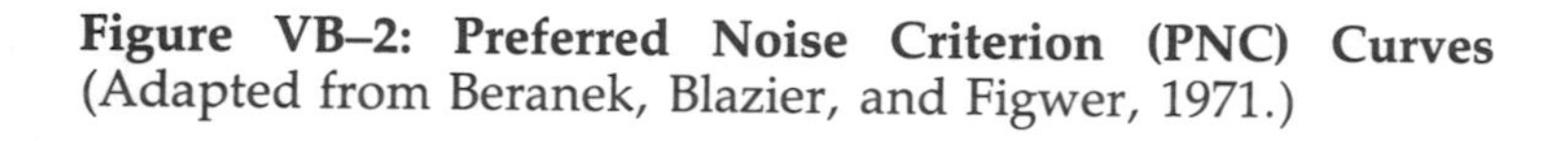

Figure VB–2: Preferred Noise Criterion (PNC) Curves (Adapted from Beranek, Blazier, and Figwer, 1971.)

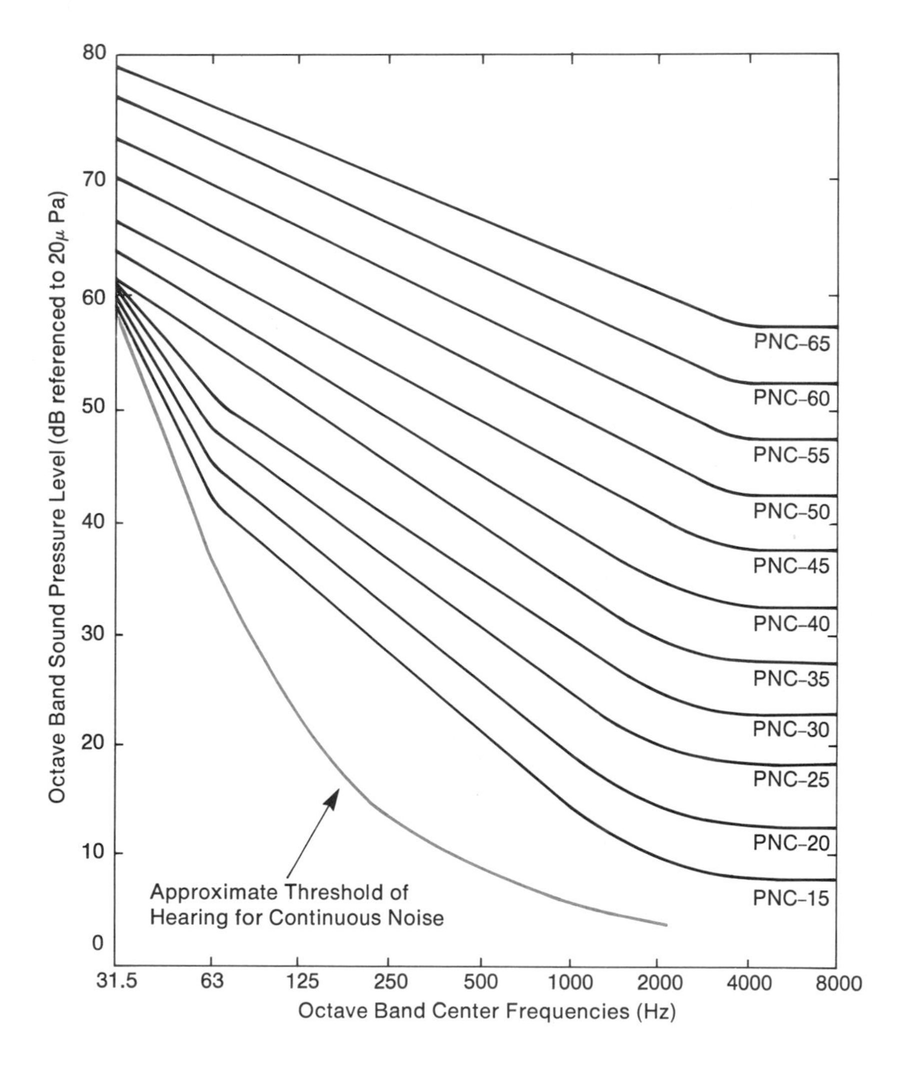

The relationships between sound-level intensity (octave band sound pressure, in decibels, dB, referenced to 20 micropascals, μPa, on the vertical axis) and frequency (represented by the nine center frequencies on the horizontal axis) are displayed for different conditions of hearing (PNC curves). The curves range from the lower threshold of hearing for continuous noise to a PNC–65 curve, where there is significant interference with communication. The higher the frequency of the noise, the lower its intensity must be to bring it to the appropriate PNC curve for hearing or communication. Descriptions of the PNC curves for different hearing conditions are given in Table VB–2.

Table VB–2: Recommended PNC Curves and Sound Pressure Levels for Several Categories of Activity (Adapted from Beranek, Blazier, and Figwer, 1971.)

Acoustical Requirements	PNC Curve*	Approximate† L_A (dBA)
Listening to faint musical sounds or distant microphone pickup used	10 to 20	21 to 30
Excellent listening conditions	Not to Exceed 20	Not to Exceed 30
Close microphone pickup only	Not to Exceed 25	Not to Exceed 34
Good listening conditions	Not to Exceed 35	Not to Exceed 42
Sleeping, resting, and relaxing	25 to 40	34 to 47
Conversing or listening to radio and TV	30 to 40	38 to 47
Moderately good listening conditions	35 to 45	42 to 52
Fair listening conditions	40 to 50	47 to 56
Moderately fair listening conditions	45 to 55	52 to 61
Just acceptable speech and telephone communication	50 to 60	56 to 66
Speech not required but no risk of hearing damage	60 to 75	66 to 80

* PNC curves are used in many installations for establishing noise spectra.
† These levels (L_A) are to be used only for approximate estimates, since the overall sound pressure level does not give an indication of the spectrum.

The PNC, or preferred noise criterion, curves (column 2) and approximate sound pressure levels (L_A in dBA, column 3) for several hearing conditions (column 1) are given. Voice sound frequencies are used to determine the approximate sound pressure levels. At higher PNC curves it becomes more difficult to hear speech or music. All of these curves represent noise exposures lower than those that may cause hearing damage.

the interference or distraction. In areas such as staff offices and quality control rooms, our experience indicates that the PNC curves may be conservative by five to ten PNC units with regard to acceptability (B. Crist, 1981, personal communication, Eastman Kodak Company).

(3) Interference with Communication

Speech interference by noise is fairly common around production machinery and business equipment. The criticalness of communications should de-

Figure VB–3: Evaluation of Pump Noise in a Foreman's Office (Adapted from B. Crist, 1978, Eastman Kodak Company.)

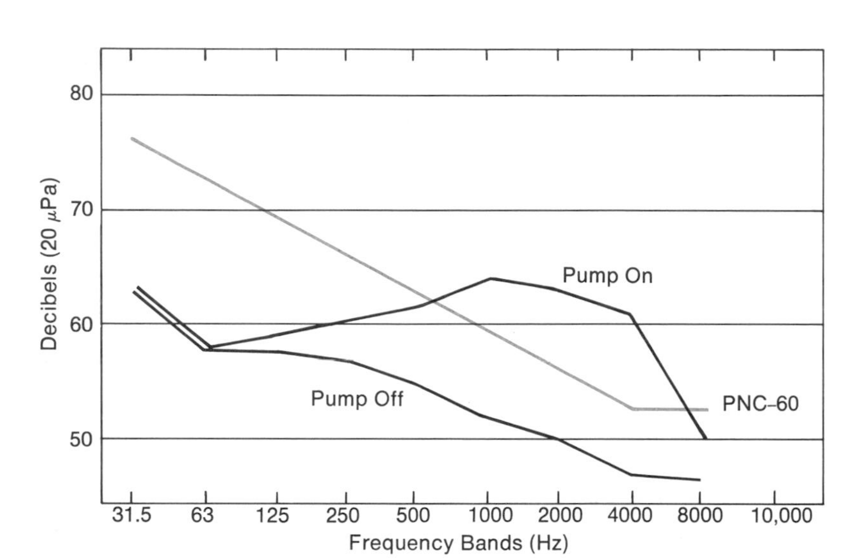

The sound pressure levels, in decibels or micropascals (μPa), in a foreman's office are shown on the vertical axis for nine different frequency bands, in hertz, on the horizontal axis. The lower curve indicates the office noise levels when the pump was turned off; the middle curve values were measured when the pump was on. The PNC–60 curve is shown for comparison. Pump noise exceeded the PNC–60 curve only in the higher frequencies. This analysis helps an engineer identify the most effective methods of noise reduction.

termine the steps to be taken to improve the noise levels in the environment. The most common method for rating the speech interference effects of noise is called the preferred speech interference level (PSIL) (Webster, 1969). This method uses the average sound pressure levels, in decibels, of the octave bands centered at 500, 1000, and 2000 Hz. Curves are drawn to estimate the distance at which intelligible communication can be made at different levels of PSIL. Figure VB–4 presents these curves organized according to the ease with which communication can take place, both for the speaker and the listener.

From this figure it can be seen that communications in the foreman's office in the earlier example (with a PSIL of 63) would require a raised voice to converse effectively at 1.8 m (6 ft). People standing more than 2.5 m (8 ft) apart will have to raise their voices to converse any time the ambient noise

Figure VB–4: Preferred Speech Interference Levels (PSIL) as a Function of Distance and Ease of Communication (Adapted from Webster, ASHA Reports 4 1969.)

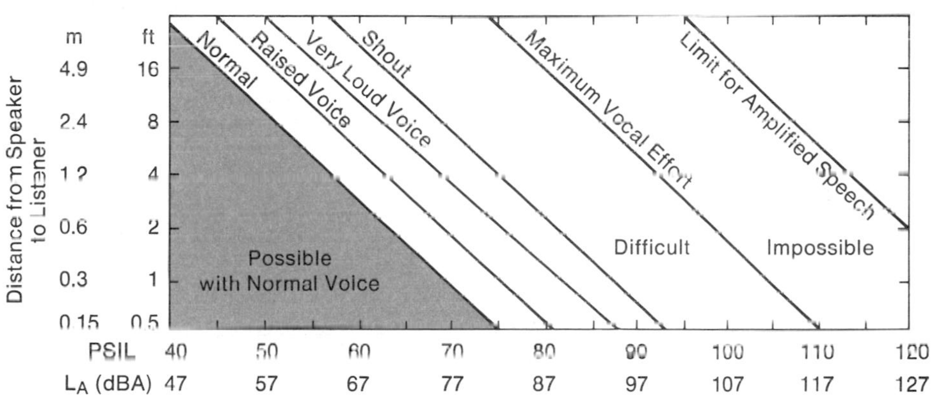

The ambient sound pressure level (L_A, in decibels, or PSIL, the preferred speech interference level) is given on the horizontal axis. The distance between the speaker and listener (on the vertical axis) will determine to how high a level the speaker must raise his or her voice in different noise conditions. The curve relating ambient noise to distance for communication for normal speaking voice levels is shown on the left. The second, third, and fourth curves illustrate the increasing voice levels needed to be heard in the ambient noise. The fifth curve (second from the right side) indicates the maximum vocal level for conversations in a noisy environment. The curve on the right shows the upper limits of ambient noise and distance for understanding amplified speech.

level exceeds 60 dBA. When the ambient noise level exceeds 85 dBA, people working for extended periods in the area should be wearing some equipment to protect their hearing, which may further reduce communications at any distance other than that of people standing right next to each other.

The ability to hear and converse effectively on the telephone is also important and can be predicted by PSIL measurements. The following guidelines for calls between telephone exchanges should be used:

PSIL	Telephone Use
60 dB or less	Satisfactory
60 to 75 dB	Difficult
Above 75 dB	Impossible

Within a single telephone exchange the values are 5 dB higher. For areas where noise cannot be reduced sufficiently to meet these levels, special telephone equipment may be used to improve communication. For example, noise-canceling transmitters are available that attach easily to the handset and help reduce background noise that would otherwise be mixed with a speaker's voice. Amplification devices are useful if the listener is in a noisy area, and enclosures or small booths may improve the situation for both listener and speaker.

b. Performance Effects of Noise

The performance effects of noise have been investigated by a number of people (Broadbent, 1957; Cohen, 1969; Kryter, 1970; Miller, 1971) with mixed results. Qualitative patterns have emerged from these studies, suggesting that certain combinations of noise variables and task types may affect performance. In addition, there are differences in the way individuals respond to the same noise/task combinations. The following general observations apply to situations where noise may affect performance:

- Paced tasks: Noise with rhythmic patterns may improve performance of a simple repetitive task by pacing the work, if the natural work pace and the noise pace are nearly synchronous.
- Information transfer: The kinds of tasks that are most likely to show degradation as a result of noise are those with mental demands, complexity, and considerable detail. Taking telephone orders which involve collection of product, pricing, billing, and shipping information through the use of computer terminals is an example of this type of task.
- Noise features: The following noise features are most likely to degrade performance:
  variability in level or content
  intermittency
  high-level repeated noises
  frequencies above approximately 2000 Hz
  any combination of the above

Degradation in performance can often be measured as a reduction in quality. In some cases average quality may not be affected, but variability of quality increases with noise (Miller, 1971). Also, tasks that directly involve verbal transmission of information will be degraded if the noise spectrum affects speech. The effects of noise on productivity are not consistent because performance on the task of interest may or may not be affected by environmental distractions. Further research is needed from before-and-after noise reduction field studies to assess the effects of noise on productivity for specific jobs.

Evaluating the possibility of performance effects of noise is based on estimates and judgments of the noise and task contents. Measurements discussed earlier, such as the octave band analyses in conjunction with the PNC and PSIL curves, are used to establish acceptability ranges for hearing and speaking. Further measures of noise characteristics—such as intermittency, level and spectral variations, and the presence of narrow-band noise, as well as measures of the task complexity and mental demands—may be made, but how they affect performance is not clear. Interviewing people to establish the specific situations in which noise is considered a problem is an effective way to determine the best approach to reducing the noise in the workplace.

c. Approaches to Reducing Noise in the Workplace

The material in this section was developed from information in Harris (1957).

There are many ways to reduce noise to desired limits in both new and existing installations. When new areas are being planned, one can obtain the power ratings of purchased equipment and estimate ratings for equipment built in-house or for older machines. Rough predictions of noise levels can be made from the ratings, the number of each kind of equipment, and the acoustical characteristics of the area.

Noise reduction methods can be applied at the noise source, in the path of noise transmission, and at the location(s) where work might be affected. Methods of reduction involve the following techniques:

- Reducing the level or altering the spectrum of the generated noise.
- Using barriers to reduce noise transmission through air or structures.
- Absorbing incident or reflected noise.

In practice, combinations of methods are usually used. Often reduction of source noise may be prohibitively expensive, and other methods must be employed. For example, one might use enclosures (barriers) with porous linings (absorbers) around engine-driven air compressors or punch presses.

Noise reduction is a specialized field and should generally be done by those with the appropriate training and experience. Some simple solutions can be found occasionally, such as the use of a screw-on muffler on the exhaust outlet of a noisy, pneumatically powered cutting machine. Such a device can effectively lower the overall noise level; it is particularly effective in reducing the high-frequency components that are characteristic of high-pressure pneumatic systems.

d. Special Considerations

(1) Music

The material in this section was developed from information in T. W. Faulkner (1969, Eastman Kodak Company) and Grandjean (1961).

Background music has been used in factories to improve the work environment. In some instances the music has been well received, while in others it has been regarded as a failure. The following observations about music in the workplace are based on a limited number of production jobs, but they have held in each case:

- There has been no conclusive proof that the presentation of music increases productivity, although there are many claims to that effect.
- Most, but not all, production workers are likely to enjoy hearing music when they work. This result is particularly true in areas where repetitive assembly tasks or heavy physical work is done. In jobs where more concentrated attention is required, music may be an undesirable distraction.
- Those who do not enjoy the music will probably complain.

If music is presented, the following guidelines should be considered:

- The employees who hear the music should have input concerning the type and the schedule; that is, whether it is presented continuously or only at specified intervals. A 15-minutes-on, 15-minutes-off schedule is often used to minimize complaints.
- Presentation should use quality systems, and usually the level should be only slightly higher than background noise.

Do not provide music if the background noise level is more than 70 dBA. The addition of music at a level high enough to be heard distinctly may make the employees regard the music as simply more noise, and the music will further interfere with oral communication.

(2) Evacuation Alarms

Evacuation alarms notify people in a work area of a situation requiring them to leave the building. The design of these alarms should take into account the ambient noise level, and they should be distinct enough to be immediately recognized. A spoken message is often desirable to reassure people and to let them know what to do. The following guidelines apply to the design of emergency alarms and public address systems (B. Crist, 1979, Eastman Kodak Company; additional information from Corliss and Jones, 1976; Fidell, 1978; Fidell, Pearsons, and Bennett, 1974):

- The evacuation tone should precede any verbal evacuation messages announced over the public address system.
- The evacuation alarm tone should have a minimum duration of 10 seconds. Coded alarms should be repeated until the building is evacuated.
- An evacuation alarm tone with a swept frequency of one octave somewhere between 500 and 2000 Hz is recommended.

- The evacuation tone should be about 10–12 dB higher than the highest one-third octave band of the ambient noise; that is, the uppermost of the three bands encompassing the tone frequencies.
- Speech should be 14 dB or more above the ambient noise level. If the resulting noise level is 70 dB or more (based on the four-band PSIL), the address system should incorporate 12 dB speech peak clipping and reamplification.

Because people with hearing loss may not recognize an audio evacuation alarm, visual alarms such as blinking lights should also be included in the emergency evacuation alarm installation.

2. VIBRATION

Vibration is the periodic motion of particles away from their position of equilibrium (*Webster's Dictionary*, 1967). It is characterized by its frequency, acceleration, and direction. The resonance band for the body is 4–8 Hz. Long-term exposure to one gravity unit (1 g) of vibration at this frequency can affect people's health (Wasserman, 1980).

The following forms of vibration are of most concern to industry:

- Whole-body vibration associated with transportation, such as truck, bus, or car driving.
- Whole-body vibration associated with the operation of large production machinery, such as presses.
- Segmental vibration associated with the operation of power tools, such as chipping hammers, chain saws, or grinders.

a. Effects on the Body

(1) Whole-Body Vibration

A person tolerates vibration slightly better while standing than while sitting. Women, on the average, experience more discomfort than men at the same vibration level (Rabineau, 1976). The acceleration and frequency interact to determine the level of discomfort; the duration of exposure will also determine vibration acceptability in the workplace. Tolerance is defined as the level of vibration at which effects on performance of a motor or visual task will be detected (Ramsey, 1975). Figure VB 5 illustrates body tolerances to vertical vibration of varying frequencies, accelerations, and durations. (Acceleration is also expressed in terms of gravity units, g. One g is equivalent to about 10 m/sec^2.)

Most heavy construction equipment, trucks, and buses produce vibrations (predominately vertical) with frequencies in the 0.1–20-Hz range and with accelerations generally less than 0.2 g but with peaks up to 0.4 g (Wasserman

Figure VB–5: Whole-Body Vertical Vibration Tolerances (Adapted from IOS, 1974; Sandover, 1979.)

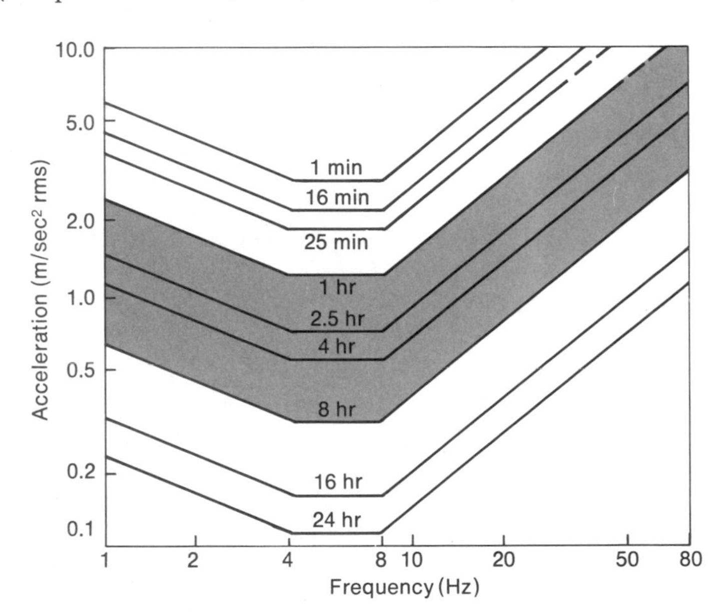

The nine curves show maximum acceptable exposure times, in minutes (min) and hours (hr), to combinations of vibration intensity, in meters per second per second (m/sec^2) of acceleration, and frequency, in hertz (Hz). The shorter the vibration exposure, the higher are the acceleration levels that can be tolerated. The least acceptable range of frequencies at all accelerations and durations of exposure is from 4 to 8 Hz.

and Badger, 1973). These values would be acceptable for an 8-hour continuous exposure, but longer exposures might be associated with discomfort and physiological responses of the sort mentioned below. The vibration characteristics of production equipment will be determined by the nature of the operation of the equipment. These characteristics can be measured by mounting accelerometers on the vibrating surface. They often are in the same range as those for vehicles.

Some of the known physiological effects of vibration in the 2–20-Hz range at 1 *g* acceleration include abdominal pain, loss of equilibrium, nausea, muscle contractions, chest pain, and shortness of breath (Beaupeurt, Snyder, Brumaghim, and Knapp, 1969; Ramsey, 1975; Whitham and Griffin, 1978). It is noteworthy that low-frequency (<1 Hz), low-acceleration (<0.3 *g*) vibration can have a relaxing effect on people (Ramsey, 1975). There is also some indication that 5-Hz, low-acceleration vibration assists in keeping people alert on long-term monitoring tasks (Poulton, 1978). Blurring of an image will occur at 10–30 Hz, resulting in some loss of visual acuity. Tracking will also exhibit performance decrements. Performance in some precision-manipulation tasks may also demonstrate decrements from vibration in the 5–25-Hz range. In these studies the accelerations were generally in the 0.2 *g*–0.3 *g* range (Ramsey, 1975).

(2) Segmental Vibration: Hands

For segmental vibration to the hands, accelerations in the range of 1.5 *g*–80 *g* and frequencies from 8 to 500 Hz are of concern. Raynaud's phenomenon is a vibration-associated circulatory disturbance of the fingers that results in stiffness, numbness, pain, and blanching of the fingers and a loss of strength. Vibration frequencies in the range of 25–150 Hz and accelerations from 1.5 *g* to 80 *g* are most commonly associated with white finger, or Raynaud's phenomenon (Ramsey, 1975). People who develop this problem are unable to do fine manipulative tasks in cool or cold environments and may lose endurance for sustained-holding tasks in warmer temperatures.

Tolerance curves for segmental vibration to the hands are under development (IOS, 1979). Table VB–3 summarizes current knowledge about the limits.

Table VB–3: Characteristics of Vibration That Are Within the Safe Range for Transmission to the Hands (Adapted from Rabineau, 1976.)

Duration	Frequency Range (Hz)	Upper Limit of Acceleration (*g*)*
30 minutes	8–15	6
	80	40
	250	125
4 to 8 hours	8–15	1.5
	80	6
	250	20

* 1 *g* of acceleration is approximately equal to 10 m/sec^2. The upper limit is the level at which performance decrements can be seen.

The upper limits of acceleration in gravity units, *g* (column 3), for vibration exposure to the hands in three frequency ranges (column 2) and for short and extended exposure durations (column 1) are given. The criterion for setting the upper limit of acceleration is a reduction in performance on a manipulation or visual task. For vibration exposure to the hand that lasts more than 4 hours, the recommended maximum accelerations drop to about a sixth of the values that are acceptable for 30-minute exposures. Lower-frequency vibration is not tolerated as well, so accelerations have to be lowered considerably below the acceptable values for higher frequencies.

Segmental vibration, primarily to the hands and arms, will vary according to the type of power tool being used. Chipping hammers, for instance, which may be used to detach caked material from metal surfaces, have accelerations of around 30 *g* and frequencies of up to several hundred hertz (Pyykkö et al., 1976; Wasserman, Taylor, and Curry, 1977). Grinders have higher frequencies and have accelerations generally under 1 *g*. The weight, size, and design of the power tool will determine its vibration characteristics (Abrams and Suggs, 1977). From this information about chipping hammer vibration characteristics and the data in Table VB–3, one can conclude that a person should not operate such a tool for a full shift. If the acceleration could be kept consistently below 20 *g*, the tool would be suitable for extended use. The chipping hammer with 30 *g* acceleration could, however, be used for less than an hour without concern about adverse effects on the operator.

b. Ways to Reduce Vibration Exposure

There are three ways to decrease a person's exposure to vibration: reduce the vibration, reduce the length of time a person is exposed to it, and isolate the person from the vibration with cushions or other isolators.

Vibrations can be reduced by engineering the equipment or workplace more effectively. For example (Ramsey, 1975):

- Mount equipment on springs or compression pads.
- Maintain equipment properly; balance and replace worn parts.
- Use materials that generate less vibration.
- Modify equipment speed, feed, or motion to change the vibration characteristics to a more suitable range.

The amount of time a person is exposed to vibration can be reduced by alternating the person between work tasks where vibration is present and those where it is negligible. As indicated above, chipping hammer use should not exceed 4 hours per person per shift if the acceleration of the tool is 20 *g* or more (Bovenzi, Petroniv, and DiMarino, 1980). Reducing exposure time is not always feasible, however, as in long-distance trucking. Thus proper engineering of seating is usually the preferred approach in such cases.

Isolating the person from the vibration source can be done in the following ways:

- By providing a spring or cushion as a vibration isolator in a seated task.
- By providing a spongy rubber or vinyl floor mat for standing operations.
- By designing tools to reduce the vibration transmitted to the hands, for example, by using isolating materials.
- By counterweighting tools to minimize the gripping forces required to operate them (see Section IIB, "Adjustable Design Approaches").

SECTION V C. ILLUMINATION AND COLOR

The amount and quality of light at a workplace and the color of the equipment, walls, and work surfaces can influence people working there and may affect their job performance. In this section guidelines are given for the design of illumination, including general, task, and special-purpose lighting, which should make visual tasks suitable for a large majority of industrial workers. A short summary of information about the use of color in the workplace concludes the section.

1. ILLUMINATION

For most jobs, vision is the main sensory channel for receiving information. Illumination, therefore, is one of the critical elements in the design of any workplace, because without adequate lighting important task elements may be incorrectly seen or not be seen at all. How adequate the lighting seems depends on its quality and quantity and on the task difficulty.

Illumination, also called illuminance, is a measure of the amount of light falling on, or incident to, a work surface or task from ambient and local light sources. It is measured with an illumination (lux or footcandle) meter, which is set directly on the work surface. The farther away the surface is from the source of light, the less the illumination will be. Luminance, on the other hand, is a measure of the light reflected off a surface and is associated with the subjective sensation of brightness. Luminance does not vary with the distance between the surface and the observer, and it is measured with a photometer located at a convenient distance from the surface and pointed toward it. For additional information, see the "Dictionary of Lighting Terms" in the *IES Lighting Handbook* (Kaufman and Christensen, 1972).

Task modification is an effective way to improve visual task performance without increasing illumination levels. Some ways to make visual tasks easier are:

- Increase the contrast of written material.
  Use black ink rather than pencil.
  Change typewriter ribbons frequently.
- Increase the size of the smallest critical detail.
  Use larger type size for printed material.
  Decrease the viewing distance.
- Alter the workplace so that the task materials are perpendicular to the operator's line of sight.

Some more specific approaches to making visual inspection tasks easier are presented in the discussion "Special-Purpose Lighting" given later in this section.

For any given workplace the designer must consider the minimum lighting requirements for each task and subtask, the types of artificial light sources that can be used to illuminate the work surface, the amount of task lighting that can be provided, and the available methods to minimize glare. These factors must be balanced with the implementation and operating costs. Only when all of these factors are carefully considered can a satisfactory design be achieved.

a. Recommended Light Levels for Different Tasks

In selecting a specific illuminance for each workplace, the designer must consider the visual difficulty (such as the contrast and spatial resolution requirements), the criticality, and the frequency of each task and subtask to be performed, as well as the time available to perform the tasks. The minimum lighting level is the level that is sufficient for people performing the most difficult and critical tasks.

(1) Quantity of Light and Productivity

Numerous studies have examined the relationship between the amount of illumination and productivity. Most have shown some increase in productivity as illumination increases, but the amount of the increase is task-dependent (Barnaby, 1980; Hopkinson and Collins, 1970; Hughes and McNellis, 1978; IERI, 1975; Ross and Baruzzini, Inc., 1975). It is not clear, however, how much of the change is attributable to improved seeing and how much to motivational factors. Typically, the productivity increases are larger when the tasks are very demanding visually or when the workers are over 45 years of age. Increases in illumination have no effect, or only a small one, on productivity for younger workers and less visually demanding tasks.

The illuminance-related increases in productivity for older workers can be attributed to several factors. With age the optic media (eye fluids, lens, cornea) can become denser, and the ability to bring close objects into sharp focus is reduced. An increase in illumination can reduce the impact of these age-related changes in visual performance. However, older persons are usually more affected by the glare of reflected light (which increases with increasing illumination level) than are younger ones. Too much light, then, can be as undesirable as too little.

Increases in illumination usually have an associated cost increase. For some applications the increase in productivity realized by higher light levels may reduce labor costs more than the costs of the additional lighting. Energy conservation needs suggest the reduction of lighting levels wherever possible. The challenge at hand, then, is to find out how to increase productivity while decreasing, or at least not increasing, the amount of energy used for lighting.

(2) Illumination Guidelines

Recommendations for illumination levels are given in Table VC–1 as ranges of illuminance for specific types of activities. These guidelines take into account the worker's age, the importance of speed and accuracy, and the reflectance of the task background. However, the responsibility for selecting a specific value is clearly left to the lighting engineer and to supervisory personnel.

Table VC–1: Recommended Range of Illuminance for Various Types of Tasks (Adapted from Flynn, 1979.)

	Range of Illuminance*	
Type of Activity or Area	**Lux**	**Footcandles**
Public areas with dark surroundings	20–50	2–5
Simple orientation for short temporary visits	>50–100	>5–9
Working spaces where visual tasks are only occasionally performed	>100–200	>9–19
Performance of visual tasks of high contrast or large size: reading printed material, typed originals, handwriting in ink, good xerography; rough bench and machine work; ordinary inspection; rough assembly	>200–500	>19–46
Performance of visual tasks of medium contrast or small size: reading pencil handwriting, poorly printed or reproduced material; medium bench and machine work; difficult inspection; medium assembly	>500–1000	>46–93
Performance of visual tasks of low contrast or very small size: reading handwriting in hard pencil on poor-quality paper, very poorly reproduced material; very difficult inspection	>1000–2000	>93–186
Performance of visual tasks of low contrast and very small size over a prolonged period: fine assembly, highly difficult inspection, fine bench and machine work	>2000–5000	>186–464
Performance of very prolonged and exacting visual tasks: the most difficult inspection, extra fine bench and machine work, extra fine assembly	>5000–10,000	>464–929
Performance of very special visual tasks of extremely low contrast and small size: some surgical procedures	>10,000–20,000	>929–1858

* The choice of a value within a range depends on task variables, the reflectance of the environment, and the individual's visual capabilities.

The ranges of illuminance, in lux (column 2) and footcandles (column 3), recommended for several types of work activities or areas (column 1) are given. These guidelines are consistent with energy conservation goals as well as with the levels needed by a diverse population doing the activities described. In general, the more demanding a visual task and the longer it is sustained, the higher the recommended illumination level will be. The upper end of the range should be used to accommodate older workers on difficult visual tasks. Local task lighting is often preferable for increasing illumination at a workplace above 1000 lux (93 footcandles).

To meet these illumination guidelines while still saving energy, the lighting engineer should be sure that all light sources and fixtures are cleaned regularly. He or she might also consider employing two other strategies:

- Use higher-efficiency (lumens per watt) light sources.
- Modify the pattern of illumination so that light levels are increased at the workplaces and decreased elsewhere.

b. Selection of Higher-Efficiency Light Sources

In the selection of artificial light sources for illuminating work areas and workplaces, the two most important considerations are efficiency, in lumens per watt (lm/W), and color rendering. Color rendering is the degree to which the perceived colors of objects illuminated by various light sources match the perceived colors of the same object when illuminated by standard light sources. Other important factors are maintenance requirements, including relamping, reballasting, and cleaning; ease of shielding and directional control; and overall system economics. Efficiency is a very important factor because it is inversely related to operating costs. Operating costs for commercial and industrial lighting systems tend to be much higher than initial costs (system design, materials, and installation costs) and maintenance costs. Use of efficient light sources also helps reduce energy consumption.

Unfortunately, the more efficient light sources are often not suitable for tasks requiring color discrimination because of their poor color rendering. Until better sources are available, the best way to determine the optimum trade-off between efficiency and color rendering is through empirical testing. Test conditions should approximate as closely as is feasible those of the tasks, workplaces, and workspaces being designed or modified.

The efficiencies and color-rendering characteristics of commonly used artificial light sources are given in Table VC–2. For additional information, see Burnham, Hanes, and Bartelson (1963), General Electric Company (1978), Hopkinson and Collins (1970), and Kaufman and Christensen (1972).

c. Direct and Indirect Lighting

Fixtures that face downward generally provide direct lighting; those that face upward give indirect (reflected) light. Some fixtures provide both types of light. With direct lighting all of the available light is directed toward the task. The reflection of light off the ceiling in indirect lighting results in some loss of energy, so it is less efficient.

There are several important differences in the quality of light for indirect and direct lighting. For example:

- With indirect lighting there is no direct glare (see below).
- There is more uniformity of the light levels in areas where indirect lighting fixtures are used.

- Veiling reflections, or reflected light that obscures the detection of a visual target, and shadowing are greater in areas with direct lighting.

Table VC–2: Artificial Light Sources (Adapted from Lum-i-neering Associates, 1979; Ross and Baruzzini, Inc., 1975.)

Type	Efficiency (lm/W)	Color Rendering	Comments
Incandescent	17–23	Good	Incandescent is a commonly used light source but is the least efficient. Lamp cost is low. Lamp life is typically less than 1 year.
Fluorescent	50–80	Fair to Good	Efficiency and color rendering vary considerably with type of lamp: cool white, warm white, deluxe cool white. Significant energy cost reductions are possible with new energy-saving lamps and ballasts. Lamp life is typically 5–8 years.
Mercury	50–55	Very Poor to Fair	Mercury has a very long lamp life (9–12 years), but its efficiency drops off substantially with age.
Metal Halide	80–90	Fair to Moderate	Color rendering is adequate for many applications. Lamp life is typically 1–3 years.
High Pressure Sodium	85–125	Fair	This lamp is a very efficient light source. Lamp life is 3–6 years at average burning rates, up to 12 hours per day.
Low Pressure Sodium	100–180	Poor	This lamp is the most efficient light source. Lamp life is 4–5 years at average burning rate of 12 hours per day. Mainly used for roadways and warehouse lighting.

The efficiency (column 2), in lumens per watt (lm/W), and color rendering (column 3) of six frequently used light sources (column 1) are indicated. Lamp life and other features are given in column 4. Color rendering is a measure of how colors appear under any of these artificial light sources compared with their color under a standard light source. Higher values for efficiency indicate better energy conservation.

To overcome the disadvantage of lower efficiency with indirect lighting, some designers have combined high-intensity discharge lamps with specially designed optics. The more efficient lamps and their fixtures cost no more than direct fluorescent lighting when the number of fixtures and the installation, maintenance, and energy implications (as affected by the heat load from the lighting sources) are taken into account. However, the use of high-intensity discharge lamps may produce complaints about the lamp color, such as "harsh blue" and "warm orange." These problems can be reduced, though, by using one of each type of lamp, metal halide and high pressure sodium, in two-lamp fixtures.

d. Task Lighting

Illuminance can be increased while energy consumption is reduced by modifying the lighting pattern. For example, traditionally, an entire large office or production area might have been lighted uniformly at a high level such as 1000–1500 lx (93–139 fc). Today, though, it may be preferable to provide a relatively low level of ambient illumination, such as 300–500 lx (28–46 fc), and use supplementary desk lamps and task illuminators at workplaces where difficult visual tasks are to be performed (General Electric Company, 1964). This technique is especially effective when a small number of workers occupy a relatively large space. However, when using the task-lighting approach, the designer must make a serious attempt to minimize direct and indirect glare, such as veiling reflections and reflected glare, or the benefit from the task lighting will be significantly reduced. Supplementary lighting should come from the left and right sides of the workplace. Desk lamps that are directly in front of the worker, often built into office furniture, are potential sources for indirect glare, especially when located in the offending zone (see below). These sources should be avoided for task lighting whenever possible.

e. Minimizing Glare

(1) General Workplaces

The quality of illumination can be improved by reducing glare. Direct glare is caused when a source of light in the visual field is much brighter than the task materials at the workplace. Indirect glare, frequently called either veiling reflections or reflected glare, depending on its severity, is caused by light reflected from the work surface. It may reduce the apparent contrast of task materials at the workplace and, unless it is carefully controlled, may affect task performance. The zones for direct and indirect glare are shown in Figures VC–1 and VC–2. Methods for controlling direct and indirect glare are given in Table VC–3.

The visual comfort probability (VCP) index is a measure that gives the percentage of individuals who would be expected to be visually comfortable at a given workplace. It is inversely related to direct glare. The VCP index should be at least 70 or higher, if practical. Methods for calculating this index

Figure VC–1: Direct- and Indirect-Glare Zones for an Operator (Adapted from Lum-i-neering Associates, 1979.)

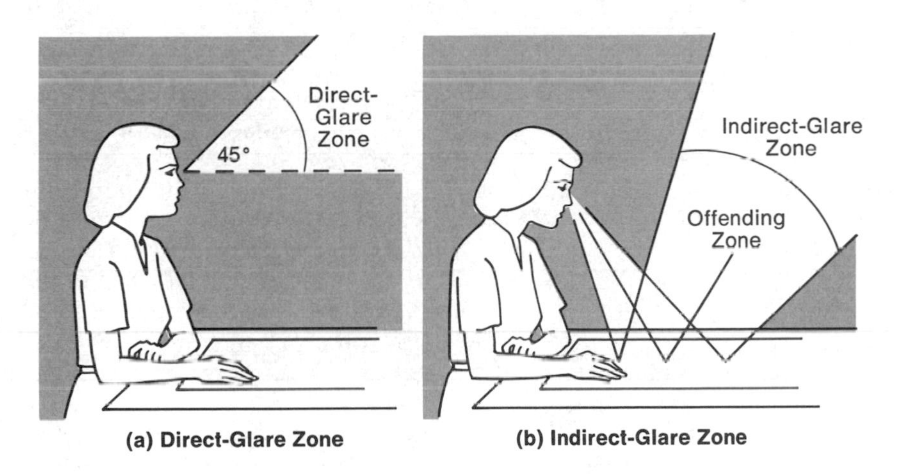

(a) Direct-Glare Zone **(b) Indirect-Glare Zone**

The direct-glare zone (part a) is shown as the region described by a 45° arc above the operator's line of sight. Indirect glare (part b) is reflected off the working surface. It reaches the operator's eyes after being reflected off the object observed or its background.

Figure VC–2: Direct- and Indirect-Glare Zones for a Luminaire, or Lighting Unit (Adapted from Lum-i-neering Associates, 1979.)

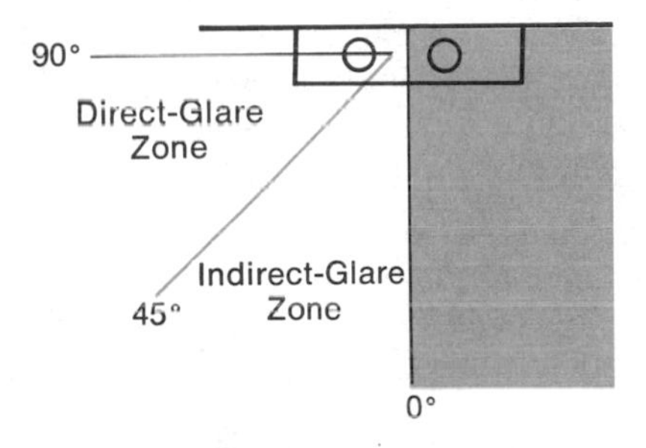

The direct-glare zone for a lighting source is in the area between the horizontal plane of the luminaire, or lighting unit, and a 45° angle downward. The indirect-glare zone is in the area between the 45° line and a vertical line drawn from the center of the luminaire to the working surface.

Table VC–3: Techniques for Controlling Glare (Adapted from Kaufman and Christensen, 1972; Morgan, Cook, Chapanis, and Lund, 1963.)

To Control Direct Glare	To Control Indirect Glare (Veiling Reflections and Reflected Glare)
Position luminaires, the lighting units, as far from the operator's line of sight as is practical	Avoid placing luminaires in the indirect-glare offending zone (see Figure VC–2)
Use several low-intensity luminaires instead of one bright one	Use luminaires with diffusing or polarizing lenses
Use luminaires that produce a batwing light distribution*, and position workers so that the highest light level comes from the sides, not front and back	Use surfaces that diffuse light, such as flat paint, non-gloss paper, and textured finishes
Use luminaires with louvers or prismatic lenses	Change the orientation of a workplace, task, viewing angle, or viewing direction until maximum visibility is achieved
Use indirect lighting	
Use light shields, hoods, and visors at the workplace if other methods are impractical	

* The effectiveness of the batwing distribution varies with the orientation of the workplace and worker. It can also be used to control indirect glare, because maximum output is in the arc between approximately 35° to 45° angles.

Examples of ways to control direct glare (column 1) and indirect glare (column 2) at the workplace are given. These methods include design approaches that can be used when installing the lighting, as well as interventions that can be made after glare has been identified in a workplace.

are given in the *IES Lighting Handbook* (Kaufman and Christensen, 1972). Luminaires that emit no light in the direct-glare zone include lighting fixtures with parabolic wedge louvers (VCP = 99+). These units are ideal for general lighting in rooms with video display units. In contrast, clear tungsten lamps have a very low VCP and are not suitable for most visual task applications.

Direct glare can also be controlled by manipulating the luminance (brightness) of the area surrounding the task. Under ideal conditions the luminance of the surround should be equal to or slightly less than the luminance of the task materials. The IES has published a list of maximum luminance ratios for specific situations. They are given in Table VC–4 and are somewhat conservative; in practice, it is often difficult to achieve luminance ratios at the workplace as low as those given.

(2) Video Display Unit (VDU) Workplaces

For VDU operators glare is luminance variation in the visual field sufficient to cause fatigue, discomfort, or annoyance. It may be generated by luminaires within the operator's field of view, daylight shining through windows, or light reflected from surfaces within the room. Reflections occurring on display screens may become annoying glare sources that reduce task visibility. To minimize the annoyance associated with these reflections, the designer might use VDUs with screens having an antireflection coating or place a non-glare,

Table VC–4: IES Recommended Maximum Luminance Ratios for Visual Tasks (Adapted from Kaufman and Christensen, 1972.)

Conditions	Luminance Ratios	
	Office	Industrial
Between tasks and adjacent darker surroundings	3 to 1	3 to 1
Between tasks and adjacent lighter surroundings	1 to 3	1 to 3
Between tasks and more remote darker surfaces	5 to 1	20 to 1
Between tasks and more remote lighter surfaces	1 to 5	1 to 20
Between luminaires (or windows, skylights) and surfaces adjacent to them	20 to 1	NC*
Anywhere within normal field of view	40 to 1	NC*

* NC means not controllable in practice.

The ratios of luminance between a task or display and its background (column 1) are given for offices (column 2) and production workplaces (column 3). These ratios are based on recommendations from the Illuminating Engineering Society (IES). No recommended ratios have been given for situations where it is not practical to try to control the brightness, as in contrasts between workplace and outdoor lighting. It is often difficult to keep the luminance ratios this low, especially in production workplaces where a multitude of tasks are done.

contrast-enhancing filter, such as a polarizing filter, directly on the screen. In some situations it may be necessary to angle the filter toward the viewer in order to reduce reflections from its surface. Add-on filters must be used with care. Some may noticeably degrade image quality and thus should be avoided.

In addition to anti-reflection screens and hoods, the following methods are suggested for reducing reflected glare at VDU workplaces:

- Lower the ambient lighting, if feasible.
- Have the operator sit with his or her back toward a dark-colored wall.
- Paint walls with dark colors, dull or matte finish only. Desk tops and other work surfaces should have matte finishes.
- Avoid placing clocks, mirrors, back-lighted displays, bulletin boards, and similar items in areas where they cause reflections that will appear on display screens.
- If all else fails, tilt the screen downward, or move it slightly to the left or right, to eliminate specific reflected images.

Glare caused by light transmitted through windows can be reduced or eliminated as follows:

- Cover windows either partially or completely with draperies, venetian blinds, woven woods, or movable louvers.
- Cover windows with neutral density filters to reduce transmittances.
- Add awnings or other devices to shield the windows from the direct rays of the sun.
- Install the VDU at right angles to the window.

f. Special-Purpose Lighting

General illumination may not be adequate for production or service operations with difficult visual tasks, such as inspection activities. Although increasing the quantity of light may improve the detectability of some visual targets, special-purpose lighting, usually local, can be far more effective and considerably less expensive to implement. Table VC–5 presents some lighting techniques that have been successfully employed in industry.

More than one type of light is needed to detect multiple defect types in a single-pass, 100 percent inspection task, as occurs in many product acceptance operations. For further information, see Section IV B, 'Product-to-Person Transfer: Visual Inspection.'' The combination of lights can be tailored to the importance and relative visibility of the defect types most likely to occur. Compromises will have to be made, favoring speed in most instances; not all defects can be optimally illuminated in these combined-lighting stations.

Table VC–5: Special-Purpose Lighting for Inspection Tasks (Developed from information in Faulkner and Murphy, 1973; Hopkinson and Collins, 1970; Hunter, 1937; Kaufman and Christensen, 1972; T. J. Murphy, 1981, Eastman Kodak Company.)

Column 1 describes fourteen improvement goals for inspection task performance. Aids that assist the inspector in detecting the defects are given in column 2; short explanations of how these aids work or descriptions of other actions that help the inspector are given in column 3. There is often more than one way to make a defect more visible; the nature of the material being inspected will help identify the most effective method. When more than one type of defect is being searched for, a combination of aids at the workplace may be appropriate.

Desired Improvement in Inspection Task	Special-Purpose Lighting or Other Aids	Techniques
1. Enhance surface scratches	Edge lighting, for a glass or plastic plate at least 1.5 mm, or 0.06 in., thick	Internal reflection of light in a transparent product; use a high-intensity fluorescent or tubular quartz lamp
	Spotlight	Assumes linear scratches of known direction; provide adjustability so that they can be aligned to one side of the scratch direction; use louvers to reduce glare for the inspector
	Dark-field illumination (e.g., microscopes)	Light is reflected off or projected through the product and focused to a point just beside the eye; scratches diffract light to one side
2. Enhance surface projections or indentations	Surface grazing or shadowing	Collimated light source with an oval beam
	Moiré patterns (to accentuate surface curvatures)	Project a bright collimated beam through parallel lines a short distance away from the surface; looking for interference patterns (Stengel, 1979); either a flat surface or a known contour is needed
	Spotlight	Adjust angle to optimize visualization of these defects

(table continues pages 236)

Table VC–5: (Continued)

Desired Improvement in Inspection Task	Special-Purpose Lighting or Other Aids	Techniques
2. Enhance surface projections or indentations (cont.)	Polarized light	Reduces subsurface reflections when the transmission axis is parallel to the product surface
	Brightness patterns	Reflection of a high-contrast symmetrical image on the surface of a specular product; pattern detail should be adjusted to product size, with more detail for a smaller surface
3. Enhance internal stresses and strains	Cross-polarization	Place two sheets of linear polarizer at 90° to each other, one on each side of the transparent product to be inspected; detect changes in color or pattern with defects
4. Enhance thickness changes	Cross-polarization	Use in combination with dichroic materials
	Diffuse reflection	Reduce contrast of brightness patterns by reflecting a white diffuse surface on a flat specular product; produces an iridescent rainbow of colors that will be caused by defects in a thin transparent coating
	Moiré patterns	See item 2 in this table
5. Enhance non-specular defects in a specular surface, such as a mar on a product	Polarized light	A specular nonmetallic surface acts, under certain conditions, like a horizontal polarizer and reflects light; non-specular portions of the surface will depolarize it; project a horizontally polarized light at a 35° angle to the horizontal

Table VC–5: (Continued)

Desired Improvement in Inspection Task	Special-Purpose Lighting or Other Aids	Techniques
5. Enhance non-specular defects in a specular surface, such as a mar on a product (cont.)	Convergent light	Project the light at a spherical mirror, reflect it off the product, and focus it at the eye; requires very rigid posture for inspectors, however; mirror should be larger than the area being inspected
6. Enhance opacity changes	Transillumination	For transparent products, such as bottles, adjust lights to give uniform lighting to the entire surface; use opalized glass as a diffuser over fluorescent tubes for sheet inspection; double transmission transillumination can also be used
7. Enhance color changes, as in color matching in the textile industry	Spectrum-balanced lights	Choose lighting type to match the spectrum of lighting conditions expected when the product is used; use 3000°K lights if the product is used indoors, 7000°K light if it is used outdoors
	Negative filters, as in inspecting layers of color film for defects	These filters transmit light mainly from the end of the spectrum opposite to that from which the product ordinarily transmits or reflects; this reversal makes the product surface appear dark except for blemishes of a different hue, which are then brighter and more apparent
8. Enhance unsteadiness, jitter	Parallel line patterns (Moiré)	Two sets of parallel lines, 3–5° offset; one set is mounted on the product and the other is stationary; this pattern can magnify the jitter 10–40 times

(table continues page 238)

Table VC–5: (Continued)

Desired Improvement in Inspection Task	Special-Purpose Lighting or Other Aids	Techniques
9. Enhance repetitive defects, as in rotating shafts or drums	Stroboscopic lighting	Adjust strobe frequency to the expected frequency of the defect
10. Enhance fluorescing defects	Black light	Use ultraviolet light to detect cutting oils and other impurities; may be used in clothing industry for pattern marking; fluorescing ink is invisible under white light, but very visible under black light
11. Enhance hairline breaks in castings	Coat with fluorescing oils	Use of ultraviolet light inspection will detect pools of oil in the cracks
12. Reduce surface glow under white light that hides defects; the surface appears to fluoresce	Complementary filter or light source, similar to a negative filter	Use a filter or light source with low transmission in wavelengths reflected by the object's surface, and high transmission in other parts of the spectrum, so as to create a gray appearance
13. Remove distracting reflections	Light shields	Place overhead or side shields on a workplace to eliminate reflections caused by room lighting
	Light traps	For VDUs mount a circular polarizer in front of the tube, set at a downward angle; the polarizer traps all incoming light from the room and allows only internally generated light back to the observer

Table VC-5: (Concluded)

Desired Improvement in Inspection Task	Special-Purpose Lighting or Other Aids	Techniques
13. Remove distracting reflections (cont.)	Reposition workplace	Rather than have operators face a wall, with ceiling lights behind them reflecting off the 45°–90° surfaces of their workpieces, have them sit with their backs to the wall so that workpieces reflect the low-luminance wall instead
14. Reduce blurring of fast-moving product, as in the printing industry	Synchronized moving images	Projected or reflected images on flat, otherwise formless webs can provide fixation points and reduction of streaming
	Stroboscopic lighting	Pulsed light above the fusion threshold, approximately 40+ Hz, will make a random spot type of defect appear as a string of pearls, even if the formless web itself is blurred (Taylor and Watson, 1972)
	Elongate the observation area	Rule of thumb: 0.3 m (1 ft) of observation area per 18.3 m/min (60 ft/min) of object speed at close inspection distances of 0.6–1.2 m (2–4 ft) allows proper fixation time, eye pursuit, and stopped images of the product (T. J. Murphy, 1981, Eastman Kodak Company)
	Group the product	For the same result, it is better to tighten the grouping and reduce the speed rather than to spread the product out and increase the speed

2. COLOR

The material in this section was developed from information in Burnham, Hanes, and Bartelson (1963); Grandjean (1980), Hopkinson and Collins (1970), Ramkumar and Bennett (1979), and Woodson (1981).

The influence of color on people in a production or office workplace has not been rigorously examined. Most studies on the effects of environmental color have been preference studies, where aesthetics are the prime consideration. From these examinations the following observations about preferences can be made:

- In Western culture the order of color preference is blue, red, green, violet, orange, and yellow. These preferences are broad and transcend racial and sexual differences. Children may prefer red-containing colors, but these preferences shift toward the blue end of the spectrum at maturity.
- Generally speaking, blue, green, and violet shades are considered cool colors, while red, orange, yellow, and brown are considered warm.
- Blue and green are considered soothing colors; orange, yellow, and brown are considered stimulating; red and violet are described as "aggressive," "alarming," "discouraging," and "disturbing."
- Color can influence a person's perception of size and distance within a closed space. Walls covered with shorter-wavelength colors, such as blues and greens, are said to recede, while walls covered with the longer-wavelength colors are said to advance. Thus blues and greens cause a room to appear larger, while reds achieve the opposite effect. Similarly, there is an interaction between brightness and color in the distance effect. Light colors tend to recede and dark colors to advance.
- The formation of a reaction to a color takes time, and the reaction, once formed, is subject to adaptation. Thus a person's initial reaction may be quite pronounced, but it will tend to diminish in magnitude with the passage of time. The end result after complete adaptation has occurred could be relative indifference.
- As the saturation (intensity) of the color is lowered to a pastel level, the perceptual and psychological effects of the color diminish.

From the above information about color, the following guidelines for application of specific colors, brightness, and saturation levels in the workplace have been developed:

- For large areas, colors that give uniform reflectivity should be chosen. Good visual contrasts can be obtained without significant brightness contrasts. For example, doors, protruding wall segments, or other barriers may be painted in a different hue of the same brightness as the overall wall space. Thus these features will be easily identifiable without unnecessarily calling attention to them or distracting the workers by using highly contrasting brightness.

- Bright, or highly saturated, primary colors should be avoided. They are undesirable because they might cause a negative afterimage, a persisting sensation after the stimulus has ceased. Pastel colors are generally preferred for walls, large room units, and tabletops or work surfaces.
- In areas where highly repetitive work is performed or where a large area of wall or floor space exists, some stimulating colors can be used to highlight a door or partition. In addition, some pieces of equipment in the area may be painted in a brighter shade of the color to which the room is keyed. Equipment such as conveyors, cabinets, shelving, and small pieces of production machinery are often painted this way. Large surfaces should not have these bright colors.
- A large area can be functionally divided by color to give identity to different groups working within it. Separate rooms can be keyed to a certain basic color to accomplish the same effect.
- In temperate climates the normal preference in the interior of buildings is for a balance of color on the warm side. Thus in windowed buildings and rooms, use poorly saturated warm colors on walls and equipment to balance the coolness of white areas of walls and the grays of metal and other equipment.
- The selection of color schemes should be coordinated with the decisions about illumination type. High-pressure sodium lighting has only fair color-rendering characteristics; subtle shadings of color that would be appropriate under white light may be lost under this type of illumination.

SECTION V D. TEMPERATURE AND HUMIDITY

The temperature in a workplace or work area can strongly influence how effectively a task is performed. Hot, humid conditions added to the demands of moderately heavy to heavy physical work may cause excessive fatigue and potential health risks by reducing a person's work capacity. Cold conditions where manipulation tasks are done, such as maintenance activities outdoors in cold weather, can lead to decreases in productivity and potentially unsafe actions because of a loss of finger flexibility. Discomfort produced by exposure to conditions outside the thermal comfort zone (described below) can distract a person from the task at hand and may increase the potential for unsafe acts. In this section the comfort, discomfort, and health risk zones for hot and cold air environments are discussed. In addition, guidelines for water immersion and contact with hot or cold surfaces are given.

Temperature and humidity extremes may occur in the following industrial operations that generate process heat:

- smelting
- drying
- boiler cleaning and maintenance
- steam cleaning
- ironing
- extruding plastics
- molding plastics
- chemical manufacturing (heat-producing reactions)

In addition, work in buildings that are not air-conditioned or work outside during the summer months in warm climates can result in significant heat and humidity exposure, especially if heavy physical effort is required.

Exposure to cold can occur in the following industrial situations:

- Refrigeration or cold storage work.
- Construction in poorly heated buildings (warehouses) or outdoors in the winter months in cold climates.
- Outdoor maintenance and service work (e.g., snow removal, trucking, and handling) in the winter months in cold climates.
- Cleaning with cold water.

1. BODY HEAT BALANCE

Body temperature is regulated very closely in order to provide the best environment for the chemical reactions essential to life. In the course of a day, if no heavy exercise is done, the variation in body core temperature is seldom more than 0.6°C (1°F) for most people (Colquhoun, Blake, and Edwards, 1968; Kleitman and Jackson, 1950). Heat balance is maintained by increasing blood flow to the skin and by sweating in hot environments, and by reducing circulation to the skin and by shivering in cold environments. Heat is generated by metabolism and muscular work and is removed from the body by convection, radiation, and evaporation (See Table VD–1). In some instances heat is lost or gained by conduction through contact with metallic or other solid surfaces. This factor is not a primary one in heat balance and will be discussed later in the section on contact with hot or cold surfaces.

As outside temperature is lowered below skin temperature, the rate of body heat loss by convection is increased. If outside temperature is raised above skin temperature, the body's net loss of heat by radiation is reduced, and evaporation and convection become the most effective ways to maintain body temperature. The heat balance equation is shown in Table VD–1. Physiological or physical factors that determine the different components of the heat balance are indicated.

Table VD–1: Body Heat Balance (Developed from information in Kamon, 1975; Leithead and Lind, 1964.)

Term	Definition	Determinants
M	Metabolic heat gain	Physical work load, or the muscular work done minus the work efficiency
C	Convective heat gain or loss	Air velocity The difference between air temperature and a person's average skin temperature
R	Radiative heat gain or loss	The difference between a person's average skin temperature and the temperature of surfaces in the environment, measured with a globe thermometer or radiometer The amount of skin exposed to the solid surface
E	Evaporative heat loss	The difference between the water vapor pressure of a person's skin and the water vapor pressure, or relative humidity, of the environment; it is indirectly related to work load and the person's sweat rate Air velocity
S	Heat storage in, or loss from, the body	Balance of the above factors Rectal and skin temperatures Body weight
Co	Conductive heat gain or loss	The area of the conductive surface The difference between the person's skin temperature and the temperature of the surface contacted

The equation describing body heat balance is given below. Each term (column 1) is defined in column 2, and its major determinants are identified in column 3. Conductive heat gain or loss (Co) has also been included at the end of the table, since this path of heat transfer is important when a hot or cold surface is directly contacted.

The heat balance equation for the body can be summarized as follows:

$$M \pm C \pm R - E = \pm S$$

Because body temperature stability is important to health, occupational tasks that consistently and continuously disturb it are not desirable. For this reason developers of heat exposure criteria documents (NIOSH, 1972b; WHO, 1969) have chosen to use a 1°C (1.8°F) rise in body core temperature as the upper limit for 1- to 2-hour exposures. This value represents a heat storage of 250 kilojoules (kJ), or 60 kilocalories (kcal).

Guidelines for acceptable durations of exposure to multiple combinations of temperature, humidity, air velocity, radiant heat, clothing ensembles, and work load can be developed from the heat balance relationships by limiting changes in body core temperature to ±1°C (1.8°F) and choosing maximum sweat rate levels for a shift or an hour of work.

2. THE COMFORT ZONE

a. Zone Definition

Under optimal comfort conditions 2.5 percent of the population is too warm and 2.5 percent is too cold (Fanger, 1970). Individual variability in assessing comfort levels is very high. The levels vary with time of day, season, diet, hormonal changes, and behavior, including clothing choices, the presence of other job stress, and cultural variables and expectations. A comfort zone has been defined on the basis of psychophysical data gathered on several thousand people (Fanger, 1970). Assuming people can alter their clothing ensembles and energy conservation goals will stimulate industry to regulate building temperatures nearer the limits of comfort, guidelines for thermal comfort in sedentary and light work have been developed; these guidelines are presented in Figure VD–1. They indicate that temperatures between 19° and 26°C (66° and 79°F) are generally considered comfortable, provided that the humidity is not too high at the upper limit and air velocity is not excessive at the lower limit.

Temperatures of 19° and 26°C (66° and 79°F) are recommended as the outer limits for thermostatic regulation in areas where sedentary or light work is done. These guidelines are for 8-hour exposures. Experience indicates that for extended exposures temperatures between 20° and 25°C (68° and 78°F) are more acceptable than the lower and higher values given above. This difference can usually be attributed to the presence of other factors, such as heat-losing surfaces, which are discussed below.

b. Factors Affecting the Feeling of Comfort Within the Zone

The following factors will affect the individual's sense of comfort within the thermal comfort zone shown in Figure VD–1:

- temperature
- humidity
- air velocity
- work load
- clothing
- radiant heat load

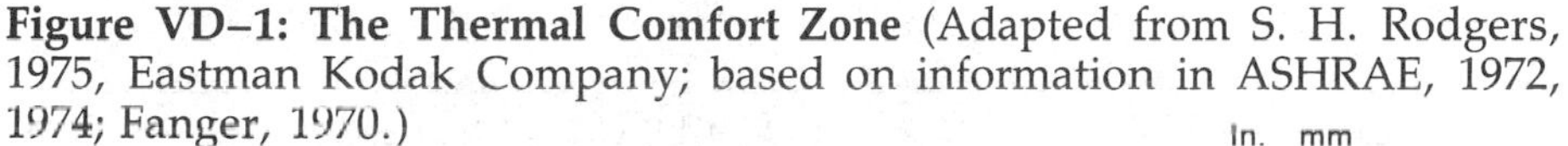

Figure VD–1: The Thermal Comfort Zone (Adapted from S. H. Rodgers, 1975, Eastman Kodak Company; based on information in ASHRAE, 1972, 1974; Fanger, 1970.)

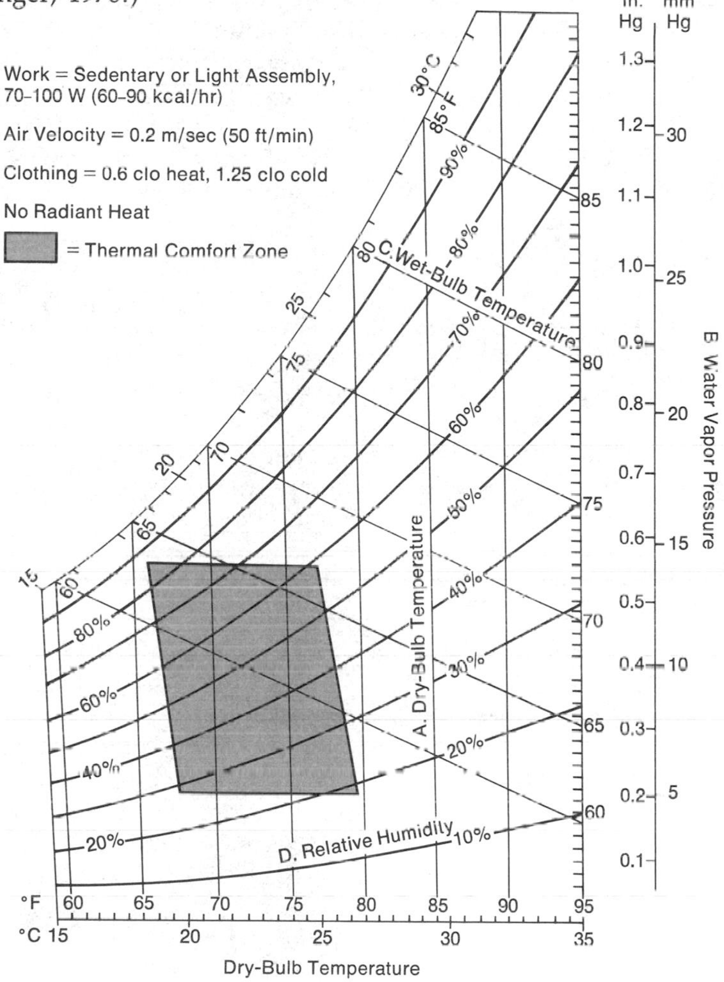

The dry bulb temperature and humidity combinations that are comfortable for most people doing sedentary or light work are shown as the shaded area on the psychometric chart. The dry bulb temperature range is from 19° to 26°C (66° to 79°F), and relative humidities (shown as parallel curves) range from 20 to 85 percent, with 35 to 65 percent being the most common values in the comfort zone. On this chart ambient dry bulb temperature (A) is plotted on the horizontal axis and indicated as parallel vertical lines; water vapor pressure (B) is on the vertical axis. Wet bulb temperatures (C) are shown as parallel lines with a negative slope; they intersect the dry bulb temperature lines and relative humidity curves (D) on the chart. In the definition of the thermal comfort zone, assumptions were made about the work load, air velocity, radiant heat, and clothing insulation levels. These assumptions are given in the top left corner of the chart.

Table VD–2 shows how changing the level of many of these factors influences thermal comfort. The upper and lower ambient dry bulb temperature limits must be adjusted to produce the same feeling of comfort when humidity, air velocity, work load, clothing insulation, and radiant heat load are varied from the levels given in Figure VD–1. Subjective discomfort is primarily related to skin temperatures greater than 34.5°C or less than 32.7°C (94° and 91°F) (Hardy, 1970) for sedentary or light work conditions. With increased work load, lower skin temperatures (<30°C, or 86°F) are needed for comfort. Brief descriptions of each of these modifiers of comfort follow.

(1) Humidity

As humidity increases, discomfort will be felt at the upper end of the thermal comfort zone. All other things being equal, raising the humidity from 50 to 90 percent at 26°C (79°F) can increase the feeling of discomfort by a factor of 4 (Fanger, 1970) for a person doing light work. Humidities below 70 percent are preferable in the summer months.

Too little humidity can produce discomfort for some people (McIntyre, 1978). Where feasible, it is recommended that humidity values above 20 percent be used for extended (more than 2 hours) exposures (ASHRAE, 1974).

(2) Air Velocity

The rate of air flow is measured as the average value at the work place. Local drafts can make a workplace feel very cool even when it is well within the comfort guidelines for dry bulb temperature (Fanger, 1977). This discomfort from local drafts increases as ambient (room) temperature falls. Increased air flow is a positive comfort factor at the upper limit of the comfort zone, but it is a major cause of complaints at the lower limit. So that complaints from drafts at the workplace are minimized, the lower limit temperatures given in Table VD–2 should be raised 3°C (5°F) at velocities greater than 0.1 m/sec and less than 0.5 m/sec (20 and 100 ft/min) and by 4°C (7°F) at higher velocities (McIntyre, 1973). Air velocities of 0.1 to 0.3 m/sec (20 to 60 ft/min) are fairly typical in the comfort zone for sedentary and light assembly work.

(3) Work Load

As work load increases, the thermal comfort zone shifts to the left (see Table VD–2), favoring cooler ambient temperatures to help rid the body of heat generated by the muscles. The heavier the work, the lower the upper limit must be in order to preserve comfort. Most moderate to heavy industrial work loads average 175 to 350 W (150 to 300 kcal per hour), with heavier work being sustained only for shorter time periods (Rodgers, 1978). Because periods of effort are interspersed with periods of light or sedentary standby activity in the same environment, one has to trade off the best environment for each and compromise on a temperature level that will provide the best heat balance opportunities. The situation will therefore often lead to relative discomfort for the worker for at least part of the work cycle.

Table VD–2: The Influence of Several Factors on the Thermal Comfort Zone Limits (Adapted from S. H. Rodgers, and K. G. Corl, 1981, Eastman Kodak Company; based on a temperature model published by Gagge, 1973, and on information in Fanger, 1970.)

		Ambient Dry Bulb Temperature*			
		Lower Limit		Upper Limit	
Factor	**Level**	**°C**	**(°F)**	**°C**	**(°F)**
Relative Humidity (%)	20	20	(68)	26	(79)
	50	19	(67)	25.5	(78)
	80	18.5	(66)	24	(76)
Air Velocity, m/sec (ft/min)	0.1 (20)	18	(65)	24	(76)
	0.25 (50)	19	(67)	25.5	(78)
	0.36 (70)	21	(70)	27	(80)
	0.51 (100)	22	(72)	28	(82)
	0.71 (140)	23	(74)	29	(84)
Work Load, 8-Hour Average, multiples of resting values	× 2	19	(67)	25.5	(78)
	× 3.5	17	(64)	23	(74)
	× 5	≃15.5	(≃60)†	20	(68)
Clothing Insulation (clo)	0.25	27	(80)	28	(83)
	1.25	19	(67)	22	(72)
	2.50	≃11	(≃52)	≃16	(≃62)
Radiant Heat, °C (°F), amount that globe temperature exceeds dry-bulb temperature	0	19	(67)	25.5	(78)
	1.1 (2)	17	(64)	24	(76)
	2.8 (5)	16	(62)	23	(74)
	5.6 (10)	13	(56)	20	(68)

* Unless otherwise noted, the following values have been used to calculate the thermal comfort zone limits: air velocity, 0.25 m/sec (50 ft/min); work load, sedentary, light assembly, up to two times resting metabolism; clothing insulation, 0.6 clo in heat, 1.25 clo in cold; no radiant heat load; humidity, 50 percent.
† The actual value depends on the work/rest pattern (see text).

The upper and lower limits of ambient dry bulb temperature (columns 3 through 6) that will provide a comfortable work environment are given for several different levels of humidity, air velocity, work load, clothing insulation, and radiant heat (columns 1 and 2). If any of these parameters is altered from the value given in the footnote without making adjustments in the ambient dry bulb temperatures at the extremes of the thermal comfort zone, people's comfort level will be affected. These guidelines are, at best, rough estimates of the magnitude of change in temperature needed to improve comfort. Other factors, such as the work and rest pattern of the job, must also be considered when selecting the temperature limits for comfort.

(4) Clothing

The amount and type of clothing worn and its insulation characteristics will determine how far the lower limit of the thermal comfort zone can be extended in the workplace as well as how much the upper limit can be modified. The insulation value of clothing is expressed in clo units. Table VD–3 shows the clo values for a number of common clothing ensembles.

Table VD–3: The Insulation Value (clo) of Various Clothing Ensembles (Adapted from Fanger, 1970.)

Clothing Ensemble	Icl (clo units)*
Sleeveless blouse, light cotton skirt, sandals	0.3
Shorts, open-neck shirt with short sleeves, light socks, sandals	0.3–0.4
Long lightweight trousers, open-neck shirt	
with short sleeves	0.5
with long sleeves	0.6
Cotton fatigues, lightweight underwear, cotton shirt and trousers, cushion-sole socks and boots	0.7
Typical light business suit; pant suit (with full jacket)	1.0
Typical light business suit and cotton coat (lab coat)	1.5
Heavy traditional European business suit, long cotton underwear, long-sleeved shirt, woolen socks, shoes; suit includes trousers, jacket, and vest	1.5–2.0

Note: A wool sweater adds approximately 0.3 to 0.4 clo of insulation to the above clothing ensembles (McIntyre and Griffiths, 1975). Clothing made from non-breathing fabrics, such as nylon, will add up to 0.6 clo to the values given above (Nevins, McNall, and Stolwijk, 1974). A necktie can produce local discomfort by increasing Icl to almost 3 at the neck (E. Kamon, 1973, Eastman Kodak Company).

* Icl = Rcl/0.18, where Icl is the insulation value, in clo units, and Rcl is the total heat transfer resistance from skin to the outer surface of the clothed body, in degrees Celsius per kilocalorie per square meter of body surface area per hour. To express Icl in degrees Celsius per watt per square meter, change the constant from 0.18 to 0.16.

Column 2 shows the insulation values, or Icl in clo units (defined in the footnote), of several clothing ensembles that are described in column 1. Insulation increases with increased layers of clothes and with fabrics, like wool, that incorporate an air layer. Artificial fabrics often have higher insulation values but may not breathe. Their low moisture permeability can limit their usefulness because they reduce evaporative cooling in hot environments and trap the moisture near the skin in cold environments.

The use of jackets, sweaters, and gloves to increase comfort in a cool environment may not always be feasible, such as in the following situations:

- When the production process requires the operator to wear a certain uniform, as in clean room operations, that reduces the options for adding insulation.
- When the additional clothing binds the arms, hands, or body and reduces the available range of motion for the tasks to be performed.
- When gloves would interfere with a task requiring fine manual dexterity.
- When the job includes periods of heavy work where sweating occurs, which reduces the quality of the insulation of the additional clothing.

At the upper limit of the thermal comfort zone some improvement can be gained by wearing loose-fitting clothing and shedding unneeded layers. This option is not feasible, however, in the following types of activities:

- When waterproof, fireproof, or chemicalproof protective clothing must be worn.
- When the operator is working around moving equipment, and loose clothing could get entangled in it.
- When a clothing regulation exists as protection for the product or as part of a professional requirement, such as whites in a clean room or a tie in some professions.

Because the opportunities for altering clothing ensembles can be limited by factors outside the employee's control, the thermal comfort envelope in Figure VD–1 assumes that insulation changes only in the range of 0.6 to 1.25 clo can be made.

(5) Radiant Heat Load

The difference between dry bulb and globe temperature defines the amount of radiant heat present in a workplace (see the section on environmental measurement protocols in Chapter VI, Appendix B). Some common sources of radiant heat are the following:

- exothermic (heat-producing chemical reactions in vessels
- furnaces
- radiant-heat lamps
- coil heaters
- the sun
- drying or cooking ovens

At the upper limit of the thermal comfort zone, radiant heat produces discomfort; at the lower limit it increases comfort. The latter property can be used to improve comfort in offices or areas of light work when local drafts or cold walls and floors make winter building temperatures uncomfortable. The impact of radiant heat on discomfort associated with temperatures above the comfort zone is discussed later in this section.

(6) Heat-Losing Surfaces

In addition to the above factors, surfaces in the workplace—such as windows, walls with inadequate insulation, and slab floors—that lose heat faster than other parts of the operator's workspace can be a source of discomfort within the lower part of the thermal comfort zone (Emmerson, 1974; Nevins and Feyerherm, 1966). These factors are a common source of complaints of cold discomfort in the workplace when ambient dry bulb temperatures are maintained in the range of 18° to 21°C (65° to 70°F). The heat-losing surface increases the net heat loss from the body. The feet, legs, arms, or hands are most commonly affected, and this local discomfort is sensed by the operator. The cold surface also results in the creation of temperature gradients of more than 1.1° to 1.7°C (2° to 3°F) between head and foot level at a seated workplace. These gradients are uncomfortable to many office or light assembly workers and can reduce productivity by distracting them. Room temperature has to be increased substantially to overcome these local discomforts, so engineering changes are preferable. Such changes include the following:

- Insulating walls and areas around window frames.
- Using curtains or blinds over windows.
- Putting rugs, mats, or other insulating surfaces over slab floors.
- Rearranging the work area to reduce exposure to the heat-losing surface.

For example, office areas are sometimes placed in buildings that were designed for other purposes, such as warehousing. Walls that were not necessarily constructed for careful internal temperature regulation become a problem during winter cold extremes. The use of additional wall insulation often abates the problem without requiring additional energy use, even saving some. In office areas where large windows predominate, curtains or blinds may reduce the impact of the heat-losing surface on people. Care must be taken, however, to avoid disturbing the air flow pattern from heaters that may be placed under the windows; a long curtain may trap the hot air by the window and not permit it to reach the rest of the room.

3. THE DISCOMFORT ZONE

Discomfort is produced when the body's physiological responses to environmental temperature and humidity exceed the usual regulatory responses. The degree of discomfort is related to the skin temperature level and to the amount of sweating. Discomfort increases as each of these increases in the heat and as skin temperature decreases and shivering increases in the cold. Circulatory collapse, radiant-heat burns, severe shivering, or frostbite should not occur in the discomfort zone. They are more possible when ambient temperature and humidity are in the "not recommended" areas of Table VD–4.

Table VD–4: Maximum Recommended Work Loads, Heat Discomfort Zone (Adapted from S. H. Rodgers, 1982, Eastman Kodak Company; based on information in Hertig, 1973; Leithead and Lind, 1964; Lind, 1963.)

Maximum Recommended Work Load					
Ambient Temperature		Relative Humidity			
°C	°F	20%	40%	60%	80%
27	80	VH	VH	VH	H
32	90	VH	H	M	L
38	100	H	M	L	NR
43	110	M	L	NR	NR
49	120	L	NR	NR	NR

Note: Assumptions include 2-hour continuous exposure, 0.6 clo insulation, air velocity less than 0.5 m/sec (100 ft/min). Higher work loads may be sustained for shorter work periods. See Figure VD–3 for further information. Definitions of work load abbreviations: VH = very heavy, 350–420 W (300–360 kcal/hr); H = heavy, 280–350 W (240–300 kcal/hr); M = moderate, 140–280 W (120–240 kcal/hr); L = light, less than 140 W (120 kcal/hr). NR = not recommended for 2 hours of continuous exposure.

The recommended maximum work loads for several combinations of relative humidity and ambient temperature (columns 1 and 2) are shown in columns 3 through 6. The abbreviations (VH, H, M, L, and NR) are defined in the footnote. A list of job tasks in each of the effort categories is given later in the section in Table VD–5. These guidelines are based on corrected effective temperatures (CETs) of 30.2°, 27.4°, and 26.9°C (86°, 81°, and 79°F) for light, moderate, and heavy work loads, respectively. If radiant heat is present, the globe thermometer reading can be used for ambient temperature. Air velocity greater than 0.5 m/sec permits heavier work to be done in each situation. Heavier work is also possible for continuous exposures shorter than 2 hours (see Table VD–7 later in the section).

Most industrial heat and cold exposures are in the comfort and discomfort zones. Although a person can tolerate discomfort in a job, it can reduce productivity and may produce undesirable physiological responses in susceptible people, such as nausea, dizziness, fatigue, or loss of finger flexibility. Interventions to increase comfort are discussed below.

a. Guidelines for Acceptable Heat Exposures in the Discomfort Zone

Table VD–4 gives the recommended maximum work loads for several combinations of ambient temperature and relative humidity in the heat discomfort zone. The following assumptions have been made in developing these guidelines:

- People working in the area will have some, but not total, acclimatization to the conditions.
- The limits are defined for 2-hour continuous exposures. The following factors limit the maximum recommended work load:
  - The person's sweat rate should not exceed 2 liters (L), or 0.5 gallon (gal), over the exposure period (Kamon, 1975).
  - The sweat rate should not be greater than 60 percent of the maximum sweat rate (Belding, 1970; Belding and Hatch, 1955).
- Clothing insulation is 0.6 clo.
- Air velocity is less than 0.5 m/sec (100 ft/min).
- If radiant heat is present, the globe temperature reading should be used instead of the dry bulb reading to define ambient temperature level (Leithead and Lind, 1964). See Chapter VI, Appendix B, for further discussion.

Increasing relative humidity reduces a person's ability to lose heat by sweating (decreases the maximum defined in Table VD–1), so less work can be sustained over a 2-hour period. Increasing ambient temperature or the amount of radiant heat will decrease the amount of heat that can be lost from the body by radiation and convection, which, in turn, reduces the recommended work load at a given humidity. The guidelines in Table VD–4 are simplified and were derived by establishing maximum corrected effective temperature limits for each work load (Lind, 1963). The wet bulb temperature for each level of dry bulb or globe temperature was read from the corrected effective temperature nomogram (Hertig, 1973). An illustration of how this information is used to assess a job in a hot environment is given in the problems section of Chapter VI, Appendix C.

The sweating associated with heat discomfort can contribute to unsafe situations in the workplace, such as the following:

- Wet hands that make holding onto objects or equipment controls more difficult and that increase the chance of losing control of a tool or load.

- Wet floors or work surfaces, resulting from people working in a hot and humid environment saturating their clothing and leaving wet areas around them, thus increasing the potential for slips.
- Sweat in the eyes, resulting in eye irritation and some visual interference with the task, which could result in an unsafe act.
- Decreased skin resistance, thereby increasing the person's vulnerability to electric shock (see Section VA, "Electric Shock").

b. Guidelines for Acceptable Cold Exposures in the Discomfort Zone

The body's ability to compensate for heat loss is less than its ability to tolerate increased heat. As ambient temperature falls, cold discomfort increases rapidly, even when more clothing is worn. The insulation of clothing slows the fall of skin temperature below 32.8°C (91°F) during light work in cold environments, and it reduces convective heat loss in high air velocities. Shivering is initiated at skin temperatures of about 30°C (86°F), and this effect further discomforts the worker. The following equation can be used to predict the insulation requirements for people working in cold environments in high air velocities (Belding, 1973):

$$\text{Icl} = 5.55 \frac{\text{Tsk} - \text{Ta}}{0.75\ \text{M}}$$

where Icl is the clothing insulation factor, as defined in Table VD–3; Tsk is skin temperature, in degrees Celsius; Ta is the ambient dry bulb temperature, in degrees Celsius; and M is the metabolic work load, in kilocalories per square meter per hour.

Since it is not always possible to adjust the clothing to tolerate the cold, one can define Icl and solve this equation for any of the other variables. This technique gives a way of approximating the thermal limits of the discomfort zone for cold. For example, a person doing light work and wearing 1.25 clo of insulation would experience some discomfort (Tsk $\leq$ 32.7°C) at an ambient temperature of 21.7°C (71°F) and more severe discomfort (Tsk = 25°C) at 14°C (57°F). Because this equation is based on studies where air velocities above 1.5 m/sec (300 ft/min) were used, the ambient temperatures will be less cold than those that could be tolerated in calmer conditions.

Local cold discomfort, most often of the hands, feet, and face, is usually the major cause of complaints in the cold discomfort zone. The hands begin to exhibit some loss of flexibility and manipulation skills at ambient dry bulb temperatures of 15.5°C (60°F) over a few hours of exposure. Figure VD–2 illustrates the performance effects of cold as a function of ambient temperature and duration of exposure. A 20 percent decrement in performance is not unusual in manual tasks at ambient temperatures of 7°C (45°F), dry bulb.

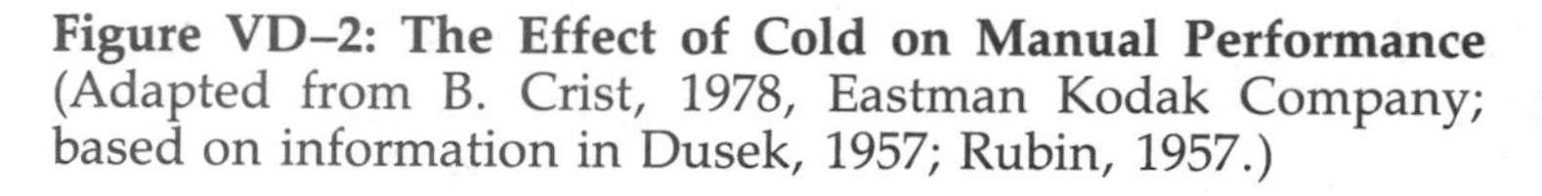

Figure VD–2: The Effect of Cold on Manual Performance (Adapted from B. Crist, 1978, Eastman Kodak Company; based on information in Dusek, 1957; Rubin, 1957.)

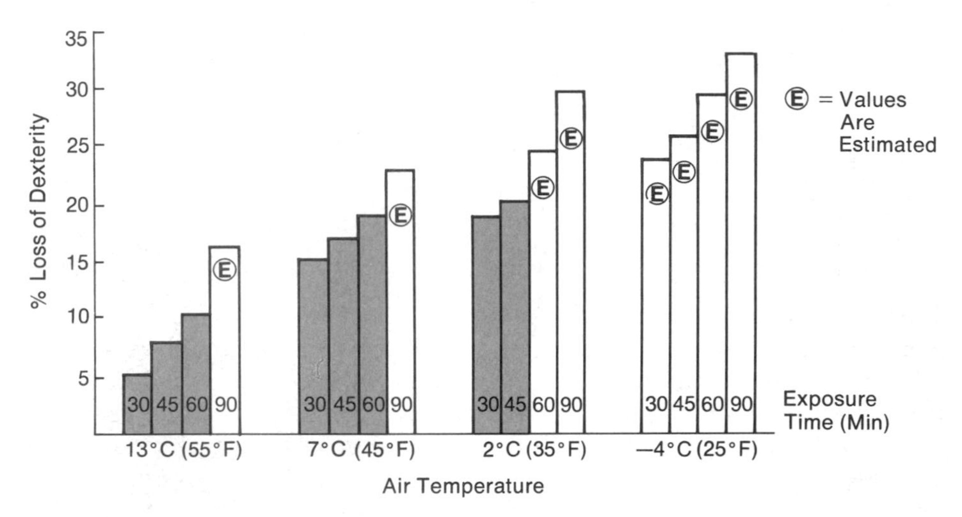

The effects of exposure to four different ambient dry bulb temperatures, −4°, 2°, 7°, and 13°C (25°, 35°, 45°, and 55°F), on people's performance on a moderately difficult manipulation assembly task are shown. Air velocity was low to moderate, similar to the levels seen inside buildings or outside on a calm day. The performance in cold conditions (shown as four groups of bar charts on the horizontal axis) is expressed as a percentage loss in dexterity (on the vertical axis) compared with performance on the same task in a 22°C (72°F) environment. For each cold condition there are four bars, representing 30, 45, 60, and 90 minutes of exposure. With increasing exposure time and cold, dexterity loss increases, providing the hands are not rewarmed in the pockets between tasks.

c. Techniques to Reduce Discomfort in the Hot and Cold Discomfort Zones

By altering temperature, humidity, air velocity, work load, clothing insulation, and radiant-heat load, one can improve comfort within the discomfort zone. Similar techniques, as well as reducing the duration of exposure, are used to bring extreme heat and cold exposures back into the discomfort zone.

(1) Heat Discomfort

(a) Decrease the Temperature

If the temperature can be reduced so that it matches the work load requirements and relative humidity, as shown in Table VD–4, there should be less risk of adverse health effects resulting from working in the heat. Humidity and radiant-heat reductions may be more effective than simple cooling in some environments, such as in the following situations:

- Areas where large amounts of air are turned over per hour, usually as part of a chemical exposure control process.
- Areas where a source of radiant heat, such as a large metallic surface of a reactor vessel or machine, provides the primary heat load.
- Localized areas where high humidity occurs, as in some chemical-making or cleaning operations where the humidity can be controlled locally.

In each situation the decision to reduce temperature or change one of the other variables should be made after each of the alternatives has been evaluated for effectiveness.

(b) Reduce the Humidity

The level of heat discomfort is strongly affected by the humidity level, because the major pathway for losing excess body heat outside the comfort zone is by evaporation of sweat from the skin. Evaporation of sweat can proceed only if the water vapor pressure of the skin is higher than the water vapor pressure of the ambient air. As humidity increases, this gradient from skin to air is reduced, and evaporative heat losses are correspondingly decreased. This result can be observed in Table VD–4; recommended maximum work loads decrease rapidly with increased humidity, particularly at ambient temperatures above 32°C (90°F).

If the humidity in an area can be decreased, higher ambient dry-bulb temperatures can be tolerated without serious discomfort or potential health risks. Water vapor in the air may be removed at the air intake side of the ventilating system or by using a dehumidifier in the work area. Whether either approach is feasible will depend on the size of the work area, the ventilation requirements (number of air changes per hour), and the seriousness of the heat and humidity problem.

(c) Increase the Air Velocity

For dry-bulb temperatures above 25°C (77°F), increasing the air velocity at the workplace improves comfort by increasing the convective and evaporative heat losses from the body (see Table VD–1). When ambient temperatures rise above 35°C (95°F), increased air velocity may be less effective in producing comfort, especially if the relative humidity is high (>70 percent). The effectiveness of air velocity in increasing heat loss reaches a plateau at about 2 m/sec (400 ft/min) in moderate to heavy work situations (Kamon and Avellini, 1979).

There may be situations in which air velocity cannot be increased sufficiently to improve comfort in a workplace, such as the following:

- High air velocities may mobilize dusts or chemical powders in the workplace, as in chemical-drying areas.
- Fans may not be able to reach the specific work areas where heat exposure takes place, as in maintenance tasks that are performed inside production machinery.
- High air velocities may disrupt work by blowing papers or tickets off a work surface.

Table VD–4 assumes an air velocity of less than 0.5 m/sec (100 ft/min). If air velocity is increased, the recommended maximum work loads at relative humidity values below 80 percent can be incremented one step.

(d) Reduce the Work Load

If it is difficult to improve the heat-exchanging conditions through changes in humidity or air velocity, people tend to reduce discomfort by altering their work load. They do so by slowing down the rate at which jobs are done or by increasing the number of breaks during which they can get away from the hot and humid environment. Most of the effects of thermal discomfort on productivity are hidden because standard production rates have been set in the hot atmosphere and already reflect the necessary work load adjustments. Improving thermal conditions in a workplace almost invariably results in increased productivity if people have been the production-limiting factors (Mackworth, 1946; Tichauer, 1962; Vernon and Bedford, 1927).

In most instances the reduction in work load is done by each individual worker in response to his or her level of discomfort. Where a machine or other external factor is pacing the operation and there is less room for individual work load adjustments, redesigning the job demands may be necessary. For instance, asking a pipe fitter to support a pipe or duct overhead for several minutes in a hot environment may be too stressful; a holding aid could be provided to reduce the muscular work.

As work load increases, more body heat is produced, and more demand is put on the circulation to get rid of that heat and keep the body core temperature steady. With increasing temperature outside the body, sweating has to increase to provide evaporative heat loss. The net effect of the ambient temperature on the cardiovascular stress experienced by a person working in the heat can be described in terms of work capacity. At temperatures between 24° and 40°C (75° and 105°F), each 1.0°C (1.8°F) rise in ambient temperature increases the physiological stress the equivalent of 1 percent of the maximum work capacity. This stress increase can be seen in a heart rate increase. Therefore a task that takes 25 percent of maximum work capacity at 24°C (75°F) may stress the system to an equivalent work load of 36 percent of maximum in a 35°C (95°F) environment (Kamon, 1975).

Table VD–5 gives the energy requirements of several industrial tasks. The actual demands of any task will vary with the skill and fitness of the person

Table VD–5: Metabolic Demands of Industrial Tasks* (Adapted from W. J. Nielsen, 1962–1982, and S. H. Rodgers, 1969–1979, Eastman Kodak Company; based on information in Davis, Faulkner, and Miller, 1969; Passmore and Durnin, 1955.)

Light	Moderate	Heavy	Very Heavy	Extremely Heavy
70–140 W (60–120 kcal/hr)	>140–280 W (>120–240 kcal/hr)	>280–350 W (>240–300 kcal/hr)	>350–420 W (>300–360 kcal/hr)	>420 W (>360 kcal/hr)
small-parts assembly	industrial sewing	making cement	shoveling (>7 kg)	stoking a furnace
typing	bench work	industrial cleaning	ditch digging	ladder or stair climbing
keypunching	filing	joining floorboards	hewing and loading coal	coal car unloading
inspecting	machine tending	plastering	handling moderately heavy (>7 kg) cases to and from a pallet	lifting 20-kg cases 10 times per minute
operating a milling machine	small-size packing	power truck operation	tree planting	
drafting	operating a lathe	handling light cases to and from a pallet	handling heavy (>11 kg) units frequently (>4 per minute)	
armature winding	operating medium-size presses	road paving		
hand typesetting	machining	painting		
operating a drill press	small-sized sheet metal work	handling operations		
desk work	electronics testing	metal casting		
small-parts finishing	plastic moulding	cutting or stacking lumber		
	operating a punch press	large-size packing		
	operating a crane			
	laying stones and bricks			
	sorting scrap			

* The values given include basal metabolism.

Several industrial jobs and tasks are listed according to their average energy demands, ranging from light (column 1) to extremely heavy (column 5) work. The range of energy expenditures is shown at the top of each column, in watts and kilocalories per hour. Tasks in the light- and moderate-effort categories (columns 1 and 2) are more likely to be done for a full 8-hour shift. Heavy to extremely heavy tasks (columns 3–5) are usually alternated with more sedentary paperwork or standby activities.

doing the job, the size and weight of the items being handled, the production pressure, and policies of management with regard to recovery break allowances. With these approximations, however, one can identify some of the less demanding tasks to which a person may be assigned in order to recover from heavier work in the heat.

The harder the work, the shorter should be the continuous duration of performing it if the person has control over the work load. The usual continuous work period for physical effort of 420 W (360 kcal/hr), for instance, is about 15 minutes.

(e) Adjust the Clothing

Propriety limits the extent to which clothing adjustments can be made in the hot workplace. Besides those clothing changes indicated in the thermal comfort discussion above, there are special suits and jackets designed to improve comfort and decrease the risk of health effects in hot environments. The two most commonly used items are ice vests (Kamon, 1980) and vortex cooling suits (Raven, Dodson, and Davis, 1979).

The ice vest uses pockets of ice or frozen gel to increase the gradient for heat loss from the torso of the worker. This effect keeps the core temperature from rising rapidly in high-heat exposures. At lower heat levels when moderate to heavy work is required, the ice vest increases comfort and reduces the load on the body's sweating mechanism. It can therefore prolong the length of time a person can work in the heat. The amount of added work time will vary with the work load, type of heat, and pattern of exposure. Figure VD–3 shows an ice vest that could be used in a maintenance task.

Some vests of this type may restrict movement of the trunk and may not be suitable for maintenance operations where awkward postures can regularly occur. Newer designs reduce the movement interference and allow flexibility in the placement of the ice packs according to the task demands.

Where whole-body protection against high heat levels (usually outside the discomfort zone but sometimes at its upper limit) is needed, vortex cooling suits have been successfully used. Figure VD–4 illustrates a vortex suit, which uses cold air to increase comfort by lowering skin temperature.

Vortex suits are not suitable for operations where a great deal of mobility is required because they restrict movement within the suit and through the tether to the air-supplying system. They are useful in some maintenance activities where continuous work, such as filing and grinding, is being done in one location in an upright posture.

In chemical operations where fabrics with low permeability are worn to protect the worker against chemical contact, reduced evaporative heat loss will increase the discomfort levels when the person is working in hot environments. If moderately heavy to heavy work is required while the person is wearing protective clothing in the heat, frequent recovery periods in cooler areas must be provided so that the accumulated heat can be lost (Mihal, 1981). The maximum work load guidelines in Table VD–4 assume a clothing insula-

Figure VD–3: Ice Vest for Improving Heat Loss in Hot Environments

(a)

(b)

The ice vest is worn under a jacket (part b) to provide a cold sink for body heat during work in high-heat environments. The vest contains multiple small pockets in which frozen water is held (part a). These ice pockets keep the trunk temperature low and permit excess heat to leave the body by facilitating conductive heat loss.

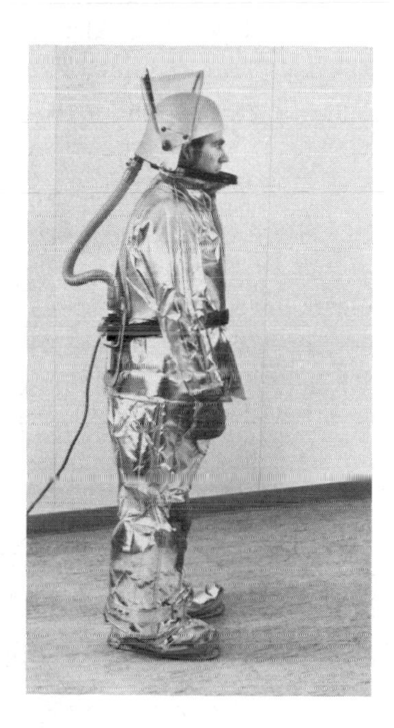

Figure VD–4: A Vortex Suit for Use in High-Heat Areas

This vortex suit covers the whole body and also has a hood to cover the head and neck. Cool air is passed through the suit via a hose connected at the back. The air increases comfort by lowering skin temperature and increasing convective—and, if the person is sweating, evaporative—heat loss. Partial suits, such as vortex jackets or hoods, can be used when heat exposure is limited to the upper body.

tion of 0.6 clo. If this value is reduced to 0.3 clo—by removing a shirt, for example—heat exchange would improve enough to add 2°C (3.8°F) to the ambient temperatures in column 1 and maintain the same work load.

(f) Provide Shields Against Radiant Heat

The discomfort produced by radiation from hot surfaces can be reduced by shielding the person from the surface (National Safety Council, 1972). This shielding can be accomplished by insulating the surface through the use of foam or polyester derivative over the metallic surface, by putting a barrier between the radiating surface and the worker (Lewis, Scherberger, and Miller, 1960), or by having the worker wear heat-reflective clothing, such as an aluminized jacket. The latter approach is not recommended for jobs where extended exposures and moderate to heavy work are required because the clothing insulation value prevents evaporative heat loss. Some effective heat barriers in the workplace are aluminized reflective curtains, water curtains, and wooden or fabric-covered standing panels.

Making a choice between coating the radiating surface or using a shield, and deciding which shield to use, depends on the work situation. If the worker must come into contact with the radiating surface to do the job, coating the surface would be preferable to using a shield; burns to the worker would not be as likely to occur (see the discussion on hot surfaces below).

(g) Work Practices for Exposure to Hot Environments

The material in this section was developed from information in Millican, Baker, and Cook (1981) and NIOSH (1972b).

Fairly substantial improvements in comfort can be achieved if the techniques just enumerated are used to reduce the heat load. In some workplaces it is not possible to provide conditions close to the comfort zone, however, so other work practices should be implemented to ensure that susceptible people will not experience heat-related health problems. The following practices have been shown to be effective:

- Workers should be provided with a cooler area in which to take work breaks. This area is often a cafeteria or break room; local, or spot, cooling areas at or near the workplace can also suffice. Spot-cooling techniques are under development as ways to conserve energy, as well as to improve comfort, in production areas (Azer, 1981; Simms, Gillies, and Drury, 1977). Since these stations usually have increased air velocity as well as cooler temperatures, they may produce discomfort for a person who has been sweating profusely during work in the heat. Temperature and air velocity should be selected to minimize this discomfort. Cool break areas should not be so cold that they reduce skin temperature enough to shut off the sweating mechanism. Temperatures of 20° to 23°C (68° to 74°F) are usually adequate to help cool off the body after heat has been accumulated.

- Drinking water should be provided in the work area. Salt should be provided in the cafeteria to use on food. Salt tablets are not recommended unless their use will be monitored strictly, since adverse effects may result from their overuse.
- Workers should be trained to recognize early signs of heat illness, such as dizziness or nausea (Leithead and Lind, 1964), and to leave the hot area when they occur. Help should be available to fill in for a person who needs to take a break to a cooler area.
- People who have been away from a hot job for at least a week should be given additional time for recovery breaks until they are reacclimatized to the heat (Givoni and Goldman, 1973).

(2) *Cold Discomfort*

The lower limit of temperature in the discomfort zone will be affected by air velocity, work load, clothing insulation, and radiant heat. Increases in air velocity result in increased discomfort, whereas increased work load and radiant-heat load improve comfort in the cold. Added clothing will eventually reduce the worker's mobility. Ways that these factors can be altered to reduce discomfort in the cold are summarized below.

(a) Reduce the Air Velocity

High air velocities increase convective heat loss from the body and increase discomfort in the cold. By lowering skin temperature, they also increase the gradient for heat transfer out of the body, some of which can be counteracted by increasing the layers of insulation over the skin. The air velocity may be difficult to alter if, for example, the person is working outdoors; shields or windproof clothing can be used to reduce the wind's influence on the worker. Indoors, air velocity problems can be solved by some of the following techniques:

- Reducing air flow through the ventilation ducts if this action is consistent with health and safety practices for adequate ventilation. Many systems have two-speed fans that permit the air velocity to be lowered in the cold season and raised in the hot season to improve comfort in the work area.
- Relocating a workplace in an area to reduce local drafts that contribute to discomfort.
- Using screens to shield the worker from the direct impact of the air velocity.

(b) Increase or Even Out the Work Load

Increasing the work load in the cold will increase comfort during the work periods (Wyndham and Wilson-Dickson, 1951); however, during recovery periods or standby time, greater discomfort may result. This discomfort is related to the amount of sweating that occurs during the physical effort. If much sweating occurs, it will reduce the effectiveness of the insulation in the clothing worn by the worker, thereby making her or him feel much colder in the ensuing recovery period (Enander, Ljungberg, and Holmer, 1979). Evaporative cooling may also occur during the recovery periods. As a general rule, increasing the work load is an effective way to counteract cold exposure only if the person can work without sweating very much. This activity corresponds to work loads below 175 W (150 kcal/hr) for most people.

Adjusting the work load so that the effort is spread more evenly over the shift, instead of alternating between periods of intense and then sedentary activity, will also reduce discomfort. However, such work patterns may be difficult to control in practice.

(c) Increase the Amount of Clothing Insulation

The most common approach to increasing comfort in the cold discomfort zone is to add layers of clothing. The degree of discomfort is related to skin temperature, so the more skin that is covered, the less is the discomfort sensed by the worker. The following clothing adjustments increase comfort in the cold:

- Use of fabrics like wool that have good insulation characteristics without being too bulky.
- Style of dress that fits closely but leaves a layer of air next to the skin and covers most of the skin.
- Pockets in which to warm the hands; pocket warmers could be used in some situations.
- Use of windproof clothing in areas where air velocity is high.

Since sweating may occur on a job being done in the cold, drying cabinets or areas should be available in break rooms to permit the clothing to dry out before the next cold exposure period. If such areas cannot be provided, more frequent breaks will be needed as the shift progresses, since cold tolerance will decrease as the clothing gets wet and has progressively lower insulating capability (Morris, 1975).

(d) Increase the Amount of Radiant Heat

In areas where it is not possible to increase the ambient temperature to increase the worker's comfort, such as working outdoors in the cold, the provision of radiant heaters may be justified. These heaters can increase comfort in a small area, similar to a spot-cooling zone in a hot environment. Such heat sources, where appropriate, can be used to heat the cabs of cranes or other heavy equipment used outdoors in cold weather. They can also be used to help dry out damp clothing, as mentioned above.

4. THE HEALTH RISK ZONE

a. Zone Definition

People with known health problems such as heart disease or some hormonal imbalances may be at risk for heat-related illness or may experience more discomfort than healthy workers in the heat discomfort zone (see Table VD–4) (Andersen et al., 1976). Most people, however, can tolerate extended exposures to these temperatures without experiencing problems. To tolerate the discomfort, though, they may do less work, so that result provides another reason to look for ways to increase comfort. Above the heat discomfort zone, however, many people may experience the symptoms of heat illness, such as nausea, dizziness, muscle cramps, and fatigue. A body temperature above 39.0°C (100.4°F), increased heart rate, a fall in diastolic blood pressure, and an increased breathing rate all signal the body's attempts to deal with the heat load and may warn of incipient failure of the heat transfer system (Leithead and Lind, 1964; NIOSH, 1972b).

Two types of exposure to hot and cold environments are of concern: long exposure to moderately low and high temperatures, and short exposures to very high or very low temperatures, known as heat and cold pulses. The former is of concern because it represents prolonged stress on the temperature-regulating system, with a life-threatening result, such as hypothermia or heat stroke, being a possible outcome. Short exposures are of concern because they represent a heat or cold load that cannot easily be accommodated by the body and may result in serious tissue damage, such as frostbite or burns (Blockley, 1964).

In the health risk zone these effects can be seen in more susceptible individuals; their occurrence is predictable enough to warrant trying to totally avoid the exposure. Where exposure cannot be controlled by engineering methods, exposure durations should be carefully regulated, and protective equipment and clothing should be provided.

b. Cold Injury

The major concern about whole-body exposure to the cold is the development of serious hypothermia and subsequent death from exposure. The body defends core temperature by intense shivering to increase metabolic heat. Exhaustion of this resource for generating heat is implied when body temperature falls below 35°C (95°F) (Burton and Edholm, 1955). Frostbite of the face or extremities may result from exposure to extreme cold, often in combination with high air velocities, or from prolonged exposure to less severe cold but with high humidity (Anon., 1970). Table VD–6 shows equivalent temperatures for several combinations of air velocity and temperature, known as the windchill index. Exposures may result in cold injury to exposed flesh at equivalent temperatures of −32°C (−36°F). For people working outdoors in the cold, body heat losses associated with high winds can be very significant. Rewarming areas should be provided to minimize the possibilities of health

Table VD–6: Equivalent Temperatures for Several Combinations of Low Temperature and Air Velocity (Anon., 1970.)

Estimated Air Velocity		Measured Dry Bulb Temperature Readings, °C (°F)									
		10 (50)	4 (40)	−1 (30)	−7 (20)	−12 (10)	−18 (0)	−23 (−10)	−29 (−20)	−34 (−30)	−40 (−40)
m/sec	mph	Equivalent Temperature, °C (°F)*									
calm	calm	10 (50)	4 (40)	−1 (30)	−7 (20)	−12 (10)	−18 (0)	−23 (−10)	−29 (−20)	−34 (−30)	−40 (−40)
2.2	5	9 (48)	3 (37)	−3 (27)	−9 (16)	−14 (6)	−21 (−5)	−26 (−15)	−32 (−26)	−38 (−36)	−44 (−47)
4.5	10	4 (40)	−2 (28)	−9 (16)	−16 (4)	−23 (−9)	−31 (−24)	−36 (−33)	−43 (−46)	−50 (−58)	−57 (−70)
6.7	15	2 (36)	−6 (22)	−13 (9)	−21 (−5)	−28 (−18)	−36 (−32)	−43 (−45)	−50 (−58)	−58 (−72)	−65 (−85)
8.9	20	0 (32)	−8 (18)	−16 (4)	−23 (−10)	−32 (−26)	−39 (−39)	−47 (−53)	−55 (−67)	−63 (−82)	−71 (−96)
11.2	25	−1 (30)	−9 (16)	−18 (0)	−26 (−15)	−34 (−29)	−42 (−44)	−51 (−59)	−59 (−74)	−67 (−88)	−76 (−104)
13.4	30	−2 (28)	−11 (13)	−19 (−2)	−28 (−18)	−36 (−33)	−44 (−48)	−53 (−63)	−62 (−79)	−70 (−94)	−79 (−109)
15.6	35	−3 (27)	−12 (11)	−20 (−4)	−29 (−21)	−37 (−35)	−46 (−51)	−55 (−67)	−63 (−82)	−72 (−98)	−81 (−113)
17.9	40	−3 (26)	−12 (10)	−21 (−6)	−29 (−21)	−38 (−37)	−47 (−53)	−56 (−69)	−65 (−85)	−73 (−100)	−82 (−116)
		LITTLE DANGER IF PROPERLY CLOTHED; REDUCED MANUAL DEXTERITY				DANGER OF FREEZING EXPOSED FLESH			VERY DANGEROUS		

* Combination of estimated air velocity and the measured dry bulb temperature readings given in the first row (above) of the table.

The equivalent temperatures for combinations of nine air velocities (columns 1 and 2) and ten dry bulb temperatures (columns 3 through 12) are shown in the body of the table. These values are known as the windchill index. Air velocity is given in meters per second (m/sec) in column 1 and in miles per hour (mph) in column 2. The values are further divided into three regions: those where there is little danger of cold injury if a person is properly clothed (the left side of the table); those where there is danger of freezing exposed flesh even in short, 1-minute exposures (the middle of the table); and those where a 30-second exposure of uncovered skin could result in frostbite (the right side of the table). Covering the exposed skin reduces the danger of cold injury.

effects to workers exposed to this cold.

Although a person may add several layers of clothing for protection against a cold environment, it is difficult to cover all of the body, especially the hands and face, and still be able to do some tasks. Therefore the usual limitation to cold exposure in very low temperatures is the time for the exposed skin to reach a temperature of 18°C (64°F) when pain is sensed (Hardy, 1970).

Cold injury to the hands can occur during work outdoors in cold climates, such as in construction, trucking, and handling operations, or work indoors in cold storage areas. This cold injury can result in frostbitten fingers or potential vibration injury syndromes (see Section VB, "Noise and Vibration"), and it can aggravate preexisting arthritic conditions. Most people who work in cold environments use their pockets to warm their hands between tasks, so long-term exposure to moderate cold is usually not a problem. Short-term exposure to cold lower than −23°C (−10°F), however, can be of concern if the task does not allow the operator to wear gloves (Fisher, 1957; Goldman, 1964). In addition, skin contact with very cold surfaces can result in tissue damage even for very short durations. Cold injuries to the hands often result in a lessening of manipulative skills of the fingers. To prevent this gloves with the tips of the fingers removed (called Miller mits) are often worn. Frequent breaks in the job schedule when the hands can be rewarmed are also helpful in preventing cold injuries.

c. Heat/Humidity Illness and Injury

The following examples identify some short duration industrial tasks that must be done in high-heat environments:

- Drawdown tasks in smelting operations.
- Fan adjustments in air-drying equipment.
- Emergency repairs in chemical-making equipment where processes produce heat.
- Plastics extruding operations.
- Forging operations.
- Oven-loading or oven-unloading tasks.
- Steam-cleaning operations.
- Utilities equipment repair and maintenance.
- Fire fighting.

The effects of exposure to heat and humidity in the health risk zone can range from dehydration to heat stroke, with the potential for burns also being of concern. Table VD–7 gives the maximum recommended temperatures for short-duration exposures to heat up to 63°C (146°F). Humidity and work load are varied; air velocity effects are shown in a separate summary in the table note. These temperature limits are based on the time it takes for body

Table VD–7: Recommended Maximum Temperatures for Short-Duration Exposures to High-Heat Environments (Up to 63°C, or 146°F) (Adapted from S. H. Rodgers and K. G. Corl, 1981, Eastman Kodak Company; based on information in Bell, Crowder, and Walters, 1971; Gagge, 1973; Hardy, 1970; Leithead and Lind, 1964; Pandolf and Goldman, 1978.)

Exposure Time (min)	Work Load*	Maximum Ambient Temperature, °C (°F)†		
		Relative Humidity 20%	Relative Humidity 50%	Relative Humidity 80%
5	L	63 (146)	56 (133)	56 (133)
	M	59 (138)	48 (118)	46 (115)
	H	57 (135)	46 (115)	42 (108)
15	L	53 (128)	45 (113)	40 (104)
	M	52 (126)	43 (110)	38 (100)
	H	51 (124)	41 (106)	36 (97)
30	L	52 (126)	44 (112)	39 (102)
	M	47 (116)	38 (100)	34 (93)
	H	41 (106)	36 (97)	30 (86)
45	L	51 (124)	43 (110)	38 (100)
	M	41 (106)	36 (97)	31 (88)
	H	36 (97)	32 (90)	27 (81)

Note: For 5-min exposure times in high air velocities (2 m/sec, or 400 ft/min), the following maximum temperatures are recommended (L = light work load, M = moderate work load, H = heavy work load):

Workload	Relative Humidity, %		
	20%	50%	80%
L	56 (133)	50 (122)	48 (118)
M	54 (129)	49 (120)	44 (111)
H	52 (126)	48 (118)	42 (103)

* Work load abbreviations: L = light, up to 140 W (120 kcal/hr); M = moderate, >140 to 230 W (>120 to 240 kcal/hr); H = heavy, >230 to 350 W (>240 to 300 kcal/hr).

† These temperatures assume the following conditions: clothing insulation = 0.6 clo; air velocity = 0.1 m/sec (20 ft/min); radiant heat = 2°C (3.6°F), which is the difference between the globe and dry bulb temperature readings. Exposures are for unacclimatized workers.

The highest recommended dry bulb temperatures are shown for continuous exposures of 5 to 45 minutes (column 1) at 20, 50, and 80 percent relative humidity (columns 3 through 5). Light (L), moderate (M), and heavy (H) work loads are indicated in column 2 for each of the exposure time and relative humidity combinations. Maximum exposure temperatures decrease with increased exposure time, relative humidity, and work load. Air velocity, radiant heat, and clothing insulation assumptions are stated in a footnote. At temperatures above 50°C (122°F), especially in high humidities, high air velocity (up to 2 m/sec, or 400 ft/min) will increase body heat storage, so lower ambient temperatures would be required. At lower temperatures increased air velocity will lengthen exposure time. A summary of recommended maximum temperatures in high air flow for 5-minute exposures is given in the table note.

core temperature and skin temperature to converge or the core temperature to rise 1°C (1.8°F). The convergence signals loss of a person's capability to transfer heat out of the body. If radiant-heat load exceeds 2°C (3.6°F), or if impermeable clothing is worn that raises Icl above 0.6, or reduces evaporative cooling, maximum temperatures will have to be lowered or the exposure time reduced.

The fall in maximum exposure temperature with increasing humidity levels is greater with increased exposure time. This fall is related to a reduction in a person's evaporative capacity at humidities above 50 percent. Increasing work load reduces exposure temperature maxima by adding to the body's heat load, thus speeding the convergence of skin and core temperatures. With radiant heat up to 5°C difference in globe and dry bulb thermometer readings, each of the temperatures would have to be lowered about 2°C (3.6°F).

Table VD–7 can be used in several ways: to describe an existing work environment and determine how long a person should be able to work there: to identify to what temperature a hot area should be cooled in order to permit a person to do a specific task, which can be described by its duration and work load requirements; and to identify the effectiveness of altering any of the variables on the exposure time or maximum temperature that can be tolerated when designing work for a hot environment. Chapter VI, Appendix C, contains a problem in which this information is applied.

It is advisable to let the workers set their own heat exposure times within the limits shown in Table VD–7, especially when the work load is moderate to heavy (see Table VD–5). Also, medical evaluation of the suitability of people for work in high-heat conditions should be made on a regular basis. Full acclimatization to this type of infrequent heat exposure may not occur (Lind and Bass, 1963). At very high sweat rates, as in heavy work in high heat, dehydration of the body can result. With dehydration the rate of rise of core temperature as a function of ambient temperature increases, making the possibility of heat stroke greater (Dukes-Dobos et al., 1966). For protection against heat illness, the following additional work practices should be enforced for people working in high-heat environments (Millican, Baker, and Cook, 1981; NIOSH, 1972b).

- Provision of protective clothing in high radiant heat.
- Use of the buddy system to ensure that each person is observed when working in the heat.
- Provision of tepid drinking water to reduce the potential for gastrointestinal symptoms associated with drinking cold water after working in the heat.
- Flexibility in the scheduling of work and rest so that the worker can self-limit exposure to the heat, as needed.

- Scheduling of the task in the cooler hours of the day, such as early morning, if outside temperatures add to the heat stress.
- Training of workers in first aid principles to aid a person who experiences heat/humidity illness.

The ability of a person to tolerate the short heat pulses shown in Table VD–7 at temperatures above 50°C (122°F) will depend on the levels of other factors. If a significant radiant-heat load is present, pain and burning of the skin can prevent a person from working in heat that he or she could theoretically tolerate on the basis of the heat balance equation. Radiant-energy burns can result from exposure to ambient temperatures that raise skin temperature to 45°C (113°F) (Hardy, 1970).

Extended exposure to air with very low humidity can produce serious discomfort for some workers and may be associated with certain pulmonary problems after the exposure. Provision of humidified air through a respirator can be used to overcome this drying effect, but the respirator may limit the person's activities and extend the time needed to do the task. In some emergency situations there may be inadequate time to put on special respiratory equipment, so people should generally not be expected to enter environments hotter than 120°C (250°F) even for a brief period. At this temperature even physically fit young people show deteriorated performance on coordination tests after 5 minutes (Pepler, 1959). If they are able to bring their own environment with them, as in a vortex suit, exposures to higher temperatures are often possible.

5. WORKING IN WET ENVIRONMENTS

Some jobs include exposure to water temperatures that range from comfortable to very uncomfortable. The water may be contacted in the process of doing a production, maintenance, or cleanup task, such as the following:

- Cafeteria work, including operation of dishwashers.
- Chemical-making operations.
- Building cleaning.
- Plumbing, heating, and refrigeration work.
- Hosing-down tasks or cleaning of facilities or equipment.
- Outside work on rainy or snowy days.
- Underwater construction work.

In most industries water contact is confined to the extremities, except for work in inclement weather or the use of safety showers in chemical or fire exposure areas.

a. Wet Clothing

The effect of working in wet clothing on tolerance times for high and low temperatures is strongly influenced by the duration of exposure, the area of contact of the skin, the work load, and the air velocity. Short contacts (less than 1 minute) with extremes of temperature are more tolerable than hours of exposure to less extreme environments.

Wet clothing in the cold reduces the insulation factor. This reduction results in a drop in skin temperature, thus increasing heat loss from the body. Evaporative heat loss will increase, especially when the air velocity is above 0.3 m/sec (60 ft/min), unless the body is dried and clothed quickly. This condition is of concern in outdoor work in inclement weather where water contact occurring on a heavy task is followed by more sedentary work.

Water contacting the skin and clothing in hot environments helps increase the evaporative heat loss without requiring sweating, providing there is air flow. Thus the discomfort zone temperature limits are higher in the heat and cold when the clothing is wet than when it is dry (Craig, 1972; Craig and Moffitt, 1974).

b. Hand and Forearm Water Immersion

Loss of finger flexibility and tissue injury may result when the hands are immersed in hot or cold water. Tasks that result in immersions of this type are cleaning operations, plumbing, and various chemical manufacturing processes. Temperatures of liquids in which a worker may immerse his or her hands are seldom below 1°C (34°F) or above 70°C (158°F). At these extremes, protective equipment is usually provided to reduce the possibility of contact and skin burns. For hand and forearm immersions in watery, nonirritating liquids, temperatures above 8°C (46°F) (Provins and Morton, 1960) and less than 36°C (97°F) (Hardy, Stolwijk, Hammel, and Murgatroyd, 1965) are recommended. If immersion time is prolonged, temperatures should be as close to 32°C (90°F) as possible to minimize heat loss or gain (Craig and Dvorak, 1966). Chemicals that penetrate the skin readily may produce burns or other tissue damage sooner at lower temperatures in the hot range and at higher temperatures in the cold range.

c. Using Safety Showers and Eye Baths

In manufacturing or laboratory areas where people are likely to come into contact with chemicals or where fire could possibly occur, safety showers and eye baths are provided to permit rapid washing of the skin or eyes. Although a person can get thoroughly wetted down under a safety shower, it is not equivalent to whole-body immersion in water of the same temperature. The parts of the body nearest the shower head get more thoroughly doused than lower ones; and the person involved is usually washing off the affected skin by rubbing it and actively moving around, thus increasing the metabolic rate. Substances that burn the skin are more comfortably washed off if the skin is lightly anesthetized with cool water; cool water also reduces blood flow to the skin, thereby reducing absorption of the contacted chemical into the body.

From interviews with people who had chemical contact incidents and from an analysis of accident data, it appears that tempering of safety shower and eye bath water temperatures in the range of 27° ± 5.6°C (80° ± 10°F) should satisfy both physiological and psychological needs (S. H. Rodgers, 1980, Eastman Kodak Company). Flow rates of 42 liters per minute (L/min), or 10 gallons per minute (gal/min), for showers and 14–21 L/min (3–5 gal/min) for eye baths for 10 minutes of washing are assumed. If air velocity is high or if air temperature falls well below the comfort zone (see Figure VD–1), safety showers should be provided with heated enclosures and with water temperatures nearer the upper recommended limit of 32°C (90°F).

6. SURFACE TEMPERATURES

The presence of a metallic surface may limit the duration and acceptability of working in a hot or cold environment. A person may be able to walk into an area for a short time to do monitoring tasks, but he or she may not be able to stand for extended periods on hot or cold floors or be able to come into contact with other surfaces that exceed the guidelines given below. For maintenance tasks in hot or cold environments, the limit on continuous working time may be related to the temperature of work surfaces that make it difficult for the mechanic to maintain the posture needed to do the task, or the time may be limited by tools that become unbearably hot or cold after a short time. Kneeling on a hot grating or pressing against a cold metal access port are examples of tasks that may limit a person's ability to work in environments that are otherwise acceptable from the standpoint of body heat balance. Using insulating materials and protective clothing will overcome some of these difficulties, but new problems, such as reduced dexterity from glove interference with finger movement, may also be created.

a. Hot Surfaces

The guidelines given below for touching hot surfaces are based on very short (1-sec) contacts. The longer the contact period, the less high the surface or substance temperature can be. This result is illustrated by the following data from studies where a finger was immersed in water (adapted from Hardy, Stolwijk, Hammel, and Murgatroyd, 1965):

1-sec contact, finger, first-degree burn: 80°C (176°F).
1-sec contact, finger, pain: 53°C (127°F).
About 15 min, fingers, pain: 45°C (113°F).

Table VD–8 summarizes the information on the recommended maximum temperatures for materials often found in the workplace. Because people may contact the workplace surface for more than 1 second, and may also have injured areas on the skin that result in higher heat transfer to underlying cells, these values should be considered maximum values; designs should not exceed them.

Table VD–8: Contact with Hot Surfaces, Maximum Temperatures (Adapted from T. W. Faulkner, 1974, Eastman Kodak Company; based on data in the literature, especially Wu, 1972.)

	Temperature Threshold	
Material	**Pain Threshold, °C (°F)**	**First-Degree Burn Threshold, °C (°F)**
Polystyrene GP	77 (171)	138 (281)
Wood (average)	76 (169)	135 (275)
ABS Resins	74 (166)	131 (268)
Phenolics (average)	60 (141)	99 (210)
Brick	59 (138)	95 (202)
Heat-resistant Glass	54 (129)	82 (180)
Water	53 (127)	80 (176)
Concrete	50 (122)	73 (164)
Steel	45 (113)	62 (143)
Aluminum	45 (112)	60 (141)

Note: Data are based on a 1-second contact by the finger.

The temperature of several materials (column 1) are shown at which pain (column 2) or a first-degree burn (column 3) will result from a 1-second finger contact. These values are maximum temperatures, so designs should not exceed them. Lower temperatures are desirable, since they accommodate longer contact times, contact with other skin such as the forearm, or the presence of breaks in the skin from cuts or abrasions.

b. Cold Surfaces

Contact with cold surfaces can become a problem if water moisture from the skin freezes to the surface, thereby resulting in adhesion of the skin to the cold surface. Trying to detach the skin from the surface may result in tearing and injury to the tissue if the cold surface is not first rewarmed. This contact can occur when a person is working with metal objects in very cold climates, such as in pipe fitting or other construction and maintenance work. In general, a safe recommendation is that unheated metallic surfaces should not be contacted with bare skin in temperatures below 5°C (41°F); plastic and wood surfaces can be touched in temperatures as low as −20°C (−4°F), providing they are dry. So that the possibility of skin adhering to the surface is reduced, however, gloves should be worn when touching these surfaces if they are 0°C (32°F) or colder.

For a reduction of the probability of a worker contacting cold or hot surfaces, the following design guidelines and actions should be considered:

- If possible, put the hot or cold components in an inaccessible location.
- If the component must be accessible, place it out of range of the normal hand and arm motion patterns needed to operate or maintain neighboring components.
- Design the hot or cold component so that it does not make a convenient place to rest an arm or hand and does not resemble controls, handles, or other frequently touched portions of the equipment.
- Label hot or cold surfaces.
- Use plastics or similar less conductive materials over hot and cold surfaces.
- Use paint or other coatings on metal surfaces to reduce their thermal conductivity.

REFERENCES FOR CHAPTER V

Abrams, C. F., Jr., and C. W. Suggs. 1977. "Development of a Simulator for Use in the Measurement of Chain Saw Vibration." *Applied Ergonomics, 8 (3):* pp. 130–134.

ACGIH. 1982. *Threshold Limit Values for Chemical Substances in Workroom Air Adopted by ACGIH for 1982.* Cincinnati: American Conference of Governmental Industrial Hygienists, pp. 82–83.

Anderson, I., P. B. Jensen, P. Junker, A. Thomsen, D. P. Wyon. 1976. "The Effects of Moderate Heat Stress on Patients with Ischemic Heart Disease." *Scandinavian Journal of Work, Environment and Health, 4:* pp. 256–272.

Anon. 1970. *Cold Injury.* TB MED 81, NAVMED P-5052-29 AFP 161-11. Washington D.C.: Departments of the Army, Navy, and Air Force, 15 pages.

ANSI. 1973. *American National Standard for Leakage Current for Appliances.* C101.1. New York: American National Standards Institute, 8 pages.

ANSI. 1981. *National Electric Safety Code.* C2. New York: American National Standards Institute.

ASHRAE. 1972, Chapter 7, "Physiological Principles" in *Handbook of Fundamentals.* New York: American Society of Heating, Refrigerating and Air-Conditioning Engineers, pages 111–126.

ASHRAE. 1974. *Thermal Environmental Conditions for Human Occupancy.* ASHRAE Standard 55-74 (ANSI B193.1-76). New York: American Society of Heating, Refrigerating and Air-Conditioning Engineers, 8 pages.

Azer, N. Z. 1981. "Design Guidelines for Spot Cooling Systems. Part 1: Assessing the Acceptability of the Environment. Part 2: Cooling Jet Model and Design Procedure." Papers submitted to ASHRAE, 62 pages.

Barnaby, J. F. 1980. "Lighting for Productivity Gains." *Lighting Design and Application, 10 (2):* pp. 20–28.

Beaupeurt, J. E., F. W. Snyder, S. H. Brumaghim, and R. K. Knapp. 1969. *Ten Years of Human Vibration Research.* DS-7888. Wichita: Boeing Company, 63 pages.

Belding, H. S. 1970. "The Search for a Universal Heat Stress Index." Chapter 13 in *Physiological and Behavioral Temperature Regulation,* edited by J. D. Hardy, A. P. Gagge, and J. A. J. Stolwijk, Springfield, Ill.: C. C. Thomas. pp. 193–204.

Belding, H. S. 1973. "Control of Exposures to Heat and Cold." Chapter 38 in NIOSH (1973), pp. 563–572.

Belding, H. S., and T. F. Hatch. 1955. "Index for Evaluating Heat Stress in Terms of Resulting Physiological Strains." *Heating, Piping and Air Conditioning,* August 1955, pp. 129–136.

Bell, C. R., M. J. Crowder, and J. D. Walters. 1971. "Durations of Safe Exposure for Men at Work in High Temperature Environments." *Ergonomics, 14 (6):* pp. 733–757.

Beranek, L. L., W. E. Blazier, and J. J. Figwer. 1971. "Preferred Noise Criterion (PNC) Curves and Their Application to Rooms." *Journal of the Acoustical Society of America, 50 (5.1):* pp. 1223–1228.

Berenson, P. J., and W. G. Robertson. 1973. "Temperature." Chapter 3 in *Bioastronautics Data Book,* 2nd edition, edited by J. F. Parker, Jr., and T. R. West, NASA SP-3006, pp. 65–148. Washington, D.C.: U.S. Government Printing Office.

Blockley, W. V. 1964. "Temperature." Section 7 in *Bioastronautics Data Book,* edited by P. Webb, NASA SP-3006, pp. 103–131. Washington, D.C.: National Aeronautics and Space Administration.

Bovenzi, M., L. Petroniv, and F. DiMarino. 1980. "Epidemiological Survey of Shipyard Workers Exposed to Hand-Arm Vibration." *International Archives of Occupational and Environmental Health, 46:* pp. 251–266.

Broadbent, D. E. 1957. "Effects of Noise on Behavior." Chapter 10 in *Handbook of Noise Control,* edited by C. Harris, pp. 10–1 to 10–34. New York: McGraw-Hill.

Burnham, R. W., R. M. Hanes, and C. J. Bartelson. 1963. *Color: A Guide to Basic Facts and Concepts.* New York: John Wiley and Sons, Inc., 249 pages.

Burton, A. C., and O. G. Edholm. 1955. *Man in a Cold Environment.* London: Edward Arnold, Ltd. Cited in ASHRAE (1972).

Carter, H. H., and S. S. Coulter. 1942. "Threshold of Stimulation of Alternating Currents." *Archives of Physical Therapy, 23:* pp. 207–213. Cited in Turner (1972).

Cohen, A. 1969. "Effects of Noise on Psychological State." In *Noise as a Public Health Hazard,* ASHA Reports 4 (February 1969), pp. 74–88; proceedings of the conference in Washington, D.C., June 13–14, 1968. Washington, D.C.: American Speech and Hearing Association.

Colquhoun, W. P., M. J. F. Blake, and R. S. Edwards. 1968. "Experimental Studies of Shift Work. I. A Comparison of Rotating and Stabilized 4-Hour Shift Systems." *Ergonomics, 11:* pp. 437–453.

Corliss, E. L. R., and F. E. Jones. 1976. "Method for Estimating the Audibility and Effective Loudness of Sirens and Speech in Automobiles." *Journal of the Acoustical Society of America, 60 (5):* pp. 1126–1131.

Craig, A. B., Jr., and M. Dvorak. 1966. "Thermal Regulation During Water Immersion." *Journal of Applied Physiology 21 (5):* pp. 1577–1585.

Craig, F. N. 1972. "Evaporative Cooling of Man in Wet Clothing." *Journal of Applied Physiology, 33 (3):* pp. 331–336.

Craig, F. N., and J. T. Moffitt. 1974. "Efficiency of Evaporative Cooling from Wet Clothing. *Journal of Applied Physiology, 36 (3):* pp. 313–316.

Dalziel, C. F. 1953. "A Study of the Hazards of Impulse Currents." *IEEE Transactions on Communications and Electronics, 72:* pp. 1032–1043.

Dalziel, C. F. 1972. "Electric Shock Hazard." *IEEE Spectrum, 9:* pp. 41–50.

Dalziel, C. F., and W. R. Lee. 1969. "Lethal Electric Currents." *IEEE Spectrum, 6:* pp. 44–50.

Davis, H. L., T. W. Faulkner, and C. I. Miller. 1969. "Work Physiology." *Human Factors, 11 (2):* pp. 157–166.

Dukes-Dobos, F. N., A. Henschel, C. Humphreys, K. Kronoveter, M. Brenner, and W. S. Carlson. 1966. *Industrial Heat Stress; Southern Phase.* Cincinnati: U.S. Department of Health, Education, and Welfare, Division of Occupational Health, 85 pages.

Dusek, E. R. 1957. *Manual Performance and Finger Temperatures as a Function of Ambient Temperature.* Tech. Rpt No. EP-68. Natick, Mass.: Headquarters Quartermaster Research and Engineering Center Environmental Protection Research Division.

Emmerson, W. E. 1974. "Don't Sit Near the Window." *ASHRAE Journal, 80:* pp. 53–56.

Enander, A., A. S. Ljungberg, and I. Holmer. 1979. "Effects of Work in Cold Stores on Man." *Scandinavian Journal of Work, Environment and Health, 5:* pp. 195–204.

Fanger, P. O. 1970. *Thermal Comfort, Analyses and Applications in Environmental Engineering.* Copenhagen: Danish Technical Press, 244 pages.

Fanger, P. O. 1977. "Local Discomfort to the Human Body Caused by Non-Uniform Thermal Environments." *Annals of Occupational Hygiene, 20:* pp. 285–291.

Faulkner, T. W., and T. J. Murphy. 1973. "Lighting for Difficult Visual Tasks." *Human Factors, 15 (2):* pp. 149–162.

Fidell, S. 1978. "Effectiveness of Audible Warning Signals for Emergency Vehicles." *Human Factors. 20 (1):* pp. 19–26.

Fidell, S., K. S. Pearsons, and R. Bennett. 1974. "Prediction of Aural Detectability of Noise Signals." *Human Factors. 16 (4):* pp. 373–383.

Fisher, F., ed. 1957. *Protection and Functioning of the Hands in Cold Climates.* Proceedings of a conference sponsored by the Quartermaster Research and Development Command, April 23–24, 1956, Natick, Mass. Washington, D.C.: National Academy of Science, 176 pages.

Flynn, J. E. 1979. "The IES Approach to Recommendations Regarding Levels of Illumination." *Lighting Design and Application, 9 (9):* pp. 74–77.

Gagge, A. P. 1973. "A TWO-NODE Model of Human Temperature Regulation in FORTRAN." Appendix to Chapter 3 "Temperature" in *Bioastronautics Data Book,* 2nd edition, edited by J. F. Parker, Jr., and T. R. West, pp. 142–146. Washington, D.C.: U.S. Government Printing Office.

General Electric Company. 1964. *Supplemental Lighting.* TP-121-R. Cleveland: Large Lamp Department, 16 pages.

General Electric Company. 1978. *Light and Color.* TP-119. Cleveland: Lighting Business Group, 31 pages.

Givoni, B., and R. F. Goldman. 1973. "Predicting Effects of Heat Acclimatization on Heart Rate and Rectal Temperature." *Journal of Applied Physiology, 35 (6):* pp. 875–879.

Goldman, R. F. 1964. "The Arctic Soldier: Possible Research Solutions for His Protection." In *Science in Alaska,* edited by G. Dahlgren, pp. 401–419; proceedings of the Fifteenth Alaskan Science Conference, August 31–September 4, 1964. College, Alaska. Washington, D.C.: American Association for the Advancement of Science, Alaska Division.

Grandjean, E. 1961. "Musik und Arbeit." *Zeitschrift für Präventivmedizin, 6:* pp. 65–70.

Grandjean, E. 1980. *Fitting the Task to the Man—An Ergonomic Approach.* 2nd edition. New York: International Publications Service, 379 pages (also London: Taylor and Francis).

Hardy, J. D. 1970. "Thermal Comfort, Skin Temperature, and Physiological Thermoregulation." Chapter 57 in *Physiological and Behavioral Temperature Regulation,* edited by J. D. Hardy, A. P. Gagge, and J. A. J. Stolwijk, pp. 856–873. Springfield, Ill.: C. C. Thomas.

Hardy, J. D., J. A. J. Stolwijk, H. T. Hammel, and D. Murgatroyd. 1965. "Skin Temperature and Cutaneous Pain During Warm Water Immersion." *Journal of Applied Physiology, 20 (5):* pp. 1014–1021.

Harris, C. M. 1957. *Handbook of Noise Control.* New York: McGraw-Hill.

Hertig, B. A. 1973. "Thermal Standards and Measurement Techniques." Chapter 31 in NIOSH (1973), pp. 413–430.

Hodgkin, B. C. 1974. "Some Consequences of Electric Shock." *Journal of the Maine Medical Association, 65:* pp. 1–3.

Hodgkin, B. C., O. Langworthy, and W. B. Kouwenhoven. 1972. "Effects on Breathing of an Electric Shock Applied to the Extremities." Paper T-72087-0, IEEE winter meeting, New York, January 30–February 4, 1972.

Hopkinson, R. G., and J. B. Collins. 1970. *The Ergonomics of Lighting.* London: MacDonald and Co., Ltd. 272 pages.

Hughes, P. C., and J. F. McNellis. 1978. "Lighting, Productivity and the Work Environment." *Lighting Design and Application, 8 (12):* pp. 32–40.

Hunter, R. S. 1937. "Methods of Determining Gloss." *Journal of Research of the National Bureau of Standards, 18:* pp. 19–39.

IERI. 1975. *Annual Report, Illuminating Engineering Research Institute.* New York: IERI, pp. 1–12.

IOS. 1974. *Guide for the Evaluation of Human Exposure to Whole-Body Vibration.* IOS 2631. Geneva: International Organisation for Standardisation.

IOS. 1979. "Guide for the Measurement and the Evaluation of Human Exposure to Vibration Transmitted to the Hand." Draft Proposal No. 5349, IOS TC/108/SC4/WG3. Geneva: International Organisation for Standardisation.

Kamon, E. 1975. "The Ergonomics of Heat and Cold." *Texas Reports on Biology and Medicine, 33 (1):* pp. 145–182.

Kamon, E. 1980. "Ergonomic Aspects of Exposure to Thermal and Other Environmental Stresses." In *Environment and Health,* edited by N. Triets, p. 305. Ann Arbor: Ann Arbor Science.

Kamon, E., and B. Avellini. 1979. "Wind Speed Limits to Work Under Hot Environments for Clothed Man." *Journal of Applied Physiology: Respiratory, Environmental and Exercise Physiology, 46 (2):* pp. 340–345.

Kaufman, J. E., and J. F. Christensen, eds. 1972. *IES Lighting Handbook.* 5th ed. New York: Illuminating Engineering Society, 726 pages.

Keesey, J. C., and F. S. Letcher. 1970. "Human Thresholds of Electric Shock at Power Transmission Frequencies." *Archives of Environmental Health, 21:* pp. 547–552.

Kleitman, N., and D. P. Jackson. 1950. "Body Temperature and Performance Under Different Routines." *Journal of Applied Physiology, 3:* pp. 309–328.

Kleronomos, C. C., and E. C. Cantwell. 1979. "Determining Maximum Impedance in a Grounding Conductor." *Electronic Engineering Times,* June 4/11, 1979, pp. 42–43.

Kouwenhoven, W. B. 1969. *Human Safety and Electric Shock.* ISA Monograph No. 112, pp. 91–97.

Kryter, K. D. 1970. *The Effects of Noise on Man.* New York: Academic Press, pp. 545–585.

Leithead, C. S., and A. R. Lind. 1964. *Heat Stress and Heat Disorders.* Philadelphia: F. A. Davis, 304 pages (also London: Cassell).

Lewis, C. E., R. F. Scherberger, and F. A. Miller. 1960. "A Study of Heat Stress in Extremely Hot Environments, and the Infra-Red Reflectance of Some Potential Shielding Materials." *British Journal of Industrial Medicine, 17:* pp. 52–59.

Lind, A. R. 1963. "A Physiological Criterion for Setting Thermal Environmental Limits for Everyday Work." *Journal of Applied Physiology, 18 (1):* pp. 51–56.

Lind, A. R., and D. E. Bass. 1963. "The Optimal Exposure Time for the Development of Acclimatization to Heat." *Federation Proceedings, 22:* p. 704. Cited in Dukes-Dobos et al. (1966).

Lum-i-neering Associates. 1979. *Lighting Design Handbook.* Final report, U.S. Government Contract No. N68305-76-C-0017 (AD A074836), Port Hueneme, Calif., Civil Engineering Laboratory, Dept. of the Navy, 198 pages.

McIntyre, D. A. 1973. "A Guide to Thermal Comfort." *Applied Ergonomics, 4 (2):* pp. 66–72.

McIntyre, D. A. 1978. "Response to Atmospheric Humidity at Comfortable Air Temperature: A Comparison of Three Experiments." *Annals of Occupational Hygiene, 21:* pp. 177–190.

McIntyre, D. A., and I. D. Griffiths. 1975. "The Effects of Added Clothing on Warmth and Comfort in Cool Conditions." *Ergonomics, 18 (2):* pp. 205–211.

Mackworth, N. H. 1946. "Effects of Heat on Wireless Operators." *British Journal of Industrial Medicine, 3:* p. 143.

Mihal, C. P., Jr. 1981. "Effect of Heat Stress on Physiological Factors for Industrial Workers Performing Routine Work and Wearing Impermeable Vapor-Barrier Clothing." *American Industrial Hygiene Association Journal, 42:* pp. 97–103.

Miller, J. D. 1971. "Noise and Performance." In *Effects of Noise on People,* NTID 300-7, pp. 118–121. Washington, D.C.: Environmental Protection Agency. Supt. of Documents.

Millican, R., R. C. Baker, and G. T. Cook. 1981. "Controlling Heat Stress—Administrative Versus Physical Control." *American Industrial Hygiene Association Journal, 42:* pp. 411–416.

Morgan, C. T., J. S. Cook III, A. Chapanis, and M. W. Lund. 1963. *Human Engineering Guide to Equipment Design.* New York: McGraw-Hill, 609 pages.

Morris, J. V. 1975. "Developments in Cold Weather Clothing." *Annals of Occupational Hygiene, 17:* pp. 279–294.

National Safety Council. 1972. "Radiant Heat Control." Data Sheet 381, Revision A, by the Glass and Ceramics Section, and the Engineering and Health Committees. Chicago: National Safety Council. Published in *National Safety News,* January 1972, pp. 77–86.

Nemecek, J., and E. Grandjean. 1973. "Noise in Landscaped Offices." *Applied Ergonomics, 4 (1):* pp. 19–22.

Nevins, R. G., and A. M. Feyerherm. 1966. "Effect of Floor Surface Temperature on Comfort. Part IV: Cold Floors." *ASHRAE Transactions, 73 (2),* 6 pages.

Nevins, R. G., P. E. McNall, Jr., and J. A. J. Stolwijk. 1974. "How to be Comfortable at 65 to 68 Degrees." *ASHRAE Journal, 80 (1):* pp. 41–43.

NIOSH. 1972a. *Criteria for a Recommended Standard—Occupational Exposure to Noise.* Cincinnati: U.S. Department of Health, Education, and Welfare, National Institute for Occupational Safety and Health.

NIOSH. 1972b. *Occupational Exposure to Hot Environments, Criteria for a Recommended Standard.* Cincinnati: U.S. Department of Health, Education, and Welfare, National Institute for Occupational Safety and Health, 104 pages.

NIOSH. 1973. *The Industrial Environment—Its Evaluation and Control.* Washington, D.C.: U.S. Department of Health, Education, and Welfare, National Institute for Occupational Safety and Health, 719 pages.

Pandolf, K. B., and R. F. Goldman. 1978. "Convergence of Skin and Rectal

Temperatures as a Criterion for Heat Tolerance." *Aviation Space and Environmental Medicine, 49 (9):* pp. 1095–1101.

Passmore, R., and J. V. G. A. Durnin. 1955. "Human Energy Expenditure." *Physiological Reviews, 35:* pp. 801–840.

Pepler, R. D. 1959. "Extreme Warmth and Sensorimotor Coordination." *Journal of Applied Physiology, 14 (3):* pp. 383–386.

Plutchik, R., and H. Bender. 1966. "Electrocutaneous Pain Thresholds in Humans to Low Frequency Square Wave Pulses." *Journal of Psychology, 62:* pp. 151–154.

Poulton, E. C. 1978. "Increased Vigilance with Vertical Vibration at 5 Hz: An Alerting Mechanism." *Applied Ergonomics, 9:* pp. 73–76.

Provins, K. A., and R. Morton. 1960. "Tactile Discrimination and Skin Temperature." *Journal of Applied Physiology, 15 (1):* pp. 155–160.

Pyykkö, I., M. Färkkilä, J. Toivanen, O. Korhonen, and J. Hyvärinen. 1976. "Transmission of Vibration in the Hand-Arm System with Special Reference to Changes in Compression Force and Acceleration." *Scandinavian Journal of Work Environment and Health, 2:* pp. 87–95.

Rabineau, G. 1976. "Whole Body Vibration." Document prepared for Bruel and Kjaer, October 22, 1976.

Ramkumar, V., and C. A. Bennett. 1979. "How Big is Dark?" In *Proceedings of the Human Factors Society, 23rd Annual Meeting, Boston,* October 24–27, 1979. Santa Monica: Human Factors Society, pp. 116–118.

Ramsey, J. D. 1975. "Occupational Vibration." Chapter 23 in *Occupational Medicine: Principles and Practical Applications,* edited by C. Zenz, pp. 553–562. Chicago: Year Book Medical Publishers.

Raven, P. B., A. Dodson, and T. O. Davis. 1979. "Stresses Involved in Wearing PVC Supplied-Air Suits: A Review." *American Industrial Hygiene Association Journal, 40:* pp. 592–599.

Rodgers, S. H. 1978. "Metabolic Indices in Materials Handling Tasks." In *Safety in Manual Materials Handling,* edited by Colin G. Drury, pp. 52–56. Cincinnati: U.S. Department of Health, Education, and Welfare, National Institute for Occupational Safety and Health.

Ross and Baruzzini, Inc. 1975. "Energy Conservation Principles Applied to Office Lighting." Conservation Paper No. 18, Contract No. 14-01-0001-1845. Washington, D.C.: Federal Energy Administration, 198 pages.

Rubin, L. S. 1957. "Manual Dexterity of the Gloved and Bare Hand as a Function of the Ambient Temperature and Duration of Exposure." *Journal of Applied Psychology, 41 (6):* pp. 377–383.

Sandover, J. 1979. "A Standard on Human Response to Vibration—One of a New Breed?" *Applied Ergonomics, 10 (1):* pp. 33–37.

Simms, M., G. Gillies, and R. Drury. 1977. "Using a Cool Spot to Improve the Thermal Comfort of Glassmakers." *Applied Ergonomics, 8 (1):* pp. 2–6.

Stengel, R. F. 1979. "Moiré Topography Records Surface Contour Intervals." *Design News, 8:* p. 126.

Taylor, E. R., and R. G. Watson. 1972. "Surface Inspection Experience Using Strobe Illumination." In *Proceedings of the 13th Mechanical Working and Steel Processing Conference,* Pittsburgh: Republic Steel Corporation, 9 pages.

Tichauer, E. R. 1962. "The Effects of Climate on Working Efficiency." *Impetus, 1 (5):* pp. 24–31 (official journal of the Chamber of Manufacturers of New South Wales, Australia).

Turner, H. S. 1972. *Human Responses to Electricity: A Literature Review.* Wright-Patterson AFB, Ohio: Aerospace Medical Research Labs (N73-10065); also Ohio State Research Foundation (NASA CR 128422), 115 pages.

Vernon, H. M., and T. Bedford. 1927. *The Relation of Atmospheric Conditions to the Working Capacity and the Accident Rate of Coal Miners.* Medical Research Council Rpt 39, Industrial Fatigue Research Board. London: His Majesty's Stationery Office, 34 pages.

Wasserman, D. E. 1980. "Occupational Vibration Studies in the U.S.—A Review." *Sound and Vibration,* October 1980, pp. 21–24.

Wasserman, D. E., and D. W. Badger. 1973. *Vibration and the Worker's Health and Safety.* Tech. Rpt No. 77. Washington, D.C.: U.S. Department of Health, Education, and Welfare, National Institute for Occupational Safety and Health, 251 pages.

Wasserman, D. E., W. Taylor, and M. Curry. 1977. *Proceedings of the International Occupational Hand-Arm Vibration Conference.* Cincinnati, Ohio, October 28–31, 1975. Publication No. 77-170. U.S. Department of Health, Education, and Welfare, National Institute for Occupational Safety and Health, Washington, D.C. U.S. Dept. of Commerce.

Watson, A. B., and J. S. Wright. 1973. "Electrical Thresholds for Ventricular Fibrillation in Man." *Medical Journal of Australia, 6* (June 16): pp. 1179–1182.

Webster, J. C. 1969. "Effects of Noise on Speech Intelligibility." In *Noise as a Public Health Hazard,* ASHA Reports 4 (February 1969), pp. 49–73; proceedings of the conference in Washington, D.C., June 13–14, 1968. Washington, D.C.: American Speech and Hearing Association.

Webster's Seventh New Collegiate Dictionary. 1967. Springfield, Mass.: G. & C. Merriam Company, p. 990.

Whitham, E. M., and M. J. Griffin. 1978. "The Effects of Vibration Frequency and Direction on the Location of Areas of Discomfort Caused by Whole-Body Vibration." *Applied Ergonomics, 9 (4):* pp. 231–239.

WHO. 1969. *Health Factors Involved in Working Under Conditions of Heat Stress.* WHO Tech. Rpt. Series 412. Geneva, Switzerland: World Health Organization.

Woodson, W. E. 1981. *Human Factors Design Handbook.* New York: McGraw-Hill, 1072 pages.

Wu, Yung-Chi. 1972. "Material Properties Criteria for Thermal Safety." *Journal of Materials, 7 (4):* pp. 573–579.

Wyndham, C. H., and W. G. Wilson-Dickson. 1951. "Physiological Responses of Hands and Feet to Cold in Relation to Body Temperature." *Journal of Applied Physiology, 4:* pp. 199–207.

Chapter VI APPENDICES

Contributing Authors

Paul C. Champney,
B.A., Psychology

Brian Crist,
M.A., Psychology

William H. Cushman,
Ph.D., Psychology

Richard L. Lucas,
Ph.D., Psychology

Suzanne H. Rodgers,
Ph.D., Physiology

Chapter Outline

Section VI A. Appendix A: Anthropometric Data

1. Limitations in the Use of Anthropometric Data in Design
   a. Defining the Population
   b. Effects of Clothing and Posture
   c. Designs Involving Several Anthropometric Variables
2. Anthropometric Measurements
   a. The Data
   b. The Definitions
3. Examples of the Use of Anthropometric Data
   a. Criteria for New Design
   b. Evaluation of Existing Design

Section VI B. Appendix B: Methods

1. Aids for Human Factors/Ergonomics Field Surveys
   a. Survey Techniques
   b. Presentation of Survey Results
2. Environmental Measurement Protocols
   a. Temperature, Humidity, and Air Velocity
   b. Noise
   c. Illumination and Luminance
3. Psychophysical Scaling Methods
   a. Types of Scales for Psychophysical Evaluations
   b. Interval Scales in Psychophysical Data Collection
   c. Rating of Perceived Exertion

Section VI C. Appendix C: Problems with Solutions

1. Workplace Design
   a. Problem 1—Seated Workplace
   b. Problem 2—Standing Inspection Workplace
2. Equipment Design
   a. Problem 1—Selecting a Pop Riveter
   b. Problem 2—Safety Shower Handle Height
3. Information Transfer
   a. Problem—Inventory Control Card Design
4. Environment
   a. Problem 1—Warning Alarm Characteristics
   b. Problem 2—Noise Levels in Office Area
   c. Problem 3—Humid Heat Exposure

Guidelines were given in the preceding chapters of this book for the design of workplaces, equipment, and environments within the capabilities and capacities of most people. Much of the work for a human factors practitioner involves evaluating industrial jobs and workplaces and making recommendations for improvements where potential health and safety problems are present and where productivity or performance is being negatively affected.

In this chapter two types of information are included to assist the practitioner in measuring and evaluating workplaces and environments: anthropometric data, including methods of solving workplace design problems; and methods for surveying workplaces, collecting information from people in a job, and measuring noise, illumination, and temperature in the workplace environment. In addition, a set of problems with solutions are included to give the human factors practitioner experience in dealing with situations likely to be encountered in the workplace.

Appendix A, "Anthropometric Data," is a compilation of the anthropometric data used by the Eastman Kodak Company Human Factors Section to develop many of the workplace and equipment design guidelines found in this book. Limitations in data use and examples of applications of some of the data in developing new design guidelines and evaluating existing workplaces are also given.

Appendix B, "Methods," includes the following discussions:

1. "Aids for Human Factors/Ergonomics Field Surveys," which presents techniques for surveying workplaces and descriptor, or reminder, lists and checklists.
2. "Environmental Measurement Protocols," which gives brief descriptions of methods to quantify noise, light, temperature, and humidity levels at the workplace.
3. "Psychophysical Scaling Methods," which describes techniques for the collection of opinion data. They are useful for structuring job and workplace data collection in the field, using the operators to identify the degree of seriousness of a perceived problem.

Appendix C, "Problems with Solutions," presents examples from field requests for human factors evaluation of workplace, equipment, and environmental design problems. Each problem is described, steps to its solution are given in relation to the material in this book, and approaches to resolving the problem are discussed. The problems are designed to assist the less experienced human factors practitioner in learning how the guidelines in this book can be applied in the field.

SECTION VI A. APPENDIX A: ANTHROPOMETRIC DATA

Anthropometrics is the science of measurement of body size (NASA, 1978). By measuring body lengths, girths, and breadths, one can describe, with frequency distributions, the population's size. The anthropometric data included in this compilation have been useful in the design of workplaces, equipment, and products relative to people's dimensions. They were used to develop the guidelines for heights, reaches, grips, and clearances presented throughout this book. The goal of human factors guidelines is to make workplaces, equipment, and product fit the capacities for reach, grasp, and clearance of most of the potential work force. That work force encompasses men and women who are large, small, old, and young, and it includes some people with chronic illnesses or conditions that may limit their range of motion or reach capability.

The frequency distributions for each measurement of population size are expressed as percentiles. The xth percentile indicates that x percent of the population has the same value or less than that value for a given measurement, such as stature, and $100 - x$ percent of the population has a higher value (see Figure VIA–1). For instance, a 25th percentile value for stature indicates that 25 percent of the population is equal to or shorter than that height and 75 percent is taller. A 95th percentile value for hand length indicates that 95 percent of the population will have shorter or the same length hands; 5 percent will have longer hands. The percentiles are derived by taking measurements on many people and developing a normal frequency distribution for statistical analysis. The z statistic is related to the variance of the values around the mean, or the standard deviation (see Figure VIA–1). It is used to determine what percentage of the population falls below a specific value (x_i) for any measurement (ASQC, 1973; Freund, 1967).

In this section the limitations of anthropometric data are discussed. Also presented are the ways the data can be used to develop new designs and to evaluate existing designs. Tables of anthropometric data and descriptions of the measurements are included as well.

1. LIMITATIONS IN THE USE OF ANTHROPOMETRIC DATA IN DESIGN

a. Defining the Population

The data from which most of the values in Tables VIA–2 and VIA–3, presented later in this section, are drawn were gathered on U.S. Air Force and Army women and men (Clauser et al., 1972; Hertzberg, Daniels, and Churchill, 1954; NASA, 1978; Randall and Baer, 1951; White, 1961; White and Churchill, 1971). Because there is some selection in the military population, the industrial population can be expected to differ from it mainly at the extremes of the distribution for each measurement. For example, the very overweight or the very small person may be seen more often in industry than in the military. This difference between the populations has its primary impact on the determination of clearances, such as the design of access ports and overhead clearances. The clearance values given in Chapter II, "Workplace Design," are, generally, somewhat greater than the 99th percentile values one can calculate from the data in Tables VIA–2 and VIA–3. This

Figure VIA–1: Percentiles and the *z*-Statistic in an Anthropometric Data Distribution (Adapted from Freund, 1967.)

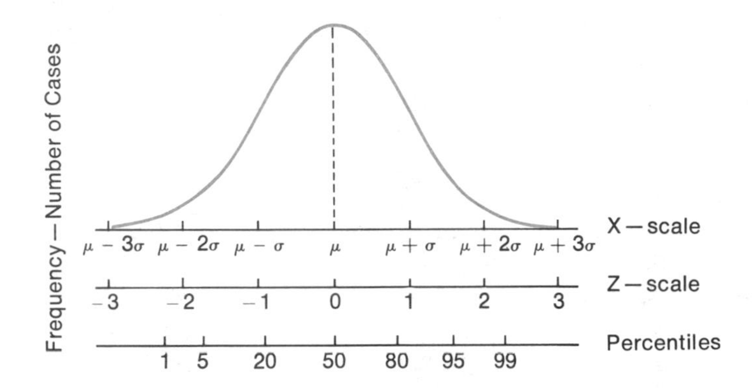

A frequency distribution for an anthropometric variable, such as stature, is given. The number of people, on the vertical axis, at each height value *(x)* along the horizontal axis forms the bell-shaped curve. The vertical line through the center of the distribution marks where half (50 percent) of the values fall on either side, or the average height of this population. The top horizontal axis is marked with one, two, and three standard deviations (σ) on either side of the mean (μ), or average, value, also designated as $\pm z$ on the second horizontal scale. The percent of the population that falls below any given value for height can be calculated by using a *z*-statistic table (ASQC, 1973). The 1st, 5th, 20th, 50th, 80th, 95th, and 99th percentiles are shown on the bottom horizontal scale. The equation for the *z* statistic is as follows:

$$z = \frac{x_i - \mu}{\sigma}$$

where

μ = mean of distribution
x_i = value of interest
σ = standard deviation of mean

change was made to accommodate the extremes seen in the industrial population. Some industrial data, usually from small populations, have been included in these tables for comparative purposes.

Since the population sample used in this section is from the United States, applying the data to other geographic regions with significant size differences may be inappropriate. Compilations of anthropometric data for specific populations are available, and, where appropriate, they should be used instead of the U.S. data (Chapanis, 1975; Grunhofer and Kroh, 1975; White, 1964). The average difference in stature of people from Mexico and the United States, for instance, is about 5 cm (2 in.); the differences, on the average, for other dimensions will vary, and the shapes of the distributions may also differ.

Consequently, it is not accurate to apply a constant correction factor to all measures (Chapanis, 1975). If anthropometric data on the population of interest do not exist, designing to accommodate the lower end of the U.S. distribution for a given measurement, the 5th to 25th percentiles, will usually reduce the probability of making serious design errors.

b. Effects of Clothing and Posture

Since military anthropometric data were gathered on minimally clothed men and women who were standing or sitting erect, some adjustments are needed to apply the data to industrial workers. Since most people in industry are usually fully clothed and stand or sit "at ease" instead of erect, increments in measures that reflect clothing and postural slump should be made. These clothing measurements are as follows:

+2.5 cm (1 in.) for standing heights.
+0.5 cm (0.2 in.) for sitting heights.
+0.8 cm (0.3 in.) for breadths.
+3.0 cm (1.2 in.) for foot length.

Postural slump must be subtracted from the measurements in Tables VIA–2 and VIA–3 (see below), since those measurements were taken in the erect posture. Slump decrements have the following average values:

−2.0 cm (0.8 in.) for standing height.
−4.5 cm (1.8 in.) for sitting height.

Since the slump decrement and the clothing increment balance each other in the standing posture, the values in Tables VIA–2 and VIA–3 can be used for standing dimensions in workplace design. The values for sitting workplaces, however, should be reduced by about 4 cm (1.6 in.).

The range of acceptable heights of displays or shelves in the workplace can be broadened by acknowledging that going up on the toes and squatting can be employed for short-duration, infrequent tasks. Forward reaches can also be extended by bending at the waist or hips in many workplaces. These allowable adjustments are estimated as follows (Damon, Stoudt, and McFarland, 1966; B. Muller-Borer, 1981, Eastman Kodak Company):

- Additional height on tiptoe, 7.5 cm (3 in.).
- Reduced height by squatting, 15 cm (6 in.).
- Extended reach, bending from waist, 20 cm (8 in.).
- Extended reach, bending from hips, 36 cm (14 in.).

These allowances are appropriate only for occasional, short-duration tasks, and only if one can bend at the waist or hips when working at a counter or around other equipment.

c. Designs Involving Several Anthropometric Variables

Although a person may be in the 50th percentile for one dimension, that person may be in the 20th or 80th percentile for another dimension. Table VIA–1 illustrates variability in hand dimensions for several people. If more than one anthropometric variable is important in design, it is difficult to define the population accommodated by the design, even though correlation coefficients and ratios are available to describe relationships between anthropometric variables (Clauser et al., 1972; NASA, 1978; Roebuck, Kroemer, and Thomson, 1975; Roozbazar, 1978). This size variability means that inaccuracy will result if dimensions that are not included in the tables and figures in this section are derived. If a measure is not available, data should be collected on a representative population.

Table VIA–1: Percentile Rankings for Several Hand Dimensions (Adapted from P. C. Champney, 1977, Eastman Kodak Company.)

Subject	Hand Length	Digit Two Length	Hand Breadth	Hand Thickness	Grip Breadth	Hand Spread Wedge
1	1st	2nd	1st	52nd	3rd	2nd
2	2nd	15th	3rd	77th	32nd	17th
3	5th	15th	19th	63rd	32nd	2nd
4	7th	48th	26th	42nd	32nd	17th
5	8th	18th	1st	8th	6th	2nd
6	8th	48th	26th	52nd	17th	38th
7	9th	56th	54th	94th	17th	17th
8	41st	48th	3rd	3rd	50th	2nd
9	47th	56th	64th	60th	32nd	84th
10	51st	66th	59th	60th	50th	84th
11	66th	44th	84th	65th	32nd	95th
12	75th	69th	81st	98th	17th	17th
13	81st	56th	3rd	1st	50th	38th
14	87th	48th	96th	60th	84th	6th

The variability in percentile rankings for six dimensions of people's hands (columns 2 through 7) is shown. For the fourteen people studied (column 1), it is apparent that there is no such thing as a fully 5th, 50th, or 95th percentile hand. This result illustrates the importance of identifying the relevant dimensions in a task and finding a design solution that satisfies most of the capacity needs. The percentiles were established from studies on the hands of Air Force men (Garrett, 1971).

The best approach for dealing with the problems of variable dimensions when designing workplaces or equipment is to simulate the task and ask people representing the extremes of the measurements of interest to work there and identify potential problems. Even if only one measurement is thought to be of concern, workplace simulation is still useful; it may identify an additional limiting dimension not previously considered.

Since most workplace designs, except for clearances, represent compromises made to accommodate the needs of people who are at the extremes of the distributions for specific dimensions, it is desirable to provide adjustability in the workplace. This factor has been discussed in Section II B, "Adjustable Design Approaches."

2. ANTHROPOMETRIC MEASUREMENTS

a. The Data

The data presented in Tables VIA–2 and VIA–3 and Figure VIA–2 are for U.S. men and women, minimally clothed and standing or sitting erect. Because of a lack of data on industrial populations, most of the data are from Air Force or Army studies with thousands of subjects. For comparative purposes some measurements from our own studies of small groups of industrial workers are also included. Since they represent few people, they should be used cautiously. The main difference between the industrial and the military data is in the range of values recorded; the industrial population's range is often wider, reflecting less selection at the extremes of the dimension's distributions.

Table VIA–2: Anthropometric Data, Centimeters (Adapted from P. C. Champney, 1979, and B. Muller-Borer, 1981, Eastman Kodak Company; NASA, 1978).*

The mean, or 50th percentile, values, plus or minus one standard deviation (S.D.) of the mean, are shown for 43 anthropometric variables (column 1). Variables 1 through 10 are standing heights, clearances, or reaches, and variables 11 through 25 are measurements for the subject seated. Data on American men (columns 2 and 3) and women (columns 4 and 5) are statistically combined to derive the 5th, 50th, and 95th percentile values for a 50/50 mix of these populations (columns 6 through 8). The data are taken primarily from military studies, where several thousand people were studied. The entries shown in parentheses are from industrial studies, where 50–100 women and 100–150 men were studied. The data in the footnote are from a study on 50 women and 100 men in industry. Figures VIA–3 and VIA–4 illustrate the measurements, which are described later in the section.

	Males		Females		Population Percentiles, 50/50 Males/Females		
Measurement	50th percentile	±1S.D.	50th percentile	±1S.D.	5th	50th	95th
STANDING							
1. Forward Functional Reach							
a. Includes body depth at shoulder	82.6 (79.3)	4.8 (5.6)	74.1 (71.3)	3.9 (4.4)	69.1 (65.5)	77.9 (74.8)	88.8 (86.5)
b. Acromial Process to Functional Pinch	63.8	4.3	62.5	3.4	57.5	65.0	74.5
c. Abdominal Extension to Functional Pinch†	(62.1)	(8.9)	(60.4)	(6.7)	(48.5)	(61.1)	(74.5)
2. Abdominal Extension Depth	23.1	2.0	20.9	2.1	18.1	22.0	25.8

* These values should be adjusted for clothing and posture.
† Add the following for bending forward from hips or waist: Male: waist, 25 ± 7; hips, 42 ± 8. Female: waist, 20 ± 5; hips, 36 ± 9.

Table VIA–2: (Continued)

Measurement	Males		Females		Population Percentiles, 50/50 Males/Females		
	50th percentile	±1S.D.	50th percentile	±1S.D.	5th	50th	95th
3. Waist Height	106.3	5.4	101.7	5.0	94.9	103.9	113.5
	(104.8)	(6.3)	(98.5)	(5.5)	(91.0)	(101.4)	(113.0)
4. Tibial Height	45.6	2.8	42.0	2.4	38.8	43.6	49.2
5. Knuckle Height	75.5	4.1	71.0	4.0	65.7	73.2	80.9
6. Elbow Height	110.5	4.5	102.6	4.8	96.4	106.7	116.3
	(114.6)	(6.3)	(107.1)	(6.8)	(98.8)	(110.7)	(123.5)
7. Shoulder Height	143.7	6.2	132.9	5.5	124.8	137.4	151.7
	(146.4)	(7.8)	(135.3)	(6.6)	(126.6)	(140.4)	(156.4)
8. Eye Height	164.4	6.1	151.4	5.6	144.2	157.7	172.3
9. Stature	174.5	6.6	162.1	6.0	154.4	168.0	183.0
	(177.5)	(6.7)	(164.5)	(7.2)	(155.1)	(170.4)	(188.7)
10. Functional Overhead Reach	209.6	8.5	199.2	8.6	188.0	204.5	220.8
SEATED							
11. Thigh Clearance Height	14.7	1.4	12.4	1.2	10.8	13.5	16.5
12. Elbow Rest Height	24.1	3.2	23.1	3.0	18.4	23.6	28.9
13. Midshoulder Height	62.4	3.2	58.0	2.7	54.5	60.0	66.5
14. Eye Height	78.7	3.6	73.7	3.1	69.7	76.0	83.3
15. Sitting Height Normal	86.6	3.8	81.8	4.0	76.6	84.2	91.6
16. Functional Overhead Reach	128.4	8.5	119.8	6.6	110.6	123.6	139.3

Table VIA–2: (Continued)

Measurement	Males		Females		Population Percentiles, 50/50 Males/Females		
	50th percentile	±1S.D.	50th percentile	±1S.D.	5th	50th	95th
17. Knee Height	54.0	2.7	51.0	2.6	47.5	52.5	57.7
18. Popliteal Height	44.6	2.5	41.0	1.9	38.6	42.6	47.8
19. Leg Length	105.1	4.8	100.7	4.3	94.7	102.8	111.4
20. Upper-Leg Length	59.4	2.8	57.4	2.6	53.7	58.4	63.3
21. Buttocks-to-Popliteal Length	49.8	2.5	48.0	3.2	43.8	49.0	53.6
22. Elbow-to-Fist Length	38.5	2.1	34.8	2.3	31.9	36.7	41.1
	(37.1)	(3.0)	(32.9)	(3.1)	(28.9)	(35.0)	(41.0)
23. Upper-Arm Length	36.9	1.9	34.1	2.5	31.0	35.7	39.4
	(37.0)	(2.5)	(33.8)	(2.1)	(28.9)	(35.0)	(41.0)
24. Shoulder Breadth	45.4	1.9	39.0	2.1	36.3	42.3	47.8
25. Hip Breadth	35.6	2.3	38.0	2.6	32.4	36.8	41.5
FOOT							
26. Foot Length	26.8	1.3	24.1	1.1	22.6	25.3	28.4
27. Foot Breadth	10.0	0.6	8.9	0.5	8.2	9.4	10.8
HAND							
28. Hand Thickness, Metacarpal III	3.3	0.2	2.8	0.2	2.7	3.0	3.6
29. Hand Length	19.0	1.0	18.4	1.0	17.0	18.7	20.4
30. Digit Two Length	7.5	0.7	6.9	0.8	5.8	7.2	8.5
31. Hand Breadth	8.7	0.5	7.7	0.5	7.0	8.2	9.3

Table VIA–2: (Concluded)

Measurement	Males		Females		Population Percentiles, 50/50 Males/Females		
	50th percentile	±1S.D	50th percentile	±1S.D.	5th	50th	95th
32. Digit One Length	12.7	1.1	11.0	1.0	9.7	11.8	14.2
33. Breadth of Digit One Interphalangeal Joint	2.3	0.1	1.9	0.1	1.8	2.1	2.5
34. Breadth of Digit Three Interphalangeal Joint	1.8	0.1	1.5	0.1	1.4	1.7	2.0
35. Grip Breadth, Inside Diameter	4.9	0.6	4.3	0.3	3.8	4.5	5.7
36. Hand Spread, Digit One to Digit Two, 1st Phalangeal Joint	12.4	2.4	9.9	1.7	7.5	10.9	15.5
37. Hand Spread, Digit One to Digit Two, 2nd Phalangeal Joint	10.5	1.7	8.1	1.7	5.9	9.3	12.7
HEAD							
38. Head Breadth	15.3	0.6	14.5	0.6	13.8	14.9	16.0
39. Interpupillary Breadth	6.1	0.4	5.8	0.4	5.2	6.0	6.7
40. Biocular Breadth	9.2	0.5	9.0	0.5	8.3	9.1	10.0
OTHER MEASUREMENTS							
41. Flexion-Extension, Range of Motion of Wrist Radians (57 degrees/radian)	2.33	0.33	2.46	0.26	1.92	2.4	2.8
42. Ulnar-Radial Range of Motion of Wrist Radians (57 degrees/radian)	1.05	0.23	1.17	0.24	0.81	1.15	1.49
43. Weight, in kilograms	83.2	15.1	66.4	13.9	47.7	74.4	102.9

Table VIA–3: Anthropometric Data, Inches (Adapted from P. C. Champney, 1979, and B. Muller-Borer, 1981, Eastman Kodak Company; NASA, 1978.)*

The data here are the same as the data in Table VIA–2, but they are expressed in inches rather than centimeters.

	Males		Females		Population Percentiles, 50/50 Males/Females		
Measurement	50th percentile	±1S.D.	50th percentile	±1S.D.	5th	50th	95th
STANDING							
1. Forward Functional Reach							
a. includes body depth at shoulder	32.5	1.9	29.2	1.5	27.2	30.7	35.0
	(31.2)	(2.2)	(28.1)	(1.7)	(25.7)	(29.5)	(34.1)
b. Acromial Process to Functional Pinch	26.9	1.7	24.6	1.3	22.6	25.6	29.3
c. Abdominal Extension to Functional Pinch†	(24.4)	(3.5)	(23.8)	(2.6)	(19.1)	(24.1)	(29.3)
2. Abdominal Extension Depth	9.1	0.8	8.2	0.8	7.1	8.7	10.2
3. Waist Height	41.9	2.1	40.0	2.0	37.4	40.9	44.7
	(41.3)	(2.1)	(38.8)	(2.2)	(35.8)	(39.9)	(44.5)
4. Tibial Height	17.9	1.1	16.5	0.9	15.3	17.2	19.4
5. Knuckle Height	29.7	1.6	28.0	1.6	25.9	28.8	31.9
6. Elbow Height	43.5	1.8	40.4	1.4	38.0	42.0	45.8
	(45.1)	(2.5)	(42.2)	(2.7)	(38.5)	(43.6)	(48.6)
7. Shoulder Height	56.6	2.4	51.9	2.7	48.4	54.4	59.7
	(57.6)	(3.1)	(56.3)	(2.6)	(49.8)	(55.3)	(61.6)

* These values should be adjusted for clothing and posture.
† Add the following for bending forward from hips or waist: Male: waist, 10 ± 3; hips, 16 ± 3. Female: waist, 8 ± 2; hips, 14 ± 4.

Table VIA-3: (Continued)

Measurement	Males		Females		Population Percentiles, 50/50 Males/Females		
	50th percentile	±1S.D	50th percentile	±1S.D.	5th	50th	95th
8. Eye Height	64.7	2.4	59.6	2.2	56.8	62.1	67.8
9. Stature	68.7	2.6	63.8	2.4	60.8	66.2	72.0
	(69.9)	(2.6)	(64.8)	(2.8)	(61.1)	(67.1)	(74.3)
10. Functional Overhead Reach	82.5	3.3	78.4	3.4	74.0	80.5	86.9
SEATED							
11. Thigh Clearance Height	5.8	0.6	4.9	0.5	4.3	5.3	6.5
12. Elbow Rest Height	9.5	1.3	9.1	1.2	7.3	9.3	11.4
13. Midshoulder Height	24.5	1.2	22.8	1.0	21.4	23.6	26.1
14. Eye Height	31.0	1.4	29.0	1.2	27.4	29.9	32.8
15. Sitting Height, Normal	34.1	1.5	32.2	1.6	32.0	34.6	37.4
16. Functional Overhead Reach	50.6	3.3	47.2	2.6	43.6	48.7	54.8
17. Knee Height	21.3	1.1	20.1	1.0	18.7	20.7	22.7
18. Popliteal Height	17.2	1.0	16.2	0.7	15.1	16.6	18.4
19. Leg Length	41.4	1.9	39.6	1.7	37.3	40.5	43.9
20. Upper-Leg Length	23.4	1.1	22.6	1.0	21.1	23.0	24.9
21. Buttocks-to-Popliteal Length	19.2	1.0	18.9	1.2	17.2	19.1	20.9
22. Elbow-to-Fist Length	14.2	0.9	12.7	1.1	12.6	14.5	16.2
	(14.6)	(1.2)	(13.0)	(1.2)	(11.4)	(13.8)	(16.2)

Table VIA–3: (Continued)

Measurement	Males		Females		Population Percentiles, 50/50 Males/Females		
	50th percentile	±1S.D.	50th percentile	±1S.D.	5th	50th	95th
23. Upper-Arm Length	14.5 (14.6)	0.7 (1.0)	13.4 (13.3)	0.4 (0.8)	12.9 (12.1)	13.8 (13.8)	15.5 (16.0)
24. Shoulder Breadth	17.9	0.8	15.4	0.8	14.3	16.7	18.8
25. Hip Breadth	14.0	0.9	15.0	1.0	12.8	14.5	16.3
FOOT							
26. Foot Length	10.5	0.5	9.5	0.4	8.9	10.0	11.2
27. Foot Breadth	3.9	0.2	3.5	0.2	3.2	3.7	4.2
HAND							
28. Hand Thickness, Metacarpal III	1.3	0.1	1.1	0.1	1.0	1.2	1.4
29. Hand Length	7.5	0.4	7.2	0.4	6.7	7.4	8.0
30. Digit Two Length	3.0	0.3	2.7	0.3	2.3	2.8	3.3
31. Hand Breadth	3.4	0.2	3.0	0.2	2.8	3.2	3.6
32. Digit One Length	5.0	0.4	4.4	0.4	3.8	4.7	5.6
33. Breadth of Digit One Interphalangeal Joint	0.9	0.05	0.8	0.05	0.7	0.8	1.0
34. Breadth of Digit Three Interphalangeal Joint	0.7	0.05	0.6	0.04	0.6	0.7	0.8
35. Grip Breadth, Inside Diameter	1.9	0.2	1.7	0.1	1.5	1.8	2.2
36. Hand Spread, Digit One to Two, 1st Phalangeal Joint	4.9	0.9	3.9	0.7	3.0	4.3	6.1

Table VIA–3: (Concluded)

Measurement	Males		Females		Population Percentiles, 50/50 Males/Females		
	50th percentile	±1S.D.	50th percentile	±1S.D.	5th	50th	95th
37. Hand Spread, Digit One to Two, 2nd Phalangeal Joint	4.1	0.7	3.2	0.7	2.3	3.6	5.0
HEAD							
38. Head Breadth	6.0	0.2	5.7	0.2	5.4	5.9	6.3
39. Interpupillary Breadth	2.4	0.2	2.3	0.2	2.1	2.4	2.6
40. Biocular Breadth	3.6	0.2	3.6	0.2	3.3	3.6	3.9
OTHER MEASUREMENTS							
41. Flexion-Extension, Range of Motion of Wrist, Degrees	134	19	141	15	108	138	166
42. Ulnar-Radial Range of Motion of Wrist, Degrees	60	13	67	14	41	63	87
43. Weight, in pounds	183.4	33.2	146.3	30.7	105.3	164.1	226.8

The grasp data shown in Figure VIA–2 are from a limited population, but they have proven useful in defining the dimensions of products or containers that are manually handled. The guidelines for handle size and shape for tools and products have been developed from these data. Section IIID on the design and selection of hand tools includes more discussion of this subject.

Figure VIA–2: Functional Hand Grasp Dimensions (Adapted from R. H. Jones, 1973, Eastman Kodak Company.)

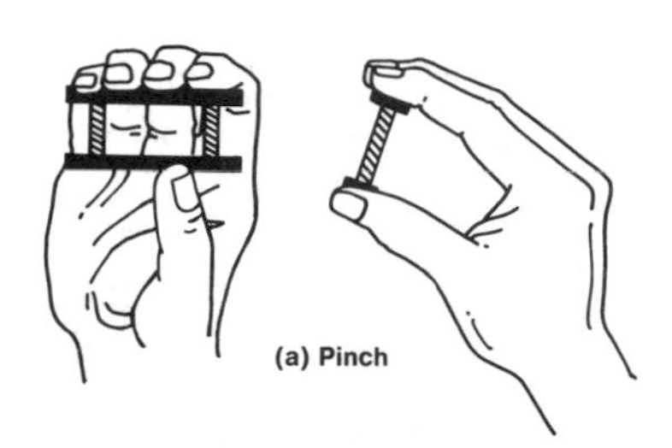

	Span, in centimeters (inches), 50/50 Male/Female Mix		
	5th Percentile	50th Percentile	95th Percentile
True	2.1 (0.8)	4.3 (1.7)	7.9 (3.1)
Maximum	10.8 (4.2)	12.5 (4.9)	15.0 (5.9)

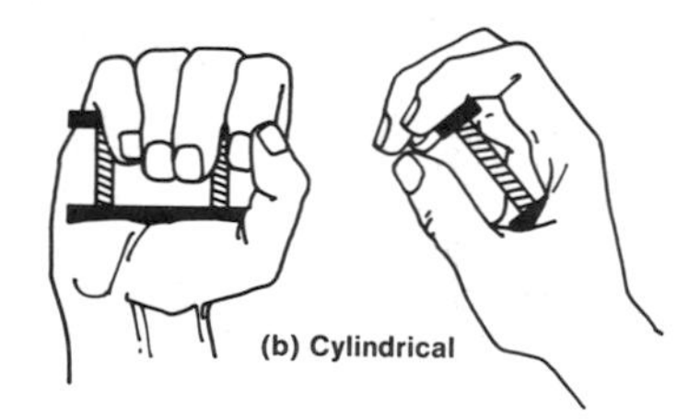

	Span, in centimeters (inches), 50/50 Male/Female Mix		
	5th Percentile	50th Percentile	95th Percentile
True	4.5 (1.8)	5.5 (2.2)	5.9 (2.3)
Maximum	9.5 (3.7)	11.0 (4.3)	13.0 (5.1)

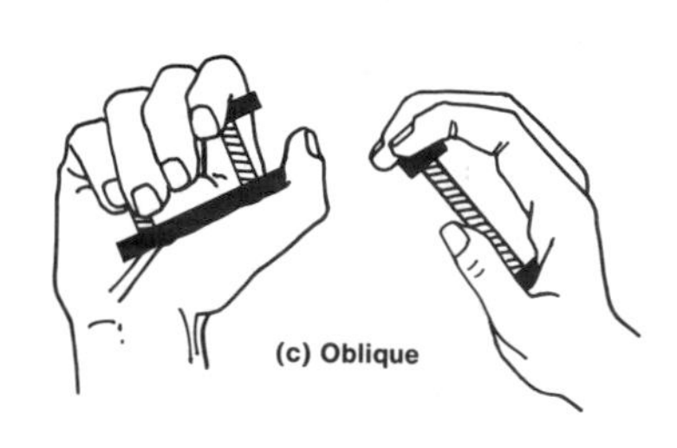

	Span, in centimeters (inches), 50/50 Male/Female Mix		
	5th Percentile	50th Percentile	95th Percentile
True	3.6 (1.4)	4.5 (1.8)	5.8 (2.3)
Maximum	9.5 (3.7)	11.0 (4.3)	13.0 (5.1)

The functional hand grasp spans for three types of grasp are shown: pinch (part a), cylindrical grasp (part b), and oblique grasp (part c). The 5th, 50th, and 95th percentile values (in the accompanying tables) are based on a 50/50 ratio of men to women from industrial studies of 46 men and 38 women. True and maximum grasp spans, which are defined in the text, are shown separately. Each value is given in centimeters, with its inches equivalent given in parentheses.

b. The Definitions

Figures VIA–3 and VIA–4 illustrate the data in Tables VIA–2 and VIA–3.

(1) Definitions of the Dimensions in Figures VIA–3 and VIA–4

The methods of measurement of the dimensions given in Tables VIA–2 and VIA–3 and shown in Figures VIA–3 and VIA–4 are described in this section. Many different techniques can be used to measure a limb length, girth, or distance; attention should be paid to how each measurement was made in order to appropriately use the data in workplace design considerations. For further information, consult NASA (1978). The Glossary near the end of this book provides additional definitions of terms used in this section.

1a. Forward Functional Reach: Body Depth at Shoulder (Standing). The subject stands erect against the wall, right arm extended forward horizontally and the tips of the thumb and index finger pressed together; shoulders should remain in contact with the wall. Functional reach is measured as the horizontal distance from the wall to the tip of the thumb. This measure could be used in defining how far forward from a fixed structure, such as a wall, a person who is restrained from leaning forward can reach for controls.

1b. Forward Functional Reach: Acromial Process to Functional Pinch (Standing). The subject stands erect against the wall, with the right arm extended forward horizontally and the tips of the thumb and index finger pressed together; shoulders should remain in contact with the wall. Functional reach is measured from the shoulder, or acromial crest, to the functional grasp. This measure identifies how far forward a person can reach into a porthole or other access hole, for example.

1c. Forward Functional Reach: Abdominal Extension to Functional Pinch (Standing). The subject stands erect against the wall, right arm extended forward horizontally and the tips of the thumb and index finger pressed together; shoulders remain in contact with the wall. Functional reach is measured from the abdominal extension point to the functional grasp. This measure indicates maximum forward reach when a person is prevented from bending forward by a machine part or workplace table, for example.

2. Abdominal Extension Depth. The subject stands erect, looking straight ahead, arms at sides, heels together, and weight distributed equally on both feet. With a beam caliper the horizontal depth of the trunk at the level of the abdominal extension landmark is measured (see Figure VIA–5). The reading is made at the end of normal expiration; the subject should not pull in his or her stomach. This measure is useful in defining whole-body clearances for confined spaces and access ports.

3. Waist Height (Standing). The subject stands erect, with heels together. Waist height is measured as the vertical distance from the floor to the upper edge (iliac crest) of the right hipbone. An anthropometer is used. The measure is a convenient marker for workplace design, because objects that have to be handled below waist level usually require some knee bending or bending at the waist.

Figure VIA–3: Anthropometric Dimensions, Standing and Sitting (Adapted from P. C. Champney, 1975, 1979, and B. Muller-Borer, 1981, Eastman Kodak Company; NASA, 1978.)

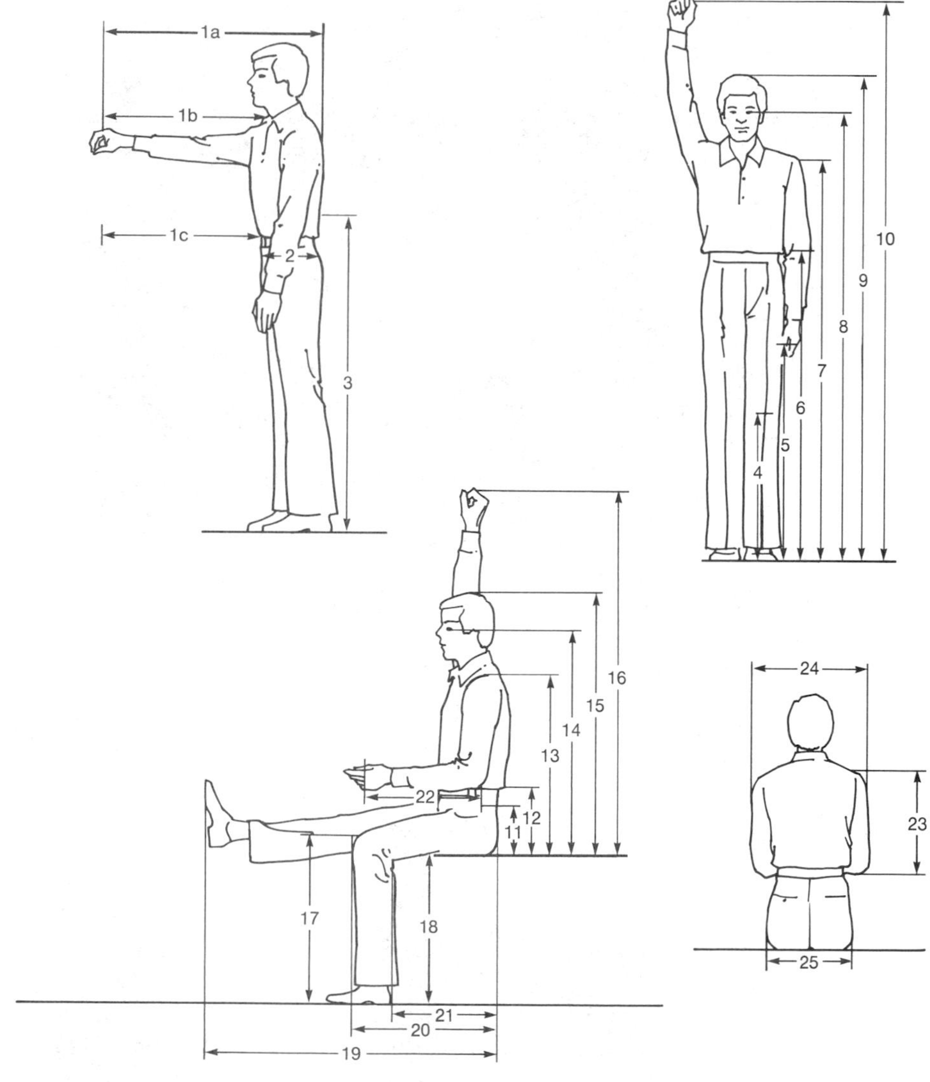

Illustrations are given of the anthropometric measures presented in Tables VIA–2 and VIA–3. Each measure is numbered to correspond to the table number and to the definitions in the text. The anatomical markers used to define the limits of each measure are described in the text definitions. Dimensions that are important in the design of standing and sitting workplaces are included in this figure.

Figure VIA–4: Anthropometric Dimensions, Hand, Face, and Foot (Adapted from P. C. Champney, 1975, 1977, 1979, and B. Muller-Borer, 1981, Eastman Kodak Company; NASA, 1978.)

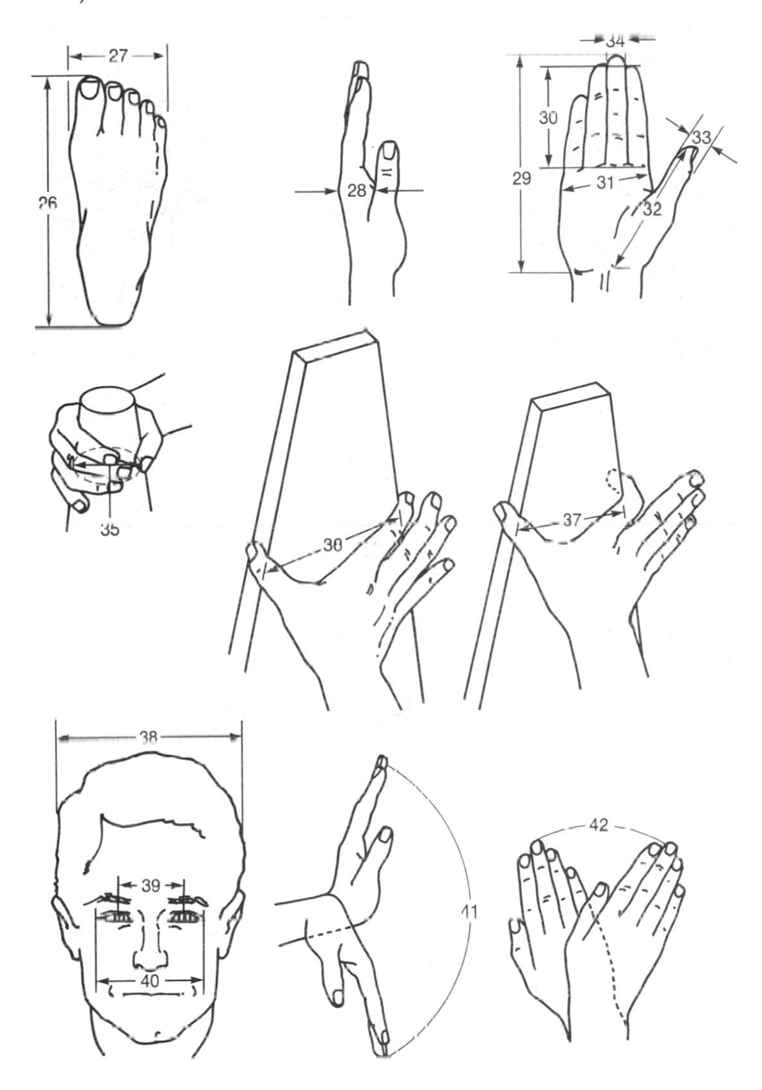

Dimensions of the hand, face, and foot, the range of motion of the wrist, and functional hand grasp dimensions are included here, illustrating the measurements given in Tables VIA–2 and VIA–3. Descriptions of these measurements are given in the text definitions. Each measure is numbered to correspond to the table number and the definition in the text. These dimensions are of importance in design of equipment, workplaces, and protective equipment such as face masks and gloves.

Figure VIA–5: Measurement of Abdominal Extension Depth

A beam caliper is used to measure the horizontal depth of the trunk. The subject stands erect, with arms at sides, heels together, and weight distributed equally on both feet. The measurement is made at the end of a normal expiration, with abdominal muscles relaxed.

4. Tibial Height (Standing). The subject stands erect, heels together and weight distributed equally on both feet. With the anthropometer the vertical distance from the standing surface to the tibial landmark on the leg is measured. This measure is a useful marker, since items below tibial height require the person to take deep knee bends to procure them; items located above this level but below knuckle height (see below) are often picked up from a stooped position.

5. Knuckle Height. The subject stands erect, heels together, arms hanging naturally at the sides, and resting equally on both feet. With the anthropometer the vertical distance to the knuckles (metacarpal-phalangeal joints) from the standing surface is measured. This height represents the lowest level from which an item can be handled or a control operated without having to bend the knees.

6. Elbow Height (Standing). The subject stands erect, arms hanging naturally at sides. With the anthropometer the vertical distance from the floor to the depression between the humerus and the radius of the right arm is measured. The depression can be seen in the lateral side of the elbow. This measure represents the work level at which a person does not have to bend at the waist or hips, thereby reducing the potential for muscle fatigue.

7. Shoulder (Acromial) Height (Standing). The subject stands erect. The distance from the floor to the marked acromial point, the outer end of the transverse spine of the shoulder blade, on the right side is measured with the anthropometer. This marker is useful for maximum height location of controls or items to be handled, since much weaker muscles are involved in lifts above this level. Routinely raising the arms to this level to operate controls or perform other assembly or packing tasks will increase fatigue of the shoulder muscles.

8. Eye Height (Standing). This height is the vertical distance from the inner angle of the eye to the floor. It is a useful marker for locating visual displays.

9. Stature. The subject stands erect, heels together, toes at a 45° angle from straight forward, back straight, and head in a horizontal plane defined by a line from the tragion, at about the top of the earlobe, to the bottom of the eye's bony orbit. The vertical distance to the standing surface from the vertex, or highest point, in the midline of the head is measured. This measure is used for vertical clearances in workplaces or in locating equipment to avoid head collisions.

10. Functional Overhead Reach (Standing). The subject stands erect, looking straight ahead, along the side of but not touching the wall-mounted scale. Holding the special pointer in the hand, the subject raises the pointer as high as possible while keeping both feet flat on the floor and both the pointer and the proximal phalanges of the fingers horizontal (see Figure VIA–6). The vertical distance from the floor to the tip of the pointer is measured on the wall scale. This measure is useful for defining how high an overhead control can be, for example.

11. Thigh Clearance Height (Sitting). The subject sits erect on a flat surface, feet on an adjustable platform, knees flexed 90°, and thighs parallel. With an anthropometer the vertical distance from the sitting surface to the highest point on the right thigh is measured. This measure is useful in the design of seated work surface heights or in the selection of the adjustability levels of a chair.

12. Elbow Rest Height (Sitting). The subject sits erect, with the feet resting

Figure VIA–6: Measurement of Functional Overhead Reach (Standing)

A wall scale is used to measure the distance from the floor to a pointer held horizontally in one hand. The hand is raised above the head as far as possible while the subject keeps both feet flat on the floor.

on a surface so that the knees are bent at about right angles, the upper arm hanging at the side, and the forearm extended horizontally. With the anthropometer the vertical distance from the sitting surface to the bottom of the elbow is measured. This measure defines the height at which there is little load on the shoulder muscles, and it is useful for the design of seated workplace height.

13. Midshoulder Height (Sitting). The subject sits erect, upper arms hanging relaxed and forearms and hands extended horizontally. With an anthropometer the vertical distance from the sitting surface to the midshoulder landmark is measured. This measure defines the height at which the forward arm reach is at its maximum value. It also defines the top of the torso for chair design.

14. Eye Height (Sitting). The subject sits erect, with the head level and with the feet resting on a surface adjusted so that the knees are bent at right angles. Eye height is measured as the vertical distance from the sitting surface

to the inner corner (internal canthus) of the eye. An anthropometer is used. This measure is used to define locations for visual displays in seated workplaces.

15. Sitting Height, Normal. The subject sits normally relaxed, hands in lap, looking straight ahead. The anthropometer is held vertically along the middle of the back, and the measuring bar is brought down into firm contact with the top of the head, in the midline. Height is measured from the sitting surface. This measure defines the vertical clearance needed for seated work.

16. Functional Overhead Reach (Sitting). This reach is measured with an anthropometer as the vertical distance from the sitting surface to a pointer held horizontally in a cylindrical grasp when the arm is extended directly upward. The subject sits erect, looking straight ahead, with the knees bent at right angles. This measure defines the maximum height at which controls should be located in overhead panels.

17. Knee Height (Sitting). The subject sits erect, with the legs bent to form right angles at the knees. The vertical distance from the surface of the footrest or floor to the top of the knee at the kneecap is measured with an anthropometer. This measure is useful in the design of seated workplace heights.

18. Popliteal Height (Sitting). The subject sits erect on a table or bench, with the legs bent to form right angles at the knees. The vertical distance from the surface of the footrest to the underside of the right knee, the tendon of biceps femoris, is measured with an anthropometer. This value is used to define seat height adjustability and footrest needs for seated workplaces.

19. Leg Length (Sitting). The horizontal distance from the wall to the bottom of the foot is measured, with the subject sitting against the wall, leg extended. This value is useful in defining maximum leg clearances for seated workplaces or for maintenance tasks where the person has to sit on the floor to do the task.

20. Upper-Leg Length (Sitting). The subject sits erect, with feet on the adjustable platform, knees flexed 90°, and thighs parallel to the floor. With a beam caliper held parallel to the long axis of the thigh, the horizontal distance from the most posterior aspect of the right buttock to the most anterior aspect of the right knee is measured. This measure is useful in defining seat depth for chairs, for example, and minimum forward clearance for seated workplaces.

21. Buttocks-to-Popliteal Length. The subject sits erect, with feet resting on a surface adjusted so that the knees are bent at right angles. Buttock-to-popliteal length is measured as the horizontal distance from the back of the buttock to the back of the knee. This value defines the minimum seat depth for a chair to ensure that pressure on the underside of the thigh is not excessive.

22. Elbow-to-Fist Length (Forearm Length). The subject holds a pointer vertically in a cylindrical grasp, with the upper arm at the side and the lower arm parallel to the floor, forming a right angle at the elbow (see Figure VIA-7). The horizontal distance from the back of the elbow, or epicondyle, to

Figure VIA–7: Measurement of Elbow-to-Fist Length (Forearm Length)

An anthropometer is used to measure the length of the forearm. The subject holds a pointer vertically in one hand, with the elbow bent to form a right angle with the upper arm. The measurement is taken from the back of the elbow to the pointer.

the pointer is measured with an anthropometer. This distance can be measured with the subject in either standing or sitting posture. This measure is useful in defining maximum forward reach in situations where elbow flexion is limited, as in glove box manipulations.

23. Upper-Arm Length. The subject stands with trunk erect, humerus, or upper arm, vertical, forearm horizontal. Upper-arm length is measured from the top of the acromial process to the bottom of the elbow. This measure is used to help define the level at which shoulder abduction, or raising the elbows, may become necessary to do a task. For maintenance tasks through access ports, this measure becomes an important clearance requirement as well.

24. Shoulder Breadth. This value is measured across the shoulders from

the large muscles on the outer side of each upper arm in the shoulder region. It is used in determining whole-body clearance requirements for access ports, for instance.

25. Hip Breadth (Sitting). This distance is measured across the widest portion of the hips. It is used in determining seat widths and whole-body access port clearances.

26. Foot Length. Foot length is measured from the heel to the tip of the most protruding toe. It is used to define foot pedal dimensions and foot clearances at standing workplaces or around production machines.

27. Foot Breadth. This value is measured at the widest part of the foot. It is used to define foot pedal dimensions and horizontal clearances at standing workplaces.

28. Hand Thickness, Third Metacarpal (Knuckle). The subject's hand is held with the palm facing downward. The fingers are extended and held together but bent downward slightly from the plane of the palm. Using light pressure and only the tips of the sliding caliper arms, one measures the thickness of the knuckle, or metacarpal-phalangeal joint, of the middle finger. The bar of the caliper may not be perpendicular to the plane of the palm. This measure is useful for designing access ports for maintenance tasks and handle widths for tools.

29. Hand Length. The subject sits with the hand and fingers extended, palm up. The length of the hand is measured from the wrist crease to the tip of the middle finger. Sliding calipers are used. This measure is useful for defining the working space of the hand in maintenance tasks and for hand tool design.

30. Digit Two Length. The subject's hand is extended, palm up. With the sliding caliper, the distance is measured along the axis of digit two from the midpoint of the tip of digit two to the level of the crotch between digits two and three. This measure is used in defining maximum depths of control buttons set inside equipment and not accessible to the full hand.

31. Hand Breadth. The fingers of the hand are extended and held together, palm up. The arms of a caliper are placed parallel to the axis of the fingers. The distance between the radial (lateral) projection of the second metacarpal's distal end and the ulnar (medial) projection of the fifth metacarpal's distal end is measured (see Figure VIA–8). This measure is useful for determining clearances for the working hand and handle dimensions of tools.

32. Digit One Length. The distance along the axis of digit one from the midpoint of the tip of digit one to the wrist crease baseline is measured. This value is used to define minimum hand length clearance with the fingers curled and clearances for the working hand.

33, 34. Breadth of Distal Interphalangeal Joint, Digits One and Three. The subject's hand is extended. With a sliding caliper, the maximum breadths of the distal interphalangeal joint of the first and third digits are measured. This measure is used in establishing minimum access port size for push buttons, for example.

Figure VIA–8: Measurement of Hand Breadth

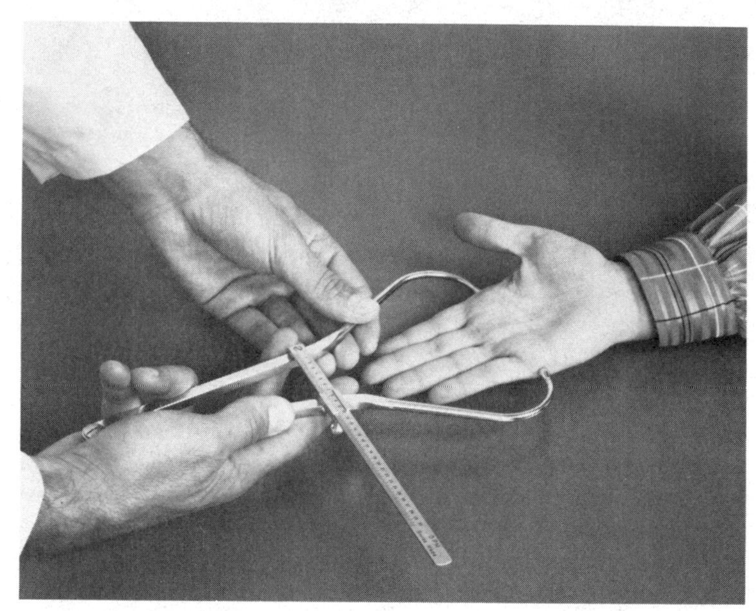

A sliding caliper is used to measure the distance across the hand. The subject holds the hand palm up, with the fingers extended and held together. The measurement is made with the caliper arms placed on the distal metacarpal projections of the second digit (index finger) and the fifth finger (little finger).

35. Grip Breadth, Inside Diameter (Thumb to Middle Finger). The subject holds a cone at the largest circumference that can be grasped with the thumb and middle fingers just touching. The diameter of the cone corresponding to this maximum circumference is recorded. This measure is used to determine the circumference of hand tools, for example.

36. Hand Spread, Digit One to Digit Two, First Phalangeal Joint. The subject places a hand on the measuring wedge so that the distal joint of the thumb is on one edge of the wedge and the *distal* joint of digit two is on the other edge of the wedge. The subject slides the hand down the sides of the wedge to the maximal spread while maintaining joint contact on the edges. A reading is taken at the last 0.635-cm (0.25-in.) line that can be completely cleared. This measure defines the maximum grip span across an object, and it can be used to limit object width for one-handed lifting tasks, for example.

37. Hand Spread, Digit One to Digit Two, Second Phalangeal Joint. The subject places a hand on the measuring wedge so that the distal joint of the thumb is on one edge of the wedge and the *middle* joint of digit two is on the other edge of the wedge. The subject slides the hand down the sides of the wedge to the maximal spread while maintaining joint contact on the edges. A

reading is taken at the last 0.635-cm (0.25-in.) line that can be completely cleared. This measure represents a more powerful grasp than measure 36. It can be used to define the dimensions of a tool or a piece of equipment that has to be grasped and controlled carefully.

38. Head Breadth. The subject sits erect, with the head level. The maximum horizontal breadth of the head is measured above and behind the ears. This measure is useful in the design of clothing and of protective equipment, such as respirators.

39. Interpupillary Breadth. The subject sits erect, with the head level. The distance between the centers of the pupils of the eyes is measured with a caliper. This measure is used in the design of eye protective equipment and in visual aids design, such as binocular microscopes for inspection tasks.

40. Biocular Breadth. The transverse distance between right and left outer canthi (outer margins of the eye socket) is measured with a caliper. This measure is used in eye protective equipment design, such as goggles.

41. Flexion-Extension, Range of Motion of Wrist. The goniometer, or joint-angle measurement device, is centered on the ulnar styloid. One arm of the goniometer is parallel to the longitudinal axis of the forearm along the ulnar border. The other goniometer arm is parallel to the longitudinal axis of the fifth metacarpal and is moved with the fifth metacarpal to measure flexion or extension (see Figure VIA–9a). This measure is useful in defining the working area of the hands in maintenance or other assembly tasks.

Figure VIA–9: Measurement of the Range of Wrist Motion

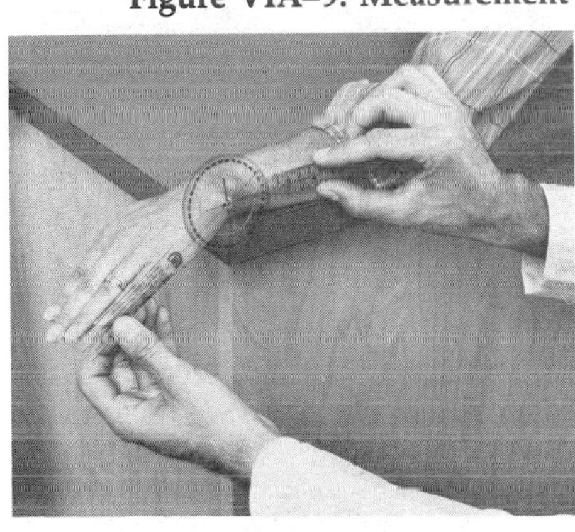

(a)

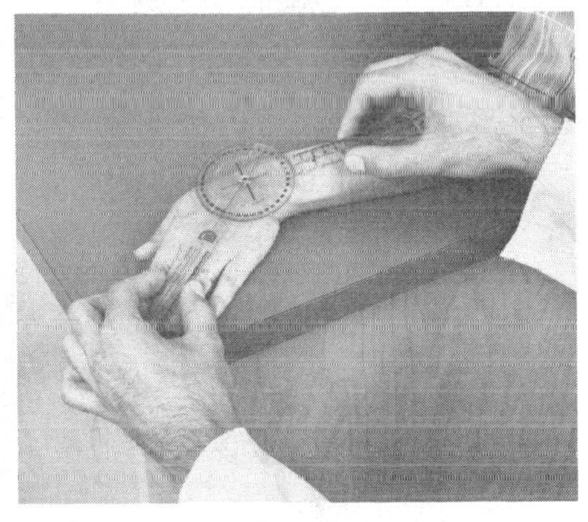

(b)

A goniometer is used to measure the angle of the wrist when moved in flexion-extension (part a) or in ulnar-radial deviation (part b). In flexion-extension the goniometer is centered on the ulnar styloid at the wrist, with one arm lying along the fifth metacarpal and moving with it. In ulnar-radial deviation the goniometer is centered on the capitate bone on the back of the wrist, with one arm lying along the third metacarpal. It is moved with the third metacarpal to measure radial and ulnar abduction.

42. Ulnar-Radial Range of Motion of Wrist. The goniometer is centered on the capitate bone; one arm of the goniometer is parallel to the transverse axis of the third metacarpal and is moved with the third metacarpal to measure radial-ulnar abduction (see Figure VIA–9b). This measure is used to define the working area of the hands in maintenance or assembly tasks.

43. Weight. The person stands on a scale, with the body weight shared equally on both feet and arms to the sides.

(2) Definitions of Functional Hand Dimensions in Figure VIA–2

The material in this section was developed from information in R. H. Jones (1973, Eastman Kodak Company).

(a) Pinch

Pinch is the grasp used when objects are held between the fingers and thumb with no contact between the object and the base of the thumb or palm (see Figure VIA–2a). The direction of pinch force is across the palm. True pinch occurs where the thumb and fingers are in direct opposition to one another. The maximum limit of true pinch is reached when the thumb can only extend and the fingers begin to drift to the ulnar side of the hand (away from the thumb) with further extension. Although the thumb starts to extend at 40 percent of maximum grasp span (hand spread, measurement 37), the limit of true pinch is at 60 percent of span. Maximum pinch force is in this 40–60 percent range. This type of grasp is commonly used in assembly tasks to procure small parts.

(b) Cylindrical Grasp

In a cylindrical grasp all but the smallest and largest objects are held between the fingers and the palm. The angle of flexion or curl for all of the fingers is similar. Contact by the thumb is not essential (see Figure VIA–2b). The limiting span of true cylindrical grasp occurs when the object loses contact with the palm and is held only by the fingers and the thumb. This limit occurs at about 45 percent of maximal grasp breadth (inside hand breadth, measurement 35). Strength falls off rapidly as palmar contact is lost. This type of grasp is used in controlling movement of a hand cart or activating a bar or wheel control.

(c) Oblique Grasp

Oblique grasp is a variant of cylindrical grasp; the object lies in the valley along the base of the thumb, and each finger assumes a different degree of flexion (see Figure VIA–2c). The index finger has the least, and the small finger has the most, flexion. The thumb often lies parallel to the long axis of the object and is used to stabilize the grasp. The limiting span of true oblique grasp is marked by the point at which the thumb can no longer be kept in this parallel position. This type of grasp is used in controlling tools, such as a pliers or a screwdriver, during assembly or repair operations.

3. EXAMPLES OF THE USE OF ANTHROPOMETRIC DATA

The two primary ways in which the anthropometric data presented above can be used in industrial workplace and equipment design are to determine criteria for new designs and to evaluate the impact of existing designs on the potential work force. These uses are each illustrated below with examples. Further examples of the use of anthropometric data are given in Appendix C.

a. Criteria for New Design

Chapter I includes a section titled "Criteria for Design: Who?" which defines the goal of new design—to accommodate the largest proportion of the potential industrial work force. An example of how this goal is accomplished follows.

(1) Problem

Choosing the height of a standing work surface where units are packed in flat boxes illustrates the need for anthropometric trade-offs. The operation is not visually demanding, but force is applied downward in the wrapping task. Thus the operator should have work surface height below his or her elbow height.

(2) Solution

The elbow heights for a 50/50, male/female population (from the military data in Tables VIA–2 and VIA–3) are as follows:

5th percentile, 96 cm (38 in.)
50th percentile, 107 cm (42 in.)
95th percentile, 116 cm (46 in.)

Since the wrapping should be done below elbow height, the work surface height should be somewhat less than 96 cm (38 in.) in order to accommodate the smaller people.

A second consideration is to provide a work surface that does not require tall people to stoop, since stooping produces rapid fatigue of the back and leg muscles. Although data on the combinations of elbow height and forearm length are not available, Tables VIA–2 and VIA–3 show that the forearm length range is from 28 to 41 cm (11 to 16 in.). It is also known that as the hand moves below the level of a 90° elbow bend, arm reach capability is reduced. Ideally, a wrapping and packing station surface should be from 8 to 20 cm (3 to 8 in.) below elbow height; it can be lower if the object to be wrapped is not large enough to require extended forward reaches.

If the work surface is located 20 cm (8 in.) below the 95th percentile elbow height, or at 96 cm (38 in.), it will be too high for less than half of the potential work force doing wrapping and packing tasks. By trading off between the too-high work surface for short people and the too-low surface for tall people, one arrives at a work surface height of 92 cm (36 in.). This height is not optimal for either extreme of the distribution of elbow heights, but it does accommodate about 85 percent of the people who fall between these extremes.

More accurate evaluations of the percentage of men or women affected can be made by using the mean and standard deviation of the measurements and calculating the z, or standardized, scores for each population (Freund, 1967), once a height or reach has been proposed.

b. Evaluation of Existing Design

Anthropometric data can be used to evaluate existing design when justification is needed to alter a workplace or when it is not clear if a problem is an individual problem or a group problem. The anthropometric data should not be used to exclude a person from a job on the basis of short arms or stature, however. In some instances job accommodations are not possible in order to help the smaller or larger person do the task, such as in manual handling of large-size product for a person with a small arm span, or entry through an access port by a very obese person. It is often difficult to validate personnel selection tests based on anthropometric characteristics with individual job performance.

An example of the use of anthropometric data to evaluate the impact of an existing design on the potential work force is given below.

(1) Problem

Individuals were having difficulty reaching a light switch in a darkroom. So that the light would not be accidentally activated as people and equipment moved in and out of the room, the switch had been placed 213 cm (84 in.) above the floor to the right of the door. But many short people had difficulty reaching the switch when light was needed.

(2) Solution

The relevant measurement is functional overhead reach, standing (see measurement 10 in Figure VIA–3 and Tables VIA–2 and VIA–3). For the 50/50, male/female potential work force, the reaches are as follows:

5th percentile, 188 cm (74 in.)
50th percentile, 204 cm (80 in.)
95th percentile, 221 cm (87 in.)

Since the light switch was placed where almost 75 percent of the potential workforce could not reach it without stretching and going on tiptoe, the problem was not confined to the smaller people. Probably more than half of the potential work force could reach the switch by standing on tiptoe; more than 40 percent had difficulty even in this posture, however, and had to find other techniques, such as jumping, to activate it.

If the switch is placed 195 cm (77 in.) from the floor, most people will be able to reach it. Accidental activation will not be a problem except, possibly, by very tall people. Protecting the switch to minimize the chance of its being turned on accidentally further reduces this possibility.

SECTION VI B. APPENDIX B: METHODS

1. AIDS FOR HUMAN FACTORS/ERGONOMICS FIELD SURVEYS

In addition to providing information for the design of new facilities and equipment, human factors guidelines can be used to develop criteria for evaluating existing workplaces and jobs. In this section a descriptor list and a checklist are presented for use as aids in identifying potential human factors/ergonomics problems in the workplace through field surveys.

A survey is commonly preceded by discussions with management and first-line supervision in the area, and it is frequently combined with medical, industrial hygiene, and safety surveys of the same jobs. There are both advantages and disadvantages to surveys; some of these are listed in Table VIB–1.

A survey team should include a member of the local supervisory, engineering, or safety staff. This person should be fully involved in the study, not just a guide. Information and suggestions from the employees should also be solicited. Neither the survey team nor the department should assume that a human factors/ergonomics survey will identify all causes of health, safety, or performance problems.

a. Survey Techniques

A successful survey should be carefully planned, and it should be perceived as useful by first-line supervision and management. The purpose of the survey should be stated positively at the outset. The surveyor should not emphasize that the survey is designed to look for problems or hazards; rather, the surveyor should state that the purpose is to identify workplace or environmental factors that might be reducing productivity or making jobs unnecessarily difficult. It is important for the survey team to be assured that plant management is committed to responding to the survey findings positively. Failing to act after problems have been identified by a third party can have a negative effect on worker morale.

Survey techniques are included in this section. They include data analysis and discussions with supervision prior to the survey, observational techniques and identification of problems during the survey, and methods for presenting survey results and prioritizing the needs for change after the survey.

(1) Data Analysis Prior to the Survey

The decision to do a survey in a work area is often associated with a concern about potential health, safety, or productivity problems. By analyzing existing data—such as accident reports, medical visits, absenteeism, and job evaluation ratings of physical effort, environmental conditions, and mental demands, if they are quantified—the survey team can develop some understanding of what areas would benefit most from human factors, health, and safety surveys. If limited resources are available, an analysis of existing

Table VIB–1: Advantages and Disadvantages of Human Factors/Ergonomics Field Surveys (Adapted from S. H. Rodgers, 1975 and 1980, Eastman Kodak Company.)

Advantages
• Usually supported by all levels of management, so major recommendations will probably be implemented, giving the survey team access to many areas that might not otherwise ask for job studies.
• Usually is an efficient way of identifying major ergonomic problems; when done as part of a health and safety team, a survey can be an effective way of prioritizing the need for workplace or job redesign.
• Gives the surveyor an appreciation of the entire manufacturing system or an entire department at one specific time, which helps in the development of recommendations and the search for workplace accommodations to resolve human factors problems.

Disadvantages
• Walk-through surveys provide only a snapshot of what is being done at that particular time. If the jobs include varied tasks, different products, or seasonal activities, the survey may not cover them. The longer a job is studied, the more representative it will be of the typical situation. In general, a job should be studied for at least one shift, preferably more, and with several people doing it, unless it is highly repetitive and unvarying.
• Some people in areas being surveyed may not understand the reason the team is there, and they may feel defensive about the need for a study. This situation can occur if upper-level supervision has not discussed the project with department personnel. This potential communications problem can be defused at the departmental level by the survey team, as is discussed later in this section.
• If the human factors/ergonomics staff is limited, the survey may take time away from more important studies with other clients who are committed to making changes in the workplace. Surveys are often done to identify areas to be improved over a number of years, or to assure management that there are no serious health and safety problems in their areas. A requirement for regular surveys of every workplace would not be an appropriate use of a small human factors/ergonomics group, although surveys of each area should be done at some point in the development of the human factors program.

Three advantages of using survey techniques to identify human factors or ergonomic problems in a manufacturing area are given at the top of the table. Three disadvantages are discussed in the lower part of the table. A decision of whether or not to survey should be based on a balancing of these factors.

data is particularly important. The advantages of reviewing the data first are improved communications with supervision about the potential value of a survey and identification of those areas of concern that will be likely to show the most improvement if changes are made.

(2) Discussions With Supervision Before the Survey

Once a survey has been requested and a commitment to implement changes has been assured, the survey team should spend some time talking to supervisory people in the work area about their perceptions of possible health, safety, or human factors problems. In the course of this conversation the team can gather information about the following work characteristics:

- The normal routes of progression through jobs.
- Hiring patterns and turnover.
- Absenteeism, accidents, and medical restrictions.
- What is done to accommodate the worker with reduced capacity, such as a person who has had a heart attack.
- Flexibility in moving people between jobs or between work groups.
- Shift-work schedules and overtime policies.
- Production pressures and perceived staffing adequacy for production goals.
- Seasonal or infrequent work demands.
- Training programs and policies.
- Specific jobs or tasks of concern from a safety or performance standpoint, such as heavy lifting or difficult visual work.

From this discussion with supervision, the survey team can identify how much time is needed in each area, what types of evaluations should be made on the first pass-through, when the survey should be made, and what options are available for recommending change if problems are identified.

(3) The Plant Survey: Information From People Doing the Job and Site Visits

Table VIB–2 presents a descriptor list of factors to consider when a team is surveying a workplace or job. This list is intended to remind the surveyor of areas to note; quantification of each factor is seldom needed. Much of the information can be collected by talking to people doing the job.

An experienced human factors/ergonomics practitioner checks the descriptor list now and then during a workplace or job evaluation to be sure all of the factors of concern have been identified. A less experienced surveyor may wish to have a more descriptive checklist as a guide in information collection in the field. Such a checklist is included in Table VIB–3. However, checklists are of limited use in surveys because they cannot always anticipate problems

Table VIB–2: Human Factors/Ergonomics Survey Descriptor List (Adapted from S. H. Rodgers, 1975, Eastman Kodak Company.)

Workplace Characteristics and Accessories

Reaches
Clearances
Crowding
Postures Required
Chairs and Footrests
Heights
Location of Controls and Displays
Motion Efficiency
Workplace Accessibility (as in moving supplies into it)

Physical Demands

Heavy Lifting or Force Exertion
Static Muscle Loading
Endurance Requirements
Work-Rest Patterns
Frequency of Handling
Repetitiveness
Grasping Requirements
Size of Articles to Be Handled: Very Large or Very Small
Sudden Movements
Stair or Ladder Use
Tool Use

Displays, Controls, Dials

Size Relative to Viewing Distance
Compatibility
Display Lighting
Labeling

Environment

Noise Level and Type
Vibration Level
Temperature
Humidity
Air Velocity
Lighting Quantity
Lighting Quality, Especially Glare
Electric Shock Potential
Floor Characteristics, Including Slipperiness, Slope, Smoothness
Housekeeping
Hot Surfaces
Protective Clothing Needed

Mental Load

Skill Requirements
Multiple Tasks Done Simultaneously
Pacing Stress
Training Time Needed
Monotony: Low Challenge
Concentration Requirements
Information-Handling Demands
Complexity of Decision Making, Defect Recognition

Perceptual Load

Visual Acuity Needs
Color Vision Needs
Space and Depth Perception Requirements
Tactile Requirements
Darkroom Vision
Auditory Demands

Six categories of descriptors are given: workplace characteristics and accessories, physical demands, environment, mental load, perceptual load, and displays, controls, and dials. Each category includes 4–12 specific descriptors of conditions that should be looked for in the survey. The list is intended to remind the surveyor to look for these potential human factors or ergonomic problems, not to define the magnitude of the problem. This approach assumes that the surveyor has some experience in this type of survey and knows what to look for.

that arise in the field. A common problem with checklists is that they are used as a substitute for observation. It is better for the less experienced surveyor to use the descriptor list and observe carefully what the job requires the person to do rather than to carefully answer the questions on the checklist without noting all that is being done.

The checklist in Table VIB–3 is not as complete as many that have been developed for ergonomics/human factors surveys (see Bainbridge and Beishon, 1964; Burger and DeJong, 1962; Woodson, 1981). It is intended to lead the surveyor into the full evaluation rather than to complete the analysis.

Table VIB–3: Survey Checklist (Adapted from S. H. Rodgers, 1980, Eastman Kodak Company.)

This checklist includes situations that may be of concern in the workplace; it is an expansion of the descriptor list given in Table VIB–2. Quantification is included where it is appropriate and available. The survey checklist is more problem-oriented than the descriptor list and, therefore, may restrict observation of situations not specifically described. The surveyor using this checklist should mark the items that exist in a given workplace but should also be attentive to conditions that may not be included. Indicate whether any of the following situations are present in the workplace.

Workplace Characteristics

____ Extended reaches, beyond normal arm reach

____ Inadequate clearance for handling or maintenance tasks

____ Inaccessible workplaces for using handling equipment

____ Poor chairs: difficult to adjust, inadequate back support, no footrests

____ Dials and displays that are difficult to read or reach

____ Inefficient work motions because of workplace layout

____ Awkward postures

____ Inadequate space for temporary storage at the workplace

____ No adjustability built into the workplace

Perceptual Load

____ Difficult-to-hear auditory signals

____ Small and difficult-to-see defects

____ Multiple types of defects to detect

____ Fine color differences to be discriminated

____ Need to judge distance rather critically, such as using the high stacking capability on a forklift truck

____ Need to detect defects or information under low light levels

____ Need to discriminate parts by touch

Table VIB–3: (Continued)

Mental Load

____ Highly complex operations

____ Need to keep track of multiple factors simultaneously

____ Demands for performance in a designated time frame, such as externally paced machine operations

____ Long training times needed

____ Highly repetitive, monotonous, unvarying activity

____ Critical tasks that involve high accountability where errors are not tolerated

____ Heavy memory (short-term) load, such as working with 9-digit codes

Environment

____ Noise that interferes with conversation or doing the job

____ Noise that is annoying and distracting

____ Vibration that is annoying

____ Temperature/humidity that is frequently uncomfortable or distracting and interferes with the job

____ Poor air circulation

____ Inadequate lighting for the job

____ Glare that interferes with reading or inspecting at the workplace

____ Wet locations that may produce electric shock hazards, such as when working with power tools

____ Floors that are uneven, with drains or pit marks

____ Slippery floors

____ Poor housekeeping, such as crowded aisles or waste on the floor

____ Hot surfaces that may increase burn hazards

____ Conditions that require protective clothing in addition to safety glasses

Displays, Controls, and Dials

____ Displays that are difficult to read from the operator's usual workplace locations: at the wrong angle, too high or low, too small

____ Displays and controls that do not follow population stereotypes or are not compatible

____ Labeling on control and display panels that is unclear

____ Displays that are not adequately lit

____ Controls that are difficult to distinguish or are not laid out consistently from one work station to another

____ Excessive number of controls needed to perform the job

____ Immoderate amount of information to be handled in a short time

Table VIB–3: (Concluded)

Physical Demands

- ____ Frequent heavy lifting (>18 kg or 40 lb, 2 hr/day)
- ____ Occasional very heavy lifting or force exertion (>23 kg, or 50 lb, 225 N, or 50 lbf)
- ____ Constant handling of materials, little variety
- ____ Handling difficult-to-grasp items
- ____ Awkward lifts or carries that are near the floor, above the shoulders, or far in front of the body
- ____ Exertion of forces in awkward positions: to the side, overhead, at extended reaches
- ____ Constant standing
- ____ Frequent daily stair or ladder climbing
- ____ High-precision movements
- ____ Sudden movements during manual handling
- Static muscle loading
- ____ High pressure on the hands from thin edges, such as pail handles or sheet metal edges
- ____ Machine pacing of handling, such as on and off of conveyors
- ____ Use of hand tools that are difficult to grasp
- ____ Short-duration heavy effort
- ____ Moderate to heavy effort sustained throughout the shift
- ____ Handling oversized objects, including two-person lifting
- ____ Lack of handling aids, such as drum carts, air hoists, scissors tables
- ____ Unavailability of help for heavy lifting or exerting forces

b. Presentation of Survey Results

In surveying a work area, a surveyor can usually find human factors problems that would benefit from redesign or changed work practices. The results of the survey should distinguish between situations that are potential contributors to health and safety problems and those that probably are not but would improve the workplace and possibly increase productivity. In many instances, production departments have recognized situations that are not optimal and are in the process of trying to resolve them before a survey is done. In others a situation may be recognized as needing improvement, but the technology is not yet available to solve it; an example of this situation is handling drums or pallets by hand.

The team's observations can be categorized for easier presentation of the survey results (P. C. Champney, 1976, Eastman Kodak Company); for example:

Category 1: Must Be Changed. These are problems that need immediate attention because they are health and safety hazards.

Category 2: Should Be Changed. These are potential health and safety problems that merit redesign or other action but do not require immediate action.

Category 3: Are Changing. These are problems that the department is aware of and trying to resolve.
Category 4: Change Next Time. These are problems that are technologically or economically difficult to resolve in existing workplaces but could be resolved in new or renovated facilities.
Category 5: Would Be Nice to Change. These are situations that would benefit from redesign in existing workplaces or change in new workplaces but that are not posing a health or safety risk. Changes would probably result in productivity increases, but this result is not obvious.

When the survey results are categorized, it is easier for priorities to be set in order to resolve problems. Some changes to category 5 situations usually can be made rapidly and inexpensively even if the priority is not high. These changes should be made, if they are worthwhile, but not at the expense of solving problems in the other, higher-priority categories. Some examples of work site survey findings in each category are given in Table VIB–4.

Most survey findings will fall into categories 2–5. Economic analyses and engineering alternatives are then developed for resolving the problems. The survey team should continue to be involved at this point to help evaluate alternatives. A cooperative effort reduces the possibility that another human factors problem will be created in the solution of the first one.

2. ENVIRONMENTAL MEASUREMENT PROTOCOLS

Nonoptimal environments often result in worker discomfort and loss of productivity in the workplace. Because it is usually possible to alter the environment, the physical environmental factors of temperature, humidity, air velocity, noise, and illumination are frequently measured in work site evaluations of job demands. In this section survey techniques are briefly discussed in terms of the measurements made, where they are taken in the workplace, and when they should be made.

a. Temperature, Humidity, and Air Velocity

(1) Instrumentation and Measurements

To assess the influence of the environment on a worker's ability to maintain body heat balance, the surveyor needs to know the dry-bulb, wet-bulb, and globe temperature readings, as well as the air velocity, at the workplace. From analyses of these data, the temperature, humidity, radiant-heat load, and air flow in the workplace can be characterized, and ways can be identified to reduce thermal discomfort or health risks. A detailed description of the comfort, discomfort, and health risk zones for heat/humidity and cold environments is given in Section V D, "Temperature and Humidity."

There are several types of instrumentation available for measuring temperature, humidity, and air velocity in industry. The choice of a given system is

Table VIB–4: Examples of Workplace Situations in Each of the Survey Result Categories (Adapted from F. C. Champney, 1976, Eastman Kodak Company.)

Category 1: Must Be Changed	Category 2: Should Be Changed	Category 3: Are Changing	Category 4: Change Next Time	Category 5: Would Be Nice to Change
Electric shock hazards Extended exposure to temperature extremes Excessively high force exertion or handling requirements	Repetitive heavy lifting (>18 kg, or 40 lb) Poor or inadequate seating Tasks that are highly machine-paced Static muscle loading Poor-quality lighting for fine visual tasks Poor grasping surfaces	Often includes handling tasks where aids are being considered or inspection tasks where lighting is being adjusted	Inadequate clearances for maintenance Awkward reaches Multiple rehandling of product	Confusing dials or displays in noncritical tasks Monotonous tasks, usually highly repetitive

Five categories of findings from a survey are given. Each category includes examples of findings from workplace surveys to illustrate possible types of problems at each response level. The surveyor can distinguish among the last three categories by working closely with a departmental or divisional representative in defining the potential problems in the workplace.

usually determined by the degree of accuracy needed to identify a heat, humidity, air velocity, or cold problem, the amount of space available for monitoring, and safety and contamination considerations. Safety considerations include limitations on the use of electric systems in areas where flammable solvents are used; contamination problems may be of concern if mercury thermometers are used in some manufacturing areas. Figure VIB–1 illustrates some of this instrumentation. A brief description of these measurements follows (Botsford, 1971; Hertig, 1973; NIOSH, 1972).

Figure VIB–1: Instrumentation for Workplace Measurements of Temperature, Humidity, and Air Velocity

(a)

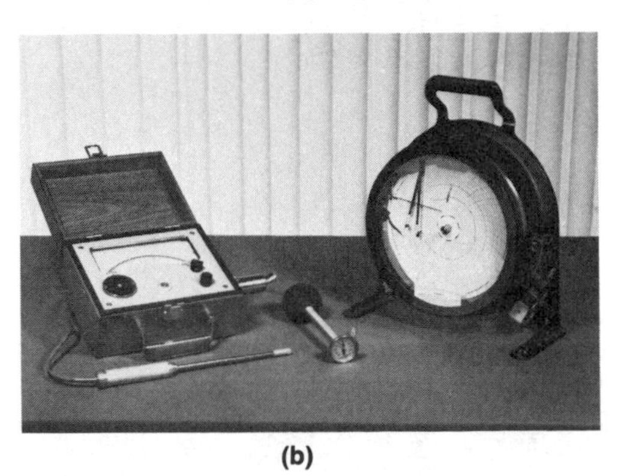

(b)

Six pieces of equipment used to measure environmental heat, humidity, and air velocity are shown. Part a shows the Christmas tree arrangement of globe, natural wet-bulb, and shielded dry-bulb thermometers used to characterize the wet-bulb/globe temperature (WBGT) index for an environment. The WBGT is related to the heat-exchanging capability of a person in the measured conditions. Part b shows a hot-wire anemometer to measure air velocity; a Botsball to measure the wet-globe temperature (WGT), which is similar to the WBGT; and a 7-day pen recorder for dry-bulb and relative humidity surveys in the workplace.

(a) Temperature

The dry-bulb temperature, shielded, is used to assess the nonradiant-heat load or the cold in the environment. It is measured with a glass thermometer or thermal probe inside a reflective shield, such as a cylinder with an aluminized outer surface.

The globe temperature is measured with a globe that has a matte black surface to absorb radiant heat. The difference between the dry-bulb and globe temperature readings gives a measure of the amount of radiant heat present.

(b) Humidity: Wet-Bulb Temperature

There are several instruments that can be used to measure the humidity in the workplace: natural wet bulb, psychrometric wet bulb, or a variety of instruments that detect humidity electronically, mechanically, or chemically and give relative humidity readings. Natural wet-bulb readings differ from psychrometric wet bulb readings by the amount of air velocity present in the workplace. Natural wet-bulb temperature is measured with the actual air velocity determining the rate of evaporation of water from a wick covering a thermometer bulb. Psychrometric wet-bulb temperature is measured with an air flow of more than 0.5 m/sec (100 ft/min) drawn past the wick. These measures are all used to estimate the evaporative capacity of the worker in the environment.

(c) Air Velocity

Air velocity measurements are important for assessing ways to reduce discomfort in hot and cold environments. A hot-wire anemometer or a mechanical velometer may be chosen, depending on the amount of flow and on safety considerations. A velometer is better when flows are under 0.1 m/sec (20 ft/min) or when the hot-wire anemometer cannot be used because open flames are not permitted. Air flow should be measured wherever people work in order to try to map out local drafts that may increase cold discomfort.

(d) Combination Measures: Heat and Humidity

Temperature, humidity, and air flow measurement equipment is available that combines the measures and prints out a weighted average or a measure to describe the heat or cold stress for the worker. Two of these measures are the wet-bulb/globe temperature (WBGT) and the wet globe temperature (WGT) readings. The former is derived by the following formulas (NIOSH, 1972, Yaglou and Minard, 1957):

Inside: WBGT = 0.7(natural wet-bulb temperature) + 0.3(globe temperature)

Outside: WBGT = 0.7(natural wet-bulb temperature) + 0.2(globe temperature) + 0.1(shielded dry-bulb temperature)

The temperatures can be expressed in either degrees Celsius or degrees Fahrenheit. WGT can be read directly from a Botsball, a globe covered with black netting that is wetted down before each measurement is made (Botsford, 1971).

(2) When and Where to Make the Measurements

In addition to identifying the types of measurements to be taken, it is important to take them in the correct places and at the appropriate times. This technique will help identify how much of the problem is related to manufacturing processes and how much to climate. It is often wise to leave a 7-day temperature and humidity chart recorder in a workplace where heat or cold is thought to be a problem; by observing the temperature and humidity trends over a week, the surveyor can often identify contributing factors, such as production equipment being turned on or the accumulation of heat associated with outside temperatures. The record also helps determine when the more thorough measurements should be made. Other factors to consider when choosing the time and place to monitor the environment are given below.

(a) When

- Determine whether the discomfort is more noticeable at any particular time, such as during the following:
  - A specific manufacturing process.
  - A particular season.
  - A certain time of day (for instance, late afternoon is usually the worst period for seasonal heat accumulation).

  Be sure to monitor during that time. Monitor at another time, as well, in order to provide comparisons.
- If possible, the worst conditions should be monitored; the results should, however, be interpreted in relation to their frequency of occurrence.
- The more data collected over time in a workplace, the better the environmental temperature and humidity can be described. Several days of monitoring helps identify the regularity of occurrence of the more extreme conditions.
- Whenever the measurements are made, conditions will probably not be as bad as they were when the problem was first identified.

(b) Where

- Find out where the exposure is occurring in the work area and how many people experience discomfort. Make measurements to determine the following:
  - How long are people exposed?

What conditions do they work in between exposures to the heat and cold?
What are the conditions in areas where breaks are taken?
How frequently do the exposures occur (times per shift, week, year)?

- Measurements in the workplace should follow the exposure intensities and durations. Preferably, 24-hour values should be recorded at each location where a person has to work for more than a few minutes.
- Some measurements should be made close to hot or cold sources in order to assess their local effects on people. For example, the surveyor may wish to assess how much radiant heat is present or how large a differential in temperature exists within the workplace.

(3) *Presentation and Interpretation of the Data*

The 24-hour recordings of temperature in a work area can be graphed against time and compared with the recommended thermal comfort and health risk levels; see Section V D, "Temperature and Humidity," for more discussion. Figure VIB–2 illustrates such a plot for a workplace where moderately heavy work was done.

The duration of temperature peaks above or below the recommended levels will indicate how important it is to make changes in the environment or work load. If the peaks are very short and infrequent, local alterations in the workplace or job are suitable. Such alterations might include the following:

- Providing shields against radiant heat.
- Wearing protective clothing for cold exposure.
- Reducing individual exposure times by rotating job responsibilities.

If the level of heat, cold, or humidity exposure is consistently above or below the recommended guidelines, interventions in the work load level or environmental conditioning may be necessary. For a more extensive description of ways to reduce heat or cold stress in the workplace, see Section V D, "Temperature and Humidity."

(4) *Additional Data for Evaluation of the Severity of a Seasonal Problem*

Many problems of temperature discomfort are seasonal. Since measurements should be made in the worst conditions as well as in average conditions, it is important to be able to evaluate the number of days of the year the worst conditions occur. If the conditions occur only a few times a year, solutions for the problem may be different from those for problems that occur more frequently. The U.S. Department of Commerce publishes weather data for specific areas of the country, summarizing the average, minimum, and maximum values for each month (Department of Commerce, 1982). Since inside temperature is often partly determined by outside temperature, these data can be

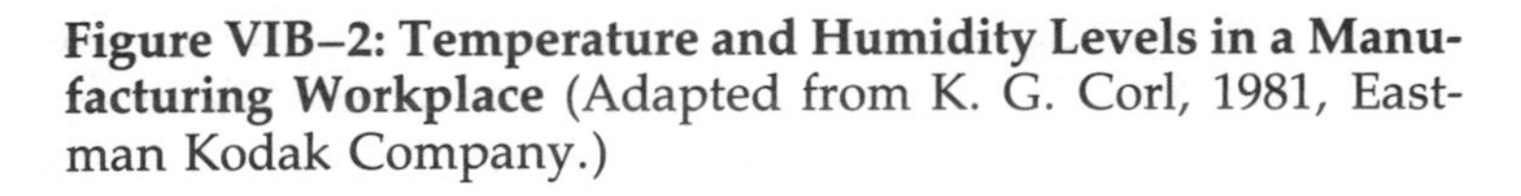

Figure VIB–2: Temperature and Humidity Levels in a Manufacturing Workplace (Adapted from K. G. Corl, 1981, Eastman Kodak Company.)

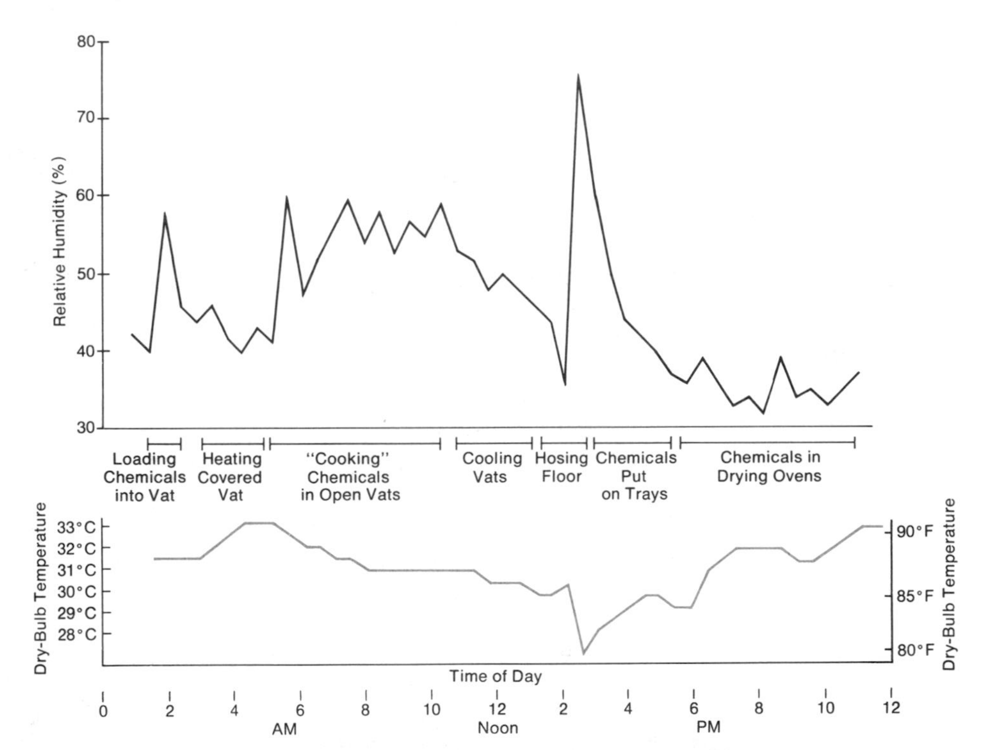

A 24-hour pattern of dry-bulb temperature (lower scale on the vertical axis) and relative humidity (upper scale) is shown for a manufacturing area during the summer. The data were taken from a chart recorder located in an area where people were monitoring a chemical-mixing process, doing cleaning operations including hosing down the floor, and performing occasional moderately heavy lifting tasks. Chart recordings help identify environmental heat and humidity loads in time (horizontal axis) over several days. By analyzing the peak loads, the surveyor can usually find effective ways to modify the environmental stress on the worker.

used to determine how typical the outside conditions are on days when environmental measurements are made in a workplace. Figure VIB–3 shows a cumulative plot of the frequency of occurrence of dry-bulb temperatures (averages, minimums, and maximums) in Rochester, New York. The data were gathered over a seven-year period by utilities engineers in a manufacturing complex. Similar curves can be constructed for other geographical locations.

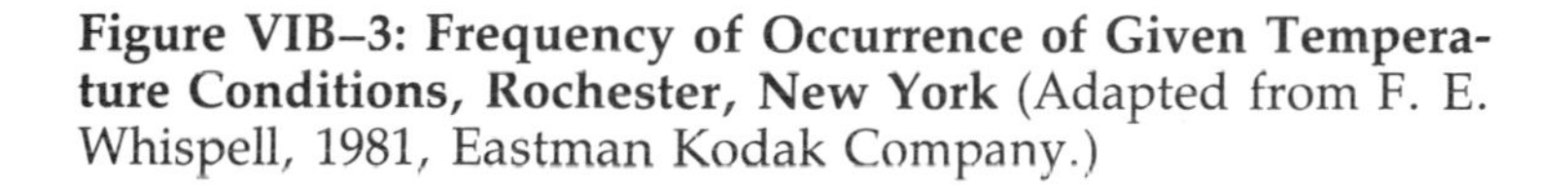

Figure VIB–3: Frequency of Occurrence of Given Temperature Conditions, Rochester, New York (Adapted from F. E. Whispell, 1981, Eastman Kodak Company.)

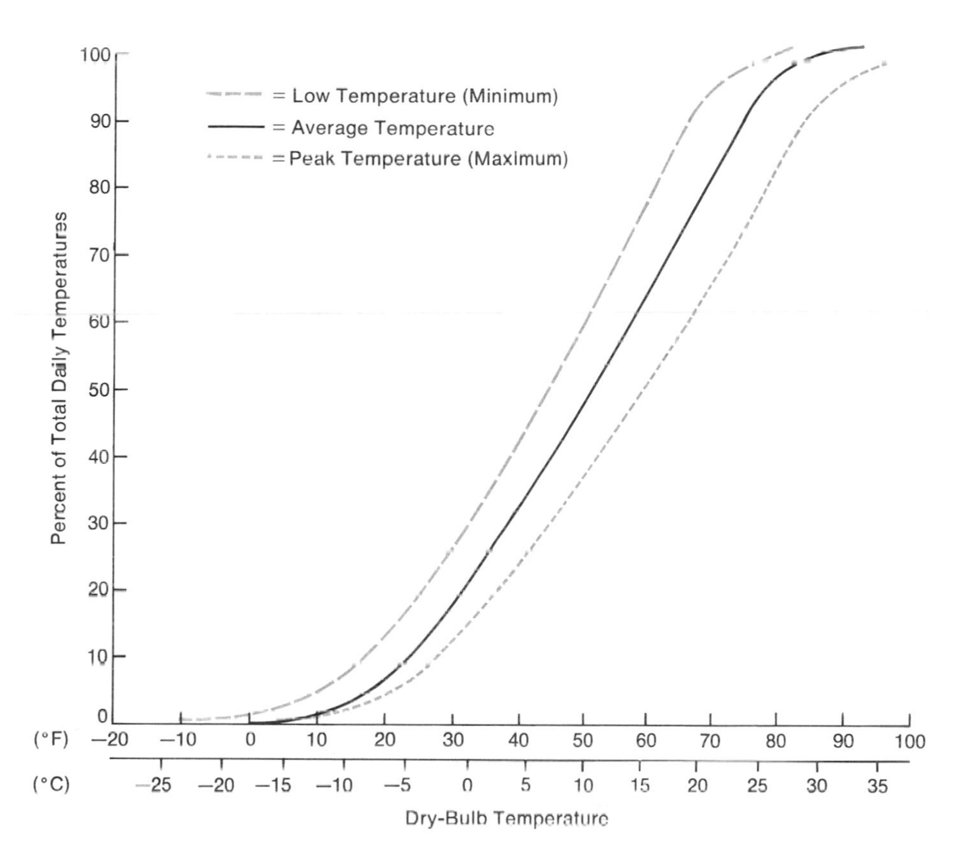

Cumulative frequency distributions for daily minimum, average, and maximum temperatures over a seven-year period (1974–1980) are shown. The horizontal axis shows dry-bulb temperature, in degrees Celsius (°C) and Fahrenheit (°F), and the vertical axis is the percentage of all daily temperatures that were recorded in the seven-year period. Average daily values were calculated from trend charts; peak and minimum values are the highest and lowest temperatures recorded during a 24-hour period. Any point on one of the curves will tell the reader the percentage of the total time that Rochester will be the same temperature as or cooler than that value on the horizontal axis. To determine how often Rochester will be the same temperature or hotter, subtract the value on the vertical axis from 100. The data represented here were collected by utilities engineers within a manufacturing complex; similar data are available from the National Oceanic and Atmospheric Administration Environmental Data Service of the U.S. Department of Commerce.

b. Noise

The operation of production equipment and other machines often introduces noise into the workplace. This noise may affect people in the workplace in one of the following ways:

- contribute to hearing loss
- interfere with communication
- annoy or distract the people nearby
- alter performance on some tasks

These concerns may be associated with the design of a new machine or facility or with the evaluation of an existing one. The ultimate goal of a noise analysis is to find ways to reduce noise to a level that does not negatively affect people or productivity.

To respond effectively to the requests from the plant, the team should include the following professionals:

- An industrial hygienist or an audiologist to measure the noise and determine whether it may affect worker health.
- A human factors specialist to evaluate the performance and distraction impacts of the noise on the worker.
- An acoustical engineer to identify engineering techniques that will eliminate some of the noise from production machinery.
- An architect with acoustical training to evaluate ways to isolate the worker from the noise sources in the workplace.
- An experienced staff engineer to identify the probable noise exposure of workers and recommend ways to reduce individual exposure to it.

With this team most noise exposure problems can be analyzed and techniques found to reduce the problem in new or existing facilities. Section V B, "Noise and Vibration," shows how to interpret noise measurements. In the following discussion, information is included about the measurement of noise and about techniques for making noise surveys, as well as the proper time and location for such surveys.

(1) Instrumentation and Measurements

The material in this section was developed from Botsford (1973), Broch (1973), and Peterson and Gross (1967).

The audible spectrum for noise is from about 20 to 20,000 Hz (Hertz). For many noise evaluations a simple sound-level meter can be used. This meter allows the surveyor to identify the sound pressure levels (in decibels on the A scale, dBA) and to classify the noise according to its potential to contribute to hearing loss over time or to interfere with communication.

A precision sound meter includes filters that permit noise to be segmented into frequency bands. By measuring the sound levels in each band, the surveyor can characterize the relative proportions of low, medium, and high frequencies, as well as the dominant frequencies. These analyses are useful to assess the effectiveness of different noise-control techniques and to help in identifying the specific machine component that needs isolation to reduce vibration or noise generation. Figure VIB–4 illustrates some noise measurement equipment.

Figure VIB–4: Instrumentation for Noise Measurement

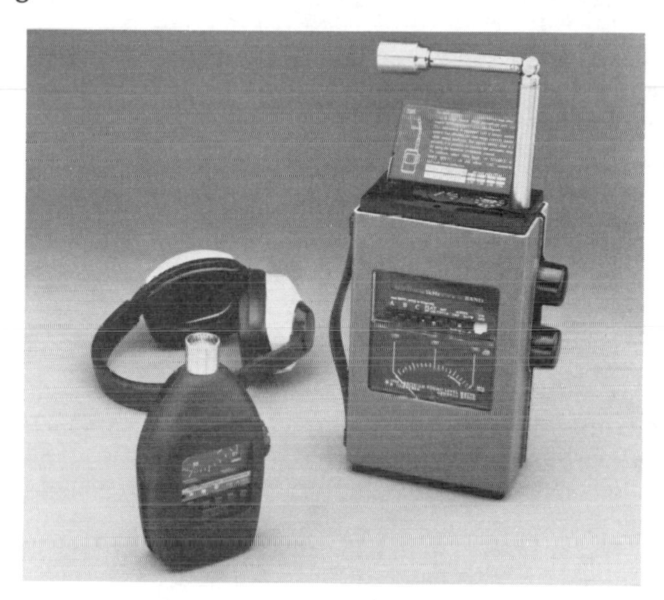

Two types of noise measurement equipment are shown. A standard sound-level meter is shown on the left; it measures the rms (root mean square) sound pressure level, in decibels. The rms value is an integration of all of the sound reaching the microphone at the top of the meter. This standard meter has three scales for noise measurement: A, B, and C. These scales represent different amounts of sound intensity attenuation in the low-frequency bands below 1000 Hz; the A band has the most attenuation. Most noise levels are expressed in decibels on the A scale (dBA). The equipment on the right is a precision sound meter with an octave-band analyzer. This meter permits the user to characterize the noise by frequency band. The sound pressure level can be plotted against frequency and compared with acceptable levels, such as the preferred noise criterion (PNC) or preferred speech interference level (PSIL) guidelines. The ear protection muffs in the background of the illustration are worn by the person making the sound measurements, if necessary, as a precautionary measure.

Noise dosimeters are available to record an individual's exposure to noise over a shift (Botsford, 1973; Metrosonics, 1981). These meters either record the amount of exposure above a specified sound level, such as 85 dBA, or provide a time history of exposure above a specified level, such as 60 dBA. The time history dosimeter can be useful in assessing the exposure of an individual to noise levels that make concentration or communication difficult.

Measurements of noise levels at a workplace are made with a sound-level meter in the A-weighted scale. An octave-band analysis is done from readings made in the following center frequencies: 31.5, 63, 125, 250, 500, 1000, 2000, 4000, and 8000 Hz. The results are plotted on a graph of frequency (in hertz) versus sound pressure level (in decibels). Preferred noise criterion (PNC) curves are plotted on the same graph to demonstrate the impact of the noise on people in the workplace. Section V B, "Noise and Vibration," contains an example of how these curves are used.

Speech interference levels are measured at the octave bands centered at 500, 1000, and 2000 Hz on the A-weighted scale; an average is taken of the three values. The average can be plotted on a graph of sound level versus the distance between speaker and listener in order to assess the ease of communication in the workplace. Such a graph is shown in Section V B.

(2) When and Where to Make Noise Measurements

As in the temperature and humidity surveys, the time to monitor the noise is when it is of concern to people in the workplace. It should be monitored where people work, rather than where the noise originates. Measuring the noise at its source may be useful, however, if its analysis can help to identify a faulty machine part or can suggest a way to reduce the problem.

If the noise is more apparent during a particular manufacturing process or at a specific machine speed, arrangements should be made to evaluate it on those occasions. In many instances the noise is not problematic until a large number of machines are operating simultaneously in the same work area. It is important to determine how frequently this condition occurs and to monitor noise during the period of noise occurrence. The data should be interpreted in terms of the frequency of occurrence when deciding what steps should be taken to reduce the noise level.

Note should be made of the communications and hearing requirements for people working on jobs in noisy work areas. Then the effect of the noise level on performance can be interpreted appropriately, assuming it is below 85 dBA, the level at which hearing may be affected. Time history dosimeter data may help identify these problems if used in conjunction with a diary of the person's job activities throughout the shift.

c. Illumination and Luminance

Manufacturing operations are sometimes carried out in areas where lighting appears to be inadequate, either because of insufficient illumination or because of excessive glare. In this section instrumentation and survey techniques for measuring illumination and luminance and for assessing direct glare are briefly discussed.

Illuminance is the amount of light falling on a work surface or object. Luminance is the brightness of a surface or an object, measured as the amount of light reflected from it. The results of a properly conducted survey may be compared with the guidelines given in Section V C, "Illumination and Color," to estimate the seriousness of a lighting problem. More detailed information on these measurements can be found in Kaufman and Christensen (1972).

(1) Measuring Illumination (Illuminance)

If people complain of difficulty in seeing assembly parts or defects, observing processes such as chemical reactions, or reading printed materials, then there may be an illumination problem. The instrument that is used for measuring the amount of light on a task or work surface is an illumination meter (see Figure VIB–5a). It usually reads directly in lux or footcandle units. The meter's sensor measures the amount of light incident upon the surface on which it is placed. The reading may then be compared with the guidelines in Table

Figure VIB–5: Instrumentation for Measuring Illuminance and Luminance

Two instruments are shown. The illumination meter on the left incorporates a hemisphere that integrates the amount of light falling on the surface on which it is laid. The meter indicates the quantity of light in either lux or footcandle units. The readings can be compared with the recommended guidelines for illumination given in Table VC–1. The photometer on the right measures the luminance, or brightness, of an object or surface. It is usually located where the head of the person who is viewing an object would be so that it measures the amount of light reflected toward the eyes. The readings from the photometer are in candelas per square meter or in footlamberts. The values can be compared to the guidelines for luminance ratios between the object being viewed and its background given in Table VC–4.

VC–1 in Section V C. To ensure that accurate meter readings are made, the surveyor should consult the instruction booklet before conducting a lighting survey.

(2) Measuring Luminance

The luminance, or brightness, of an object can be measured with a photometer (see Figure VIB–5) that reads directly in candelas per square meter or footlamberts. Photometers that have a small aperture measuring 1° or less are the most versatile. Because the luminance of an object does not vary with viewing distance, a photometer can be placed at any convenient distance and pointed toward the luminous surface to be measured. However, if the luminance of the object varies with the direction from which it is viewed, as is the case with directional light sources and illuminated surfaces that do not diffuse light, then the photometer must view the luminous object from the same direction as the worker. Also, for satisfactory results the luminance of the area on the object being measured must be uniform unless an integrated average reading is desired. It must subtend an angle greater than that of the photometer's sensor. Detailed information for using a photometer can be found in its instruction manual.

The potential for direct glare may be assessed by measuring the luminance of the task and the surrounding objects in the worker's field of view. If the luminance ratios exceed those given in Table VC–4 in Section V C, workers may experience visual discomfort.

(3) Survey Techniques

When and where to measure the lighting levels at the workplace is determined by: the type of lighting—for example, whether window lighting is involved—whether the perceived problem is unique to certain tasks or products; and how many people are affected.

Some additional factors to consider in lighting surveys are the following:

- If outdoor, or natural, lighting is used for a job, people should be interviewed first to establish whether the perceived problem occurs at specific times or under certain conditions.
- The lighting levels of adjacent workplaces in a manufacturing complex may differ, or appear to differ, because lights with different spectral characteristics are used; this difference may be a factor in one group's concern about workplace lighting levels. Sample readings in work areas adjacent to the one of interest can help to explain these apparent differences.
- Where lighting quality cannot be easily improved by altering the existing general illumination characteristics, local task lighting may be needed. More discussion on improving lighting at the workplace is given in Section V C, "Illumination and Color."

3. PSYCHOPHYSICAL SCALING METHODS

Probably the most comprehensive source of information about job demands in industry is the body of people in the workplace. Objective measures and outside observers may be able to demonstrate what is being done and how it affects the person's heart, breathing, or muscles, but only the worker can integrate all of the actions and environmental influence to assess total job demands. Each person brings his or her own set of expectations, personality, and physical fitness to such an integration; what is very stressful for one may be no problem for another. Psychophysical scaling methods can be used to collect people's opinions about job demands in a quantitative and statistically controlled manner. The results are subjective and must be carefully interpreted, especially when extrapolating from the opinions of one group to the population at large. When many people work on the same, or very similar, jobs, some of the pitfalls of subjective data collection can be reduced by gathering opinions from a large number of people.

In this section some techniques for gathering opinions about physical, environmental, mental, and perceptual work demands are discussed. The interested reader should consult the references for more extensive treatment of the specific methods. Discussion will center on the development and use of the interval scale to measure differences between conditions and the use of the Rating of Perceived Exertion (RPE) scale for physical work.

a. Types of Scales for Psychophysical Evaluations

Three types of scales are used in psychophysical scaling: ordinal, interval, and ratio. Functions and examples of each are given in Table VIB–5.

Table VIB–5: Examples of Measurement Scales (Adapted from R. T. Kintz, 1980, Eastman Kodak Company; based on information in Stevens, 1975.)

Scale	Function	Examples
Ordinal	Ranking	Product quality level: excellent, good, fair, poor
Interval	Differences or distances between situations or items	Thermometer scale: degrees Fahrenheit or Celsius Calendar dates
Ratio	Relationship between two situations or items, as in fractions, multiples	Elapsed time from zero Relative weight compared with a standard

Three types of scales are shown in column 1. The function that each serves in psychophysical scaling is given in the second column. Familiar examples of each type of scale are given in the third column. Interval scales are most commonly used for the collection of subjective data in human factors studies.

Of these scales, the one of most practical use in industry is the interval scale. Even though people are better at making relative, rather than absolute, judgments, ratio scales are more difficult to derive and are less amenable to handling large numbers of comparisons. Ordinal scales do not indicate how far apart two conditions are and, therefore, only lend themselves to statistical analysis of the median (center of the values) or the percentiles (fraction of the values below a given level). An interval scale can be summarized or categorized into an ordinal scale once the initial data have been collected; an ordinal scale cannot be made into an interval scale without collecting additional data.

b. Interval Scales in Psychophysical Data Collection

(1) Development of the Scale

(a) Word Scales

Interval scales have to be developed for opinion data collection in the plant. To accurately develop interval scales, the surveyor should run scaling experiments with people in the plant prior to the actual study. Scales have been developed that use words or phrases to describe an opinion or sensation. Risk assessment scales for the seriousness of an injury or its likelihood of occurrence use words familiar to most people, such as *minor injury, catastrophic, very unlikely,* and *rare.* The person administering the scale should be sure that people interpret these words similarly. Otherwise, the study results may be confounded, and the final risk ratings may show great variability when applied by different people to the same work situation.

Scaling experiments may involve asking respondents to locate items of interest along a dimension that is clearly defined, such as a line marked from 0 to 100 or from low to high. They should be able to place several items at the same point on the line if they wish. Figure VIB–6 illustrates the results of scaling experiments to develop the consequences scale in a risk assessment study.

(b) Paired-Comparison Technique of Developing a Scale

Another use of psychophysical methods is to create an interval scale that will help define the relationships among several situations or items. The best way to develop an interval scale when small differences must be detected among several items is to take each item and compare it with every other item two at a time. This method is called the paired-comparison technique (Torgeson, 1958). An example of its use in establishing the acceptability of several different mail routes is shown in Figure VIB–7. From the analysis it was possible to identify the routes that were considered more difficult than others on an overall basis. This information was then combined with ratings of effort and time pressure to identify the best approach for reducing the perceived difficulty of the jobs.

The paired-comparison technique can also be used to develop an attitude scale for use in field evaluations of job demands. It is the easiest method for developing interval scales because only one judgment has to be made at a time and no absolute judgments are required.

Figure VIB–6: Example of a Scaling Experiment (Adapted from S. H. Rodgers and J. Shealy, 1980, Eastman Kodak Company; original scale based on information in Fine, 1971.)

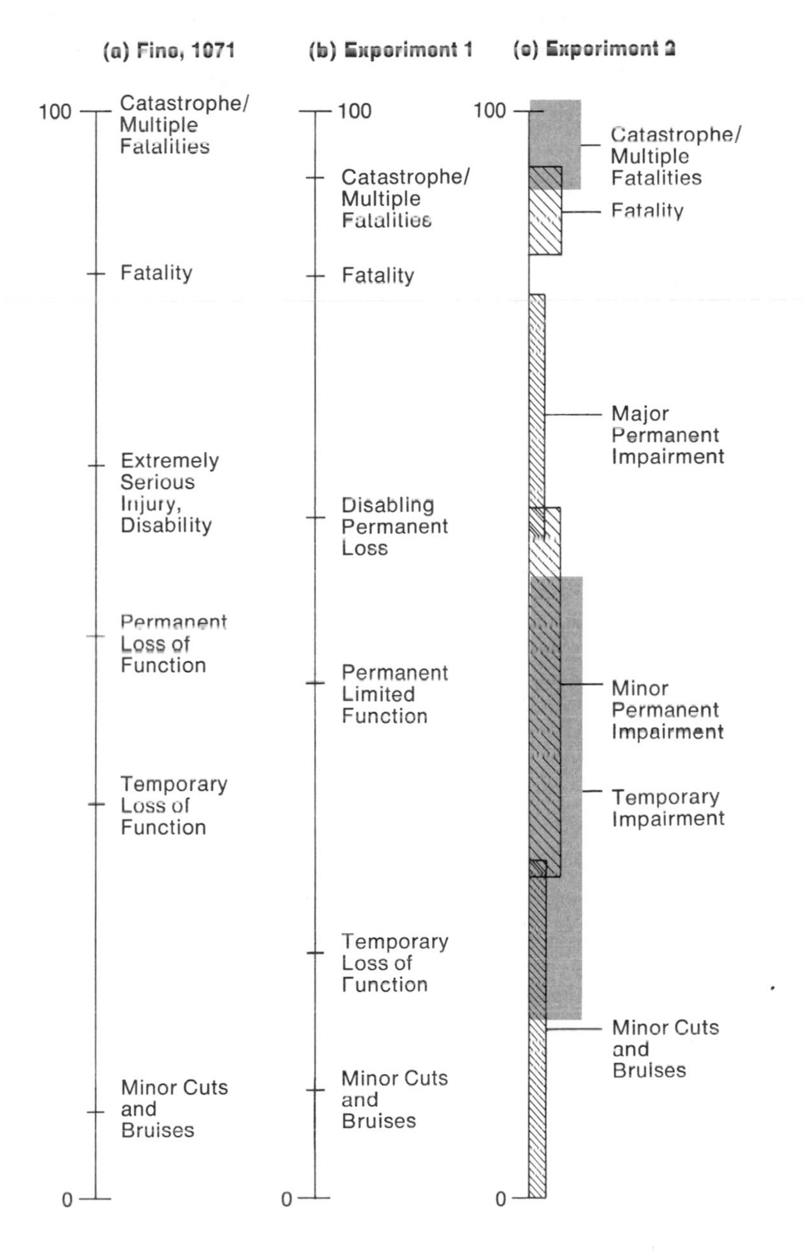

see next page for explanation

Figure VIB–6, cont.
Three interval scales are shown, representing three stages in the development of a consequences scale for a risk assessment study. The scale in part a is based on one published by Fine (1971). In this scale the word descriptors are spaced equally, except for the "Minor Cuts and Bruises" category. Because people develop their own assumptions of how different, or far apart, two descriptions are, 20 engineers and technicians were asked to locate each word descriptor on a vertical scale marked from 0 to 100 (low to high). The average responses for the group are shown in the scale in part b. This scale differs from Fine's scale primarily at the ends of the range; those descriptors are placed closer together in the new scale. From the results of the first scaling experiment (part b), some of the descriptors were changed and another experiment was done with 18 safety engineers who each located the descriptors on a scale. The average values and their standard deviations, a measure of response variability shown as shading around each descriptor's average position, are given in the scale in part c. There is still considerable overlap in different people's placement of "Temporary Impairment" and "Minor Permanent Impairment." Scaling experiments like these can help to define the significance of differences in subjective opinion when more than one person is using a scale to rate a job or situation.

Figure VIB–7: Paired-Comparison Study of Mail Routes (Adapted from S. H. Rodgers and F. E. Whispell, 1979, Eastman Kodak Company.)

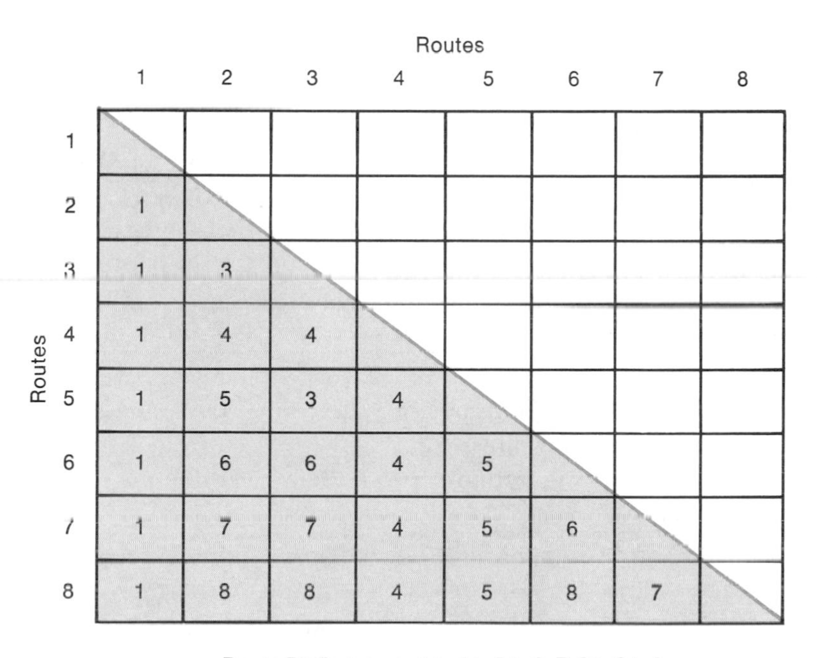

Routes	1	2	3	4	5	6	7	8
1								
2	1							
3	1	3						
4	1	4	4					
5	1	5	3	4				
6	1	6	6	4	5			
7	1	7	7	4	5	6		
8	1	8	8	4	5	8	7	

Route Preferences: 1 > 4 > 5 > 6, 7, 8 > 3 > 2

Eight routes in a mail delivery operation were compared, two at a time, and the most acceptable one was recorded in a two-way table. Routes are listed across the top and down the side of this table. When route 5 was compared with route 3 for overall demands, for example, route 3 was preferred, as indicated in the box formed by the intersection of row 5 and column 3. The overall rating for job effort level was analyzed, as well as the physical effort and time pressure, in three separate, paired rating trials. The pairing of routes for preference ratings helps define very small differences between them. An interval scale can be constructed from these data. When similar comparisons for overall effort, time stress, and muscular effort are used, approaches for reducing stress on the routes can be identified.

(2) *Use of the Scale in the Plant*

Once the interval scale has been developed, it should be administered in the plant in a consistent manner. This is more probable if the following is done:

- Write out instructions for using the scale, and read them to each person doing the scaling.
- Try out the instructions on a small group beforehand to anticipate any questions that may arise. Incorporate the answers to these questions in the revised instructions.
- Clearly mark the limits of the scale—that is, the lowest and highest values—and define them. Be consistent and follow population stereotypes in the assignment of scale values (see Figure VIB–8a). Higher values should be at the top of a vertical scale and at the right of a horizontal scale.
- Have the words move in a positive direction. For example, on an attitude scale one should move from pain to pleasure, not from pleasure to pain, when moving from the low to the high end of the scale (see Figure VIB–8b).
- If the scale is to measure sensations (such as loudness, brightness, pain, or force), define the extremes with examples or demonstrations. For instance, if a level of force is being quantified, have the person do a maximum handgrip test to identify the top of the scale and relax fully to simulate the low end.
- Ask the person doing the scaling to write down any assumptions he or she made if there is some question about how to interpret the scale.

With interval scales the surveyor can get group average responses and variability (mean ±one standard deviation) and can statistically determine the significance of the differences among two or more situations or items. The data can be displayed to show percentiles and can be correlated with other characteristics of the workplace or worker. For instance, the differences in acceptability of several manual-handling tasks can be evaluated in relation to each other as well as in relation to the rater's height, weight, age, or years of experience on the job (Snook, Irvine, and Bass, 1970).

c. Rating of Perceived Exertion

The subjective perception of physical work has been studied by using scales developed by Borg (Borg, 1962, 1970, 1980; Borg and Linderholm, 1967); this scale is called the Rating of Perceived Exertion (RPE). It is based on absolute heart rate levels for dynamic work, ranging from 60 to 200 but scaled from 6 to 20. A more recent psychophysical scale with ratio properties has been devel-

Figure VIB–8: Population Stereotypes in Scales (Adapted partly from McCormick and Sanders, 1982.)

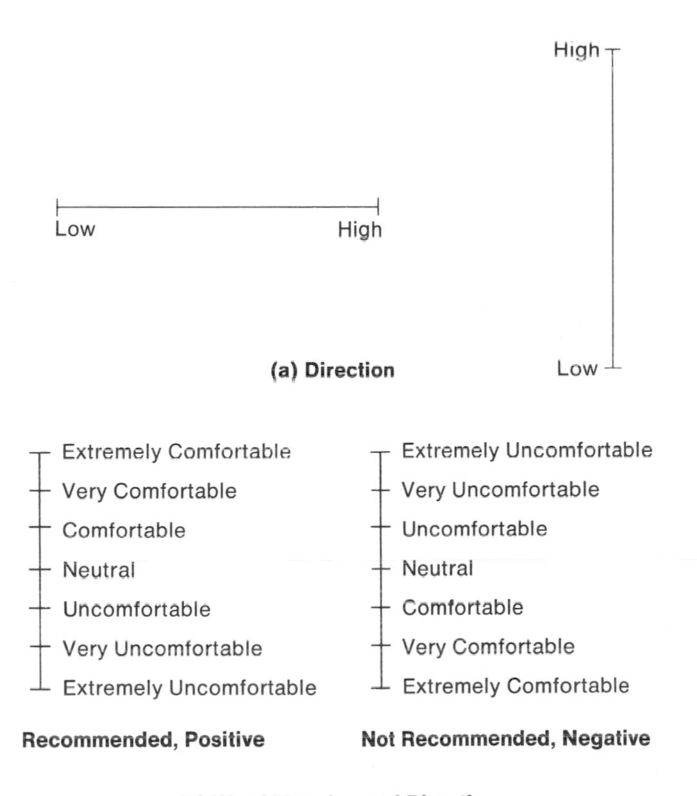

Two types of stereotypes for scales are shown. Direction for quantitative scales is shown in part a, in horizontal and vertical orientation and marked from low to high. Word meaning and direction for vertical scales are shown in part b. These scales should move in a positive direction, with the highest values being associated with the most desirable sensations, as shown on the scale at the left.

oped by Borg (1980) to make interindividual comparisons in physical effort ratings. Figure VIB–9 illustrates the earlier RPE scale (part a) and the newer category scale with ratio properties (part b). The category scale has been tested and found to be a reliable method for differentiating work loads with large muscle groups.

The subjective ratings of the scales are most closely correlated with heart rate, as intended. The subjective rating integrates both the local muscle and

Figure VIB–9: The Rating of Perceived Exertion (RPE) Scales (After Borg, 1980.)

(a) Rating of Perceived Exertion, 1962; Whole-Body Effort	(b) Category Scale for Rating of Perceived Exertion, 1980; Large-Muscle-Group Activity
20	
19 Very, very hard	· Maximal
18	10 Very, very strong (almost max)
17 Very hard	9
16	8
15 Hard	7 Very strong
14	6
13 Somewhat hard	5 Strong (heavy)
12	4 Somewhat strong
11 Fairly light	3 Moderate
10	2 Weak (light)
9 Very light	1 Very weak
8	0.5 Very, very weak (just noticeable)
7 Very, very light	0 Nothing at all
6	

(a) Rating of Perceived Exertion, 1962; Whole-Body Effort

(b) Category Scale for Rating of Perceived Exertion, 1980; Large-Muscle-Group Activity

These scales are designed for use in collecting opinion data about the level of effort in whole-body effort (part a) and in large-muscle-group activity (part b). The scale values in part a, when multiplied by 10, are roughly equivalent to an average heart rate for a healthy person doing that subjective level of work. The scale values in part b, when multiplied by 10, are roughly equal to the percentage of maximum muscular strength (%MVC) that a given effort requires. The descriptors in scale b relate to muscle work, with whole-body-effort descriptors included in parentheses. Scales of perceived exertion can be used to quantify subjective opinions about the level of effort a given task requires. By subjectively quantifying the effort of several different muscle groups, the surveyor can identify which group is likely to fatigue first by looking at the level of the rating given. The scales shown here have been changed from Borg's original ones only by making them conform to the vertical scale stereotype for word scales shown in Figure VIB–8(a). Borg's scales move from high values at the bottom of the scale to small values at the top.

joint and the central, or systemic, stress sensations. Recent research has been devoted to trying to determine the relative importance of each (Cafarelli, 1977; Pandolf, 1978; Sargeant and Davies, 1973). In one technique the person doing the task is asked to maintain a constant level of effort; for example, a person is asked to maintain a heavy-effort handgrip while an objective measure of the force exerted is recorded (Figure VIB–10). The resulting falloff of strength illustrates the relationship between intensity of effort and duration of holding.

Figure VIB–10: Force Changes with Constant RPE Handgrip
(Adapted from Cain, 1973.)

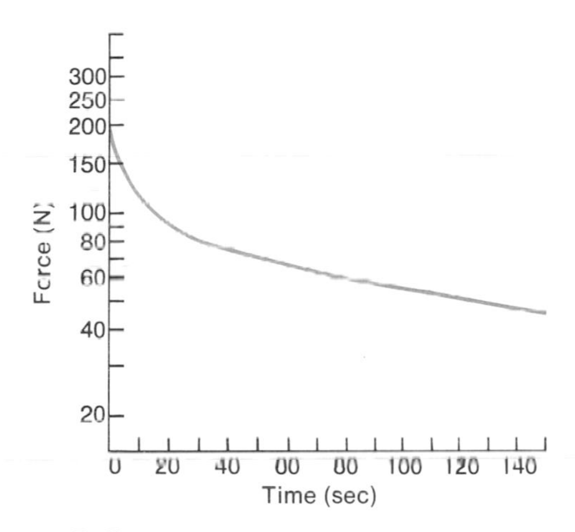

The measured change in force, in newtons on the vertical axis, over time, in seconds on the horizontal axis, is shown for people maintaining a constant level of subjective effort (RPE, or rating of perceived exertion) on a handgrip task. The averages of the responses of eight young men are shown. As the grip muscles fatigue, the subject feels that he is working just as hard, even though considerably less force is being exerted on the handgrip dynamometer than was exerted at the beginning of the task. The rating of perceived exertion for muscular work is related to more than simple muscle force levels; circulatory changes and joint receptors are probably also involved.

SECTION VI C. APPENDIX C: PROBLEMS WITH SOLUTIONS

A productive application of human factors/ergonomic principles to the design or redesign of workplaces, jobs, and working environments requires creative application of information in the literature. The more experience one gathers in field applications, the more apparent it becomes that each problem has to be analyzed on its own merits and that many solutions are possible. The usual limit to what can be done to resolve a problem is imagination. This must be tempered by the very real modifiers of cost, management commitment, expected life of the workplace or task, and individual accommodation possibilities in order to come up with the best solution for the problem.

In this section problems typical of those seen in industry are analyzed by using information contained in previous chapters. The primary human factors principle of concern is identified after each problem is described, and the methods for evaluating the human capacity or capability data in relation to the job demands are outlined. Although solutions for the problems are given, similar problems in different conditions may suggest other approaches. The reader should use these problems to practice human factors/ergonomics analysis and to become familiar with the book's contents.

A large number of the questions received from manufacturing areas involve workplace design and environmental stresses. Consequently, this section includes more problems for these areas than for any others. The problems encompass both new design and redesign of existing workplaces or jobs. For new design the ergonomics practitioner can often use directly the guidelines given in this book. For existing design the book's guidelines may be used to evaluate how close the situation is to the recommended design. To determine how many people will be affected by that design, or to balance the cost of redesign with the possible impact of the situation on people's safety and health, the human factors practitioner must combine information about people's capabilities with evaluations of job demands. This task is complicated, but this chapter provides information and methods to accomplish it. The problems in this section show how anthropometric data can be used to evaluate workplaces needing redesign.

1. WORKPLACE DESIGN

a. Problem 1—Seated Workplace

How can a bottle inspection and bottle-packing workplace be modified to reduce complaints of sore arms and shoulders by operators working for a full shift?

(1) Background

In a bottle inspection and bottle-packing job, operators are lined up on each side of a conveyor belt from which they remove product, inspect it, put it in trays, and remove the defective items. The conveyor mechanism fixes the height of the belt at 100 cm (39 in.) above the floor. The belt is recessed in a 5-cm (2-in.) guardrail to keep the bottles aligned properly. Bench height is 85 cm (33 in.) above the floor; forward reach to procure a bottle is 51 cm (20 in.). Figure VIC–1 illustrates the workplace, showing dimensions and clearances.

Operators have complained of sore arms and shoulders, which they associate with having to reach forward and up to the conveyor each time they procure a bottle. They have the option of either sitting or standing at the workplace; most choose to sit, except for the occasional procuring of supplies and handling of finished trays of product, when it is more convenient to stand.

Figure VIC–1: Existing Design of a Bottle Inspection and Bottle-Packing Workplace

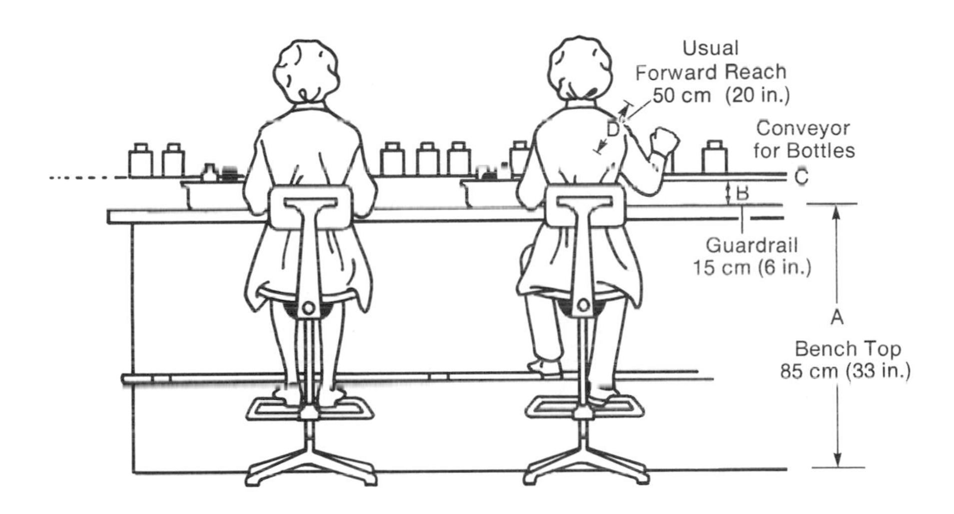

A bottle inspection and bottle-packing line is shown prior to its redesign. The operators are seated at an 85-cm (33-in.) high bench (A) that is 30 cm (12 in.) wide. They reach forward to the conveyor line (B), which is 15 cm (6 in.) above the bench top and about 30 cm (12 in.) wide, to remove bottles for inspection and packing. The usual forward reach is about 51 cm (20 in.). A 5-cm (2-in.) high guardrail (C) keeps the bottles from falling out of the conveyor. The bench top is 5 cm (2 in.) thick, leaving a vertical leg clearance of 80 cm (31 in.). Forward clearance for the legs is 35 cm (14 in.). The chairs can be adjusted up to 63 cm (25 in.) above the floor. Bottles come in at the right side of the line and exit to the left, moving at a rate of six per minute past the inspector. The operators may either sit or stand at the workplace; most prefer to sit. Many of the operators have complained of sore arms and shoulders.

(2) Human Factors Principles and Information Used to Solve the Problem

The sore arms and shoulders suggest that static loading of those muscle groups is occurring. This static loading is probably associated with the need for the operators to raise each bottle out of the conveyor at 18 cm (7 in.) above elbow height and at arm's reach. The discussions of sitting workplaces and sit/stand workplaces in Chapter II, "Workplace Design," indicate that the 85-cm (33-in.) work surface height is too high for a sitting workplace without a footrest and too low for a sit/stand workplace. With the restrictions on thigh clearance and forward leg room in the workplace, the operators cannot easily raise the chair height in order to reduce the load on their arm and shoulder muscles during this activity.

To improve the workplace for the inspection and tray-filling activity, the human factors practitioner can use one of two basic approaches (see Section IIB, "Adjustable Design Approaches"): raise the operator or lower the conveyor. Since the conveyor cannot be lowered, the operator has to be raised; the edge of the work surface makes it impossible to simply raise chair height, so a platform should be provided to raise the workplaces on either side of the conveyor to 100 cm (39 in.). A full floor platform is needed because of the multiple work stations; a low guardrail should be provided to guard against people stepping off of the edge inadvertently.

(3) Conclusion

The following workplace modifications can be made to improve comfort at the bottle inspection and bottle-packing workplace:

- Use a platform to raise the work surface to 100 cm (39 in.).
- Provide a footrest on the chair or workbench to reduce discomfort of the legs.
- If there is enough space for the task, cut a semicircular opening in the workbench at each station to reduce the arm reach requirements.

b. Problem 2—Standing Inspection Workplace

How should an inspection and sampling workplace be designed for a paper-web manufacturing system so that inspector fatigue is minimized? Where should the moving web of paper be located in relation to the inspector?

(1) Background

An inspector stands near the end of a papermaking machine and scans the full 90-cm (35-in.) width of the paper web as it moves by at 0.6 m/sec (120 ft/min). A 50-by-90-cm (20-by-35-in.) sample is cut from the web every 15 minutes, and the two ends of the web are spliced to keep the web continuous.

The work station is located near the end of the papermaking machine so that the web moves across the inspector's field of view from right to left. Paper orientation can be changed as the paper comes out of the machine; it can be raised as high as 190 cm (75 in.) above the floor and then brought down to the windup height of 90 cm (35 in.) at any angle that would be suitable for the visual and sample-cutting demands of the inspection job.

A sample is cut from the web when paper speed has been momentarily slowed to 0.15 m/sec (30 ft/min). The inspector runs a knife across the web two times to remove the 50-cm (20-in.) long sample and splices the two ends of the web together, taking about 3 to 4 seconds. A flat surface is needed to do these tasks.

(2) Human Factors Principles and Information Used to Solve the Problem

In designing the inspector's work station, the practitioner must consider both the visual inspection needs and the sample-cutting and web-splicing requirements. The following sections of the book include information that can be used in developing the design:

"Guidelines to Improve Inspection Performance" in Section IV B.
"Dimensions for Visual Work" in Section II A.
"Standing Workplace Height" in Section II A, subsection 1b.
"The Standing Work Area" in Section II A, subsection 1b.
"Appendix A: Anthropometric Data" in this chapter.

This information indicates that the inspector's work station should include the following design features to reduce the potential for fatigue on both the inspection and the sampling tasks:

- The angle of viewing the web should be downward. There should be 30 cm (1 ft) of viewing distance for every 0.3 m/sec (60 ft/min) of paper-web speed, or a total of 60 cm (24 in.) of viewing area, in this design.
- The information on eye height indicates that the web should be no more than 145 cm (57 in.) above the floor at the point where inspection is done; this height ensures that even the shorter operators will be looking downward in the inspection task. It is also advisable to keep the inspector's downward visual angle at less than 45°. The practitioner can do geometric and trigonometric analyses of the visual requirements, assuming that the inspector can stand about 50 cm (20 in.) away from the web. An example of such an analysis is given below, and an illustration of the visual angle calculation is shown in Figure VIC–2(b).
- A work surface 91 cm (36 in.) above the floor is desirable so that the inspector can exert the necessary force to cut samples in the quality control task. The paper web is 91 cm (36 in.) wide, making the inspector bend from the hips or waist to reach its far side. A work surface for sample-cutting and splicing with a height of 91 cm (36 in.) will allow the inspector to make this bending reach and will also be a convenient height for the splicing operation where force is applied.

Figure VIC–2: Design of a Paper-Web Inspection Workplace

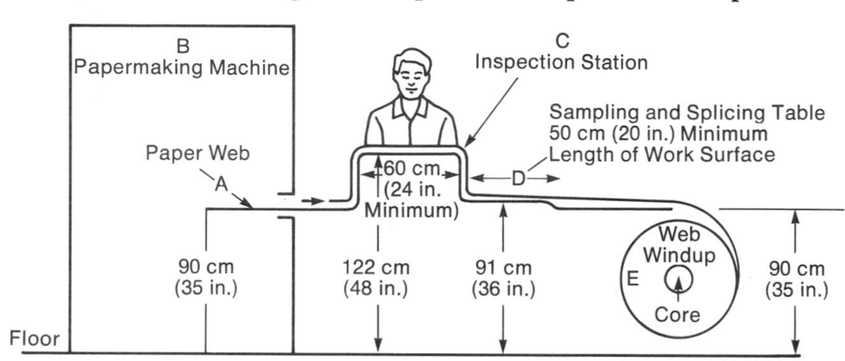

(a) Work Station

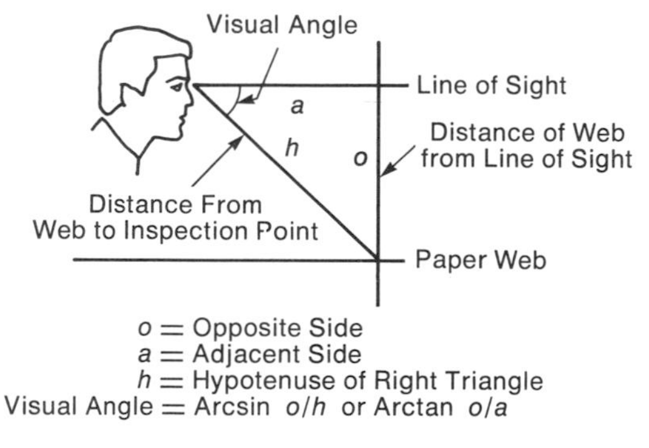

o = Opposite Side
a = Adjacent Side
h = Hypotenuse of Right Triangle
Visual Angle = Arcsin o/h or Arctan o/a

(b) Visual Inspection Demands

In part a the paper web (A) emerges from the papermaking machine (B) at a height of 90 cm (35 in.) above the floor. It is directed up to 122 cm (48 in.) above the floor at the inspection station (C), where at least 60 cm (24 in.) of the web's length is displayed as it moves by the inspector at 0.6 m/sec (120 ft/min). On leaving the inspection station, the paper web is directed back to 91 cm (36 in.) above the floor at the sampling and splicing table (D), where quality control samples are taken every 15 minutes. This work surface is at least 50 cm (20 in.) long. The web moves from the sampling area to the windup area (E), where it feeds onto a core and forms a large roll. The inspector stands and scans the paper web as it moves from right to left across the inspection station. Part b, the viewing angle, illustrates the visual angle calculation described in the text. The sides of a triangle that describes, geometrically, visual inspection of the paper web are defined, using the assumptions given in this design; the opposite side *o,* adjacent side *a,* and hypotenuse *h* of the right triangle are shown. Trigonometric calculations of the visual angle are indicated as the arcsin and arctan functions of the ratios between the sides of the triangle.

There are several ways that this inspection and sampling workplace can be configured. One design is shown in Figure VIC–2.

Since the heights of the paper web as it leaves the machine and at the paper windup are both 90 cm (35 in.) above the floor, the sampling and splicing work surface can easily be placed at that level. Some change in web height is warranted at the inspection station, however. If the inspector is standing about 50 cm (20 in.) away from the web, horizontally, and inspects the paper as it moves across the work station at 90 cm (35 in.) above the floor, the downward visual angle will be almost 60° for tall people. This angle is uncomfortable and could result in degraded performance on the inspection task as the shift progresses. Although the taller inspectors would be most affected if the web height were left at 90 cm (35 in.) above the floor, even shorter operators will exceed the 45° visual angle guideline.

Visual angles can be calculated if some distances are known or assumed (see part b of Figure VIC–2). The distance from the inspector to the web, horizontally, is the adjacent side a of a right triangle; the distance from the line of sight to the web being inspected is the opposite side o of the triangle; the triangle's third side, or distance from the eye to the web, can be calculated by using the formula $h = (a^2 + o^2)^{0.5}$, the Pythagorean theorem. Trigonometry can be used to calculate the visual angle by taking the arcsin of o/h, the ratio between the distance o from the line of sight to the web and the distance h from the eye to the web. The visual angle can also be calculated by using the arctan of o/a or the ratio of the distance from the line of sight to the web, o, and the distance from the inspector to the web, a.

To find the best design for this visual inspection task, the practitioner can define the tangent of a 45° downward visual angle as the maximum ratio between the opposite and adjacent sides of the triangle described in Figure VIC–2b. The tangent of a 45° angle is 1, thus the distance o of the web from the line of sight should be the same as the inspector's horizontal distance from the web, a. Taller people will be most affected by this design, so the calculated distance for o of 50 cm (20 in.) is subtracted from the 95th percentile value for standing eye height to estimate where the web should be placed for inspection. A location 122 cm (48 in.) above the floor is the recommendation. For more than 95 percent of the inspectors the visual angle will be less than 45° if this design is used.

The practitioner might consider displaying the web in the vertical plane, instead of the horizontal, in this inspection task. Because the paper is moving past the inspector at 0.6 m/sec (120 ft/min), at least 60 cm (24 in.) of web length will have to be viewed at one time. Thus the taller inspector's downward visual angle will exceed 45° when the bottom of the web is scanned. If the paper web is displayed at an angle from the true vertical or horizontal planes, a smaller visual angle is needed to scan at the lowest end of the web, thus providing a range of design solutions.

(3) Conclusion

Raise the paper web to a height of 122 cm (48 in.) above the floor for the inspection task and leave at least a 60-cm (24-in.) viewing area, horizontally, for the 90-cm (35-in.) wide web. Drop the web back to 91 cm (36 in.) above the floor on a flat surface for the sampling and the splicing tasks.

2. EQUIPMENT DESIGN

a. Problem 1—Selecting a Pop Riveter

What criteria should be used to select the manual pop riveters needed by personnel in a metal fabrication shop? In the past, many people have experienced sore hands and wrists after frequent use of the tool.

(1) Background

Pop riveters are used to join two pieces of sheet metal together. Figure VIC–3 illustrates several models currently available commercially.

To join two pieces of sheet metal, the mechanic has to insert the rivet through a drilled hole, then squeeze the pop riveter's handles together to expand the end of the rivet. Finally, a large force (often in the range of 225–340 N, or 50–75 lbf) is required to snap the end of the rivet off so that a smooth surface is formed. People with small hands and low hand grip strength have difficulty using many of the pop riveters currently available.

(2) Human Factors Principles and Information Used to Solve the Problem

Section III D, "Hand Tool Selection and Design," includes information from which to develop the needed criteria. Since the goal is to provide personnel with pop riveters that can be used conveniently by people with large and small hands, the following information is relevant:

- handle length
- handle span
- handle shape
- force requirements
- tool weight

The longer the handles, the better the leverage for force exertion will be. If the distance between the handles (span) exceeds 8 cm (3 in.), however, people with small hands or low strength will be at a disadvantage. Handle span should be evaluated both at the initial separation of the handles and at the separation when the heaviest forces must be exerted.

Figure IIID–8 (Section III D) shows how strength is affected at different handle spans. The figure can be used to evaluate the impact of a given pop riveter on the user population. The initial distance between the handles can be similarly evaluated by comparing the hand spread data (in measurements 36 and 37 of Tables VIA–2 and VIA–3) with the initial spans of several pop

Figure VIC–3: Examples of Pop Riveters

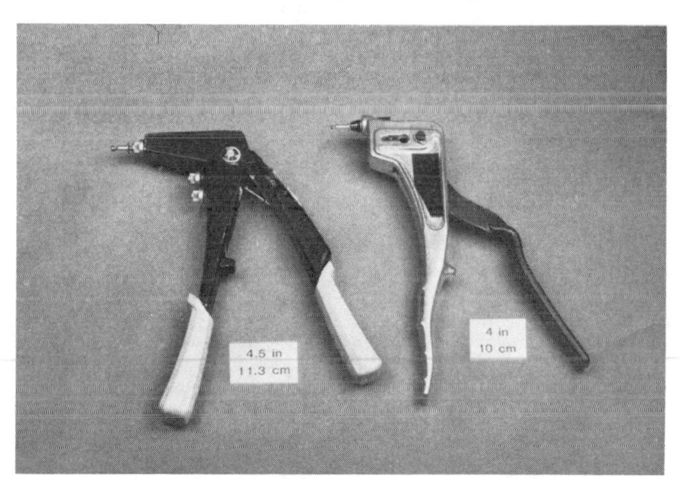

(a)

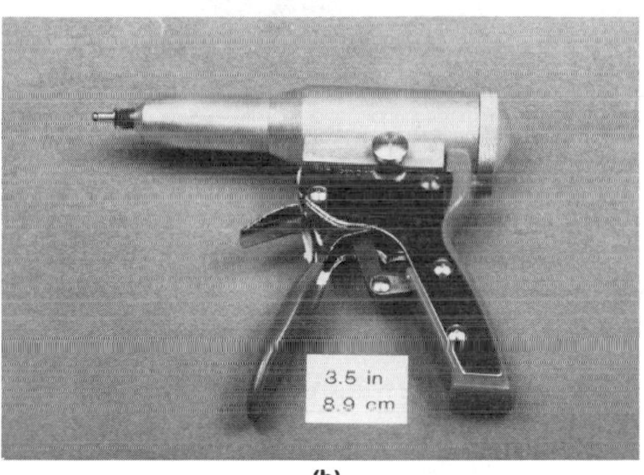

(b)

Three examples of commercially available pop riveters are shown. The position of their handles indicates the hand span needed when force is first applied to drive a rivet into place on two pieces of metal. The pop riveter on the left in part a requires the greatest hand span, while that in part b is close to the recommended grip span for force exertions that will accommodate most people. Handle shapes differ greatly, from the finger recesses on the forward handle of the riveter on the right in part a to the pistol grip of the riveter in part b. Both riveters in part a have stops on the forward handle to reduce the risk for finger pinch injuries. The pop riveter on the right in part a is the lightest of the three shown but the finger recesses in its forward handle are not recommended. Although none of these riveters have all of the desirable features, the one in part b is best for people with small hands; they would have to use both hands to operate the other pop riveters because of the wide handle spans.

riveters. The latter are frequently greater than 10 cm (4 in.); this span exceeds the comfortable hand spread of about half of the potential work force.

Safety stops (shown in Figure IIID–6) are desirable additions to any two-handled, force-requiring tool. Handle surfaces should not be too slippery, nor should they have sharp edges or finger recesses that affect the way a person can grasp them for maximum force exertion.

(3) Conclusion

The criteria for selection of pop riveters include the following:

- Handle span at maximum force exertion should be as close as possible to 8 cm (3 in.).
- Initial handle span should not exceed 12 cm (5 in.) and preferably should be close to 8 cm (3 in.).
- Handles should be curved to fit the palm. Handle length should be no less than 10 cm (4 in.).
- Handle texture should provide some frictional resistance but not have sharp edges or finger recesses.
- Safety stops should be provided between the handle parts to protect the hand and fingers from pinching.
- Tool weight should be as low as is consistent with durability and stability needs (usually less than 0.5 kg, or 1 lb).

Since it is unlikely that any one pop riveter will be found that meets all of these criteria, the nearest match usually has to be selected.

b. Problem 2—Safety Shower Handle Height

How high should a safety shower handle be if it is located in an aisle? Shorter workers have had difficulty reaching the handles now in use.

(1) Background

In areas where people are working with chemicals, it is desirable to have safety showers as close as possible to the work station. Locating the shower in an open aisle near a work station ensures that there will be easy access to it. Also, other people in the area will be immediately aware when a person has had a chemical contact incident and can come to his or her aid. If the shower is to be located in an aisle, the handle that activates it should be low enough to be reached by people with short overhead reach capability but high enough not to strike people walking in the aisle in the head or eyes.

(2) Human Factors Principles and Information Used to Solve the Problem

Anthropometric data, specifically overhead reach, stature, and standing eye height, can be used to find the best safety shower handle height. The overhead reach data of interest is that for the low end of the distribution, and the clearances, or stature and eye height, are those for the high ends of the distributions, as described in Chapter I in the section titled "Criteria for Design: Who?" In addition, the practitioner should consider what a person can do, such as standing on tiptoe, to accommodate a design that is not optimal for all people.

From information in Tables VIA–2 and VIA–3 of Appendix A the practitioner can calculate the 1st and 99th percentiles by using the standard deviations (S.D.) of the mean values; this calculation is done by subtracting the standard deviation from the 5th percentile value and adding it to the 95th percentile value. Then the following data can be assembled:

Measurement	Percentiles (percentages of population with measurements the same as or less than the values given in the table)		
	1	5	50 (the Mean)
1. Functional Overhead Reach, Cm (In.)			
a. Feet Flat on the Ground, cm (in.)	180 (71)	188 (74)	204 (80)
b. On Tiptoe, cm (in.)	188 (74)	196 (77)	212 (83)

	Percentiles		
	50 (the Mean)	95	99
2. Stature, cm (in.)	168 (66)	183 (72)	190 (75)
3. Eye Height, Standing, cm (in.)	158 (62)	172 (68)	178 (70)

From this analysis the following observations can be made:

- All but 1 percent of the population will be accommodated if the handle is located 180 cm (71 in.) above the floor. The shorter workers will have to stand on tiptoe to reach the handle.
- A handle that is 180 cm (71 in.) high represents a potential collision hazard to the eyes of 1 percent of the population. More than 5 percent of the population would be likely to strike it with their head if they walked straight into it.

People who need to activate the shower handle may have received chemical splashes on the face, restricting their vision in some cases. Thus it is more important to have the handle within an easy grasping height than to raise it to reduce the potential for collision incidents. There are other ways to deal with the collision hazard, such as the following:

- Use a longer and horizontal linkage for the shower activation valve so that the vertical handle can be moved out of the aisle (see Figure VIC–4). Then the handle can be located at about 150 cm (59 in.) from the floor.
- Locate the handle 204 cm (80 in.) above the floor. Attach a 90-cm (35-in.) cord, such as a piece of clothesline, to the handle. This extension permits the shower to be activated at any height from about 115 to 205 cm (45 to 80 in.) above the floor. The cord should be long enough to prevent its becoming an eye collision hazard, and it should be marked for easy detection. It should not interfere with the movement of handling equipment or other vehicles used in the area.
- In new facilities the safety shower should be integrated into the total workplace design so that it does not have to be located in an aisle.

Figure VIC–4: Safety Shower Handle Height

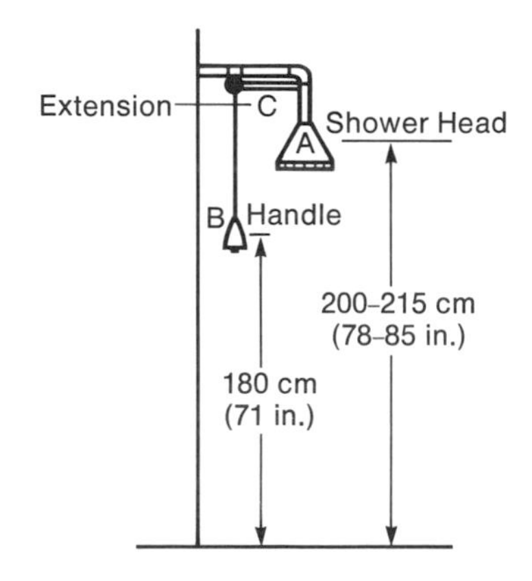

A safety shower installation is shown. The shower head (A) is located from 200 to 215 cm (78 to 85 in.) above the floor. The handle to activate the shower (B) is shown with a horizontal extension (C) to move it over toward the wall and out of the aisle. Handle height is set at 180 cm (71 in.) above the floor, but it could be lower if it is located well out of the traffic pattern in the aisle.

(3) Conclusion

Locate the handle at a height of 180 cm (71 in.) above the floor, and mark it very clearly as a potential collision hazard in the aisle. Place the handle as far as possible from the heavy-traffic parts of the aisle.

3. INFORMATION TRANSFER

a. Problem—Inventory Control Card Design

How can the presentation of information on an inventory control computer card be improved to reduce the opportunity for making errors in an order-picking task?

(1) Background

Inventory control computer cards are used by warehouse order pickers to fill small orders. The information on each card is formatted for computer needs. It includes information on the manufacturing history of the product, which is important for inventory control but is not needed by the people who do the order picking. Figure VIC–5 illustrates the original control card.

Errors have occurred in which one product is substituted for a similar one during order filling. The warehouse order pickers have indicated that it is difficult to read the control cards and to remember the important codes when they are looking for a product among many similar ones.

(2) Human Factors Principles and Information Used to Solve the Problem

The two strings of numbers across the top of the card reduce reading reliability by demanding the order picker to handle too much information. The discussions on coding, forms, and labels and signs in Section IV A should be consulted for information pertinent to this problem. For example:

- Product code information should be separated into two 4-digit numbers to aid the order picker. The lower line should be divided according to the uniqueness of each set of numbers.
- The space between the lines of numbers should be increased, and unnecessary preprinted lines should be removed.
- A dark, distinctive type of font should be used.
- Any unnecessary information should be removed.
- Lowercase letters should be used on preprinted words, where possible.
- Keypunch holes should not obliterate the coded information.
- The numbers used for order picking should be large enough to be read by the order picker while in the process of procuring a product from a shelf so that the number can be checked, if necessary. A number height of 7 to 13 mm (0.2 to 0.5 in.) should be adequate for most situations.

Figure VIC–5: Example of An Inventory Control Computer Card Used for Order Picking

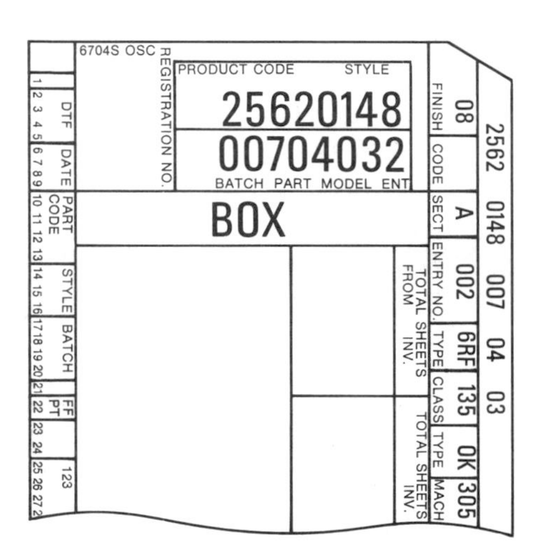

The existing, or current, format of an inventory control card is shown. The numbers in the boxes at the top of the card indicate the product code, style, batch, part, model, and entry values. These numbers describe the product and permit accurate order filling. The same information is found on the right edge of the card. The second column on the right gives additional information that is used for inventory control but not for order picking. The left margin of the card includes computer-coded descriptions of the product; it also is not used in order picking. The close packing of the numbers at the top of the card make it difficult to find and remember the pertinent information when a person is filling orders from these cards.

(3) Conclusions

A recommended improvement to the inventory control card is shown in Figure VIC–6. This card should reduce the opportunity for errors in order picking since it has the following features:

- An uncluttered appearance.
- Dark, easy-to-read numbers.
- Distinct numbers that are not easily confused.
- Shorter strings of numbers to remember.

Figure VIC–6: Recommended Changes to Control Card Used in Order Picking

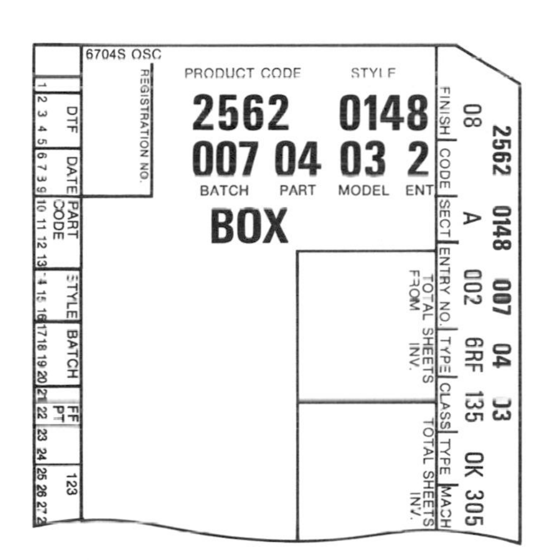

The recommended, or proposed, changes in format for the inventory control card in Figure VIC–5 are shown. In comparison with the original card, this one has a more open format, with fewer lines around the data at the top and on the right side. The numbers of interest to the order picker are now separated by spaces, and no more than four digits are grouped together. This arrangement improves the comprehensibility of the card information and reduces the demands on the order picker to get the proper information without making an error by reading the wrong set of numbers. The information of less importance in order picking has not been altered in this proposed format.

4. ENVIRONMENT

a. Problem 1—Warning Alarm Characteristics

How loud should a warning alarm be to indicate when a packaging machine is about to start up?

(1) Background

A packaging machine occasionally has to be shut down in order to clear a jam or make an adjustment. The shutdown may take only a few seconds or several minutes, depending on what has to be done. So that the equipment is not turned on when someone is working on it, the machine has a safety interlock on the start button. An auditory warning signal is being considered as an additional safety device. Injuries to the hands, caused by being caught in or between parts of the equipment, may be prevented if an effective warning alarm is provided.

(2) *Human Factors Principles and Information Used to Solve the Problem*

The auditory warning signal should be loud enough to be heard above the ambient noise level, but it should not be so loud or so similar in its sound characteristics that it can be confused with an emergency evacuation alarm for the area. In addition, its tone should not be annoying, since it will be activated frequently during a shift.

Section III B on displays indicates that an auditory warning signal should be discriminable from, and about 10 dB above, the ambient noise level. In a survey of the work area along the packaging machine when it was shut down, the following sound levels were measured:

Loading end: 72 dBA.
Cartoning machine area: 75–80 dBA.
Packing area: 72–75 dBA.

Since the signal will have to be heard all along the machine, warning systems should be mounted at several locations, especially those where operators are likely to be after a shutdown. The system should have the capability of producing a signal of up to 90 dB at the listener's location, so that it can be at least 10 dB above the background noise.

The discussion of VDU terminal alarms in Section III B indicates that signal frequency should be in the range of 200–1000 Hz; a warble of 1–8 Hz can be added to improve detectability. The signal should last no longer than a few seconds so that it is not confused with the 10-second or more duration of a building evacuation alarm (see the discussion on evacuation alarms in Section V B, subsection 1d).

(3) *Conclusion*

The warning signal characteristics should be in the following ranges:

Frequency: 200–1000 Hz, with an optional 1–8 Hz warble.
Sound level: 90 dB at 5 meters (16 ft) between the alarm and the listener.
Duration: up to 5 seconds.

Warning devices should be provided near each place where people are likely to be when the packaging machine is shut down.

b. Problem 2—Noise Levels in Office Area

People working in an office adjoining a precision-grinding production area have complained that the noise level is excessive and makes their work difficult. What noise levels would be appropriate for the office workers?

(1) *Background*

The office is a partitioned space on the edge of a precision-grinding production area. The people in the office are primarily engaged in developing computer software for the production process. They have frequent meetings and phone calls during the day.

Measurements of the noise levels were made with a precision sound-level meter, using octave-band analysis. These measurements were taken on a typical production day for the precision-grinding department. The office noise levels were as follows:

Center Frequency of Octave Band (Hz)	Sound Pressure Level (dB)
31.5	68
63	70
125	76
250	74
500	74
1000	72
2000	67
4000	63
8000	59

The nonweighted overall noise level is 80 dB.

(2) Human Factors Principles and Information Used to Solve the Problem

The discussion on noise in Section V B contains information needed to evaluate the problem. Because the noise level is below 85 dB, hearing loss is not of concern. The problem is more one of distraction, annoyance, and communications interference from the adjacent room's noise. Table VB–2 indicates that noise levels around 80 dB are not acceptable for telephone or other conversations. An overall noise level of 50 dB would be more appropriate for the type of work being done in the office area. This level would be roughly the same as a preferred noise criterion (PNC) curve of 55. A comparison of the PNC–55 sound levels, taken from Figure VB–2, and those measured in the office area follows.

Center Frequency of Octave Band (Hz)	Measured Sound Levels (dB)	PNC–55 (dB)	Measured PNC–55
31.5	68	73	−5
63	70	70	0
125	76	66	+10
250	74	62	+12
500	74	59	+15
1000	72	55	+17
2000	67	51	+16
4000	63	48	+15
8000	59	48	+11

Except in the lowest frequency bands, the measured noise exceeds the PNC–55 curve; this curve should not be exceeded in offices where people have to communicate frequently.

For further definition of the degree of interference of the ambient noise level with communications, the Preferred Speech Interference Level (PSIL) of the office area noise can be calculated; it is the simple average of the sound levels in the 500-, 1000-, and 2000-Hz bands, or (74 + 72 + 67)/3 = 71 dB. From Figure VB–4 it is clear that a PSIL of 71 dB (the upper horizontal scale), with a distance between speaker and listener (vertical scale) of 1.2 m (4 ft), requires the speaker to use a very loud or a shouting voice in order to be heard. Telephone conversations are also very difficult to hear.

For an improvement of hearing conditions in the office area, the noise should be reduced to a PSIL of about 55, or approximately 62 dB. This level is in keeping with the PNC–55 guidelines for noise in the frequency bands below 1000 Hz. The methods for accomplishing this reduced level might include engineering approaches that increase the attenuation of high-frequency noise on the precision-grinding equipment and the use of additional noise insulation materials or techniques to further isolate the office from its adjoining production area.

(3) Conclusion

The office noise level is too high for effective communication during normal grinding operations in the adjacent area. Local noise insulation techniques should be used to reduce the overall sound level to achieve an office PNC of 55. Attenuation of some of the high-frequency noise associated with the grinding equipment could also be investigated so that the noise levels can be reduced in both areas.

c. Problem 3—Humid Heat Exposure

Oil and dirt are cleaned from industrial vehicles, such as forklift and pallet trucks, by using a steam wand. During the cleaning an operator spends 15 to 30 minutes in the hot and very humid room. Is the steam-cleaning room too hot for people to work in? What is an acceptable exposure time?

(1) Background

The cleaning area is open at one end and has one overhead ventilation duct. The three walls do not have ventilation openings. In the summer months conditions are more uncomfortable because initial room temperatures are higher. Although no one operator is in the area for more than 2 hours during a shift, and exposure is broken into periods of 15 to 30 minutes, heat and humidity accumulate and increase operator discomfort as more trucks are cleaned.

Measurements of the heat and humidity during steam cleaning were made in the fall (see the discussion of temperature, humidity, and air flow measurements in Appendix B). The results are shown following:

Time of Day	Phase of Cleaning Operation	Temperature		% Relative Humidity
		°C	°F	
8:30 A.M.	Outside of Cleaning Room	22	72	50
9:00	Before Cleaning	24	75	50
9:30	After 15 Minutes of Cleaning	30	86	≈100
10:00	30 Minutes After Cleaning Done	28	83	80
after four cleaning sessions				
3:00 P.M.	Before Cleaning	27	80	60
3:30	After 15 Minutes of Cleaning	33	92	≈100
4:00	30 Minutes After Cleaning Done	31	88	85

Air velocity in the cleaning room was measured at several locations; it ranged from 0.1 to 0.2 m/sec (20 to 40 ft/min). Workload was estimated to be about 140 W (120 kcal/hr).

(2) Human Factors Principles and Information Used to Solve the Problem

Tables VD–4 and VD–7 (Section V D) help in the assessment of the severity of the heat and humidity load. Table VD–4 shows the maximum work load for a 2-hour continuous exposure. It indicates that at about 32°C (90°F) and high humidity (more than 80 percent), only light work can be tolerated in the cleaning room. Above this temperature and humidity, a 2-hour exposure would not be recommended. Table VD–7 should be consulted to find the maximum duration of exposure. The analysis is summarized below.

Temperature		% Relative Humidity	2-Hr Maximum Work Load (From Table VD–4)	Acceptable Duration (From Table VD–7)
°C	°F			
22	72	50	Very Heavy	>2 hr
24	75	50	Very Heavy	>2 hr
30	86	100	Light to Moderate	Up to 2 hr
28	83	80	Moderate to Heavy	>2 hr
27	80	60	Very Heavy	>2 hr
33	92	100	Not Recommended to Light	About 1 hr
31	88	85	Light	>2 hr

The work load in this job was between light and moderate. Once the cleaning room temperature exceeds 33°C (92°F) at relative humidities above 80 percent, exposure time has to be shortened to reduce the potential for severe heat discomfort. Because it took only 15 to 30 minutes per truck, exposure times seldom came near the limits.

Since the readings were taken in the fall, and discomfort is greater in the summer months, techniques to reduce the heat and humidity load were evaluated even though the health risks were very small. Because of the nature of the operation, temperature and humidity cannot be reduced during the cleaning operation. In the discussion on techniques to reduce discomfort in the hot and cold discomfort zones in Section V D, the effectiveness of altering several environmental factors on the level of heat discomfort is described. In this case, increasing the air flow is the most effective way to reduce discomfort; when more outside air is pulled through the room during the cleaning periods, the humidity effect is lessened, and the rate of recovery of the temperature and humidity after cleaning can also be improved.

(3) Conclusion

Although the risk of health effects for people doing the steam-cleaning operation for 15 to 30 minutes is very small, increasing the flow of outside air through the room by increasing air velocity from 0.1 to 0.5 m/sec (20 to 100 ft/min) should reduce the heat discomfort considerably.

REFERENCES FOR CHAPTER VI

ANSI. 1982. "Human Factors." Chapter 13 in *Z94.0: Industrial Engineering Terminology*—New York: American Society of Mechanical Engineers (ASME), American Institute of Industrial Engineers (AIIE), in preparation.

ASQC. 1973. *Glossary and Tables for Statistical Quality Control.* Milwaukee: American Society for Quality Control, Statistics Technical Committee, Table 1, pp. 50–53.

Bainbridge, L., and R. J. Beishon. 1964. "The Place of Checklists in Ergonomics Job Analysis." In *Proceedings of the 2nd I.E.A. Congress, Dortmund,* Ergonomics Proceedings Supplement. London: Taylor & Francis, 579 pages.

Borg, G. A. V. 1962. *Physical Performance and Perceived Exertion.* Lund, Sweden: Gleerups.

Borg, G. A. V. 1970. "Perceived Exertion as an Indicator of Somatic Stress." *Scandinavian Journal of Rehabilitation Medicine, 2:* pp. 92–98.

Borg, G. A. V. 1980. "A Category Scale with Ratio Properties for Intermodal and Interindividual Comparisons." Paper presented at the International Congress of Psychology, Leibig, West Germany, 1980.

Borg, G. A. V., and H. Linderholm. 1967. "Perceived Exertion and Pulse Rate During Graded Exercise in Various Age Groups." *Acta Medica Scandinavica, 187:* pp. 17–26.

Botsford, J. H. 1971. "A Wet Globe Thermometer for Environmental Heat Measurement." *American Industrial Hygiene Association Journal, 32:* pp. 1–10.

Botsford, J. H. 1973. "Noise Measurement and Acceptability Criteria." Chapter 25 in NIOSH (1973), pp. 321–331.

Broch, J. T. 1973. *Acoustic Noise Measurements, The Application of Bruel and Kjaer Measuring Systems.* 2nd ed. Published by Bruel and Kjaer. Naerum, Denmark. 199 pages.

Burger, C. E., and J. R. DeJong. 1962. "Aspects of Ergonomics Job Analyses." *Ergonomics, 5:* pp. 185–201.

Cafarelli, E. 1977. "Peripheral and Central Inputs to the Effort Sense During Cycling Exercise." *European Journal of Applied Physiology, 37:* pp. 181–189.

Cain, W. S. 1973. "Nature of Perceived Effort and Fatigue: Roles of Strength and Blood Flow in Muscle Contractions." *Journal of Motor Behavior, 5(1):* pp. 33–47.

Chapanis, A., 1975. *Ethnic Variables in Human Factors Engineering.* Based on a symposium held at Oosterbeck, The Netherlands, June 19–23, 1972, under the auspices of the Advisory Group on Human Factors, NATO. Baltimore: Johns Hopkins University Press, 290 pages.

Clauser, C. E., P. E. Tucker, J. T. McConville, E. Churchill, L. L. Laubach, and J. A. Reardon. 1972. *Anthropometry of Air Force Women.* AMRL-TR-70-5. Wright-Patterson AFB, Ohio: Aerospace Medical Research Labs, 1157 pages.

Damon, A., H. W. Stoudt, and R. A. McFarland. 1966. *The Human Body in Equipment Design.* Cambridge, Mass.: Harvard University Press, 360 pages.

Department of Commerce. 1982. *Local Climatological Data–Rochester, N.Y.* Washington, D.C.: National Oceanic and Atmospheric Administration Environmental Data Service, monthly reports.

Edholm, O. G. 1967. *The Biology of Work.* New York: McGraw-Hill, pp. 228–237 (developed by the International Ergonomics Association, 1965).

Fine, W. T. 1971. "Mathematical Evaluations for Controlling Hazards." *Journal of Safety Research, 3:* pp. 157–166.

Freund, J. E. 1967. *Modern Elementary Statistics.* 3rd ed. Englewood Cliffs, N.J.: Prentice-Hall, pp. 172–178.

Garrett, J. W. 1971. "The Adult Human Hand: Some Anthropometric and Biomechanical Considerations." *Human Factors, 13 (2):* pp. 117–131.

Grunhofer, H. J., and G. Kroh, eds. 1975. *A Review of Anthropometric Data of German Air Force and United States Air Force Flying Personnel 1967–1968.* AGARDograph No. 205. Advisory Group for Aerospace Research and Development, North Atlantic Treaty Organization.

Hertig, B. A. 1973. "Thermal Standards and Measurement Techniques." Chapter 31 in NIOSH (1973), pp. 413–429.

Hertzberg, H. T. E., G. S. Daniels, and E. Churchill. 1954. *Anthropometry of Flying Personnel, 1950.* WADC Tech. Rpt. 52-321. Wright-Patterson AFB, Ohio: Wright Air Development Center, 134 pages.

Kaufman, J. E., and J. F. Christensen, eds. 1972. *IES Lighting Handbook.* 5th ed. New York: Illuminating Engineering Society, 726 pages.

McCormick, E. J., and M. S. Sanders. 1982. *Human Factors in Engineering and Design.* 5th ed. New York: McGraw-Hill, 512 pages.

Metrosonics. 1981. *Metrologger System, db-301/651.* Brochure from Metrosonics, Inc., Box 18090, Rochester, N.Y., 14618.

NASA. 1978. *Anthropometric Source Book.* Volumes I–III (Reference Publication 1024). Edited by the staff of the Anthropology Research Project, Webb Associates. Yellow Springs, Ohio: NASA Scientific and Technical Information Office, 1167 pages.

NIOSH. 1972. *Occupational Exposure to Hot Environments, Criteria for a Recommended Standard.* Cincinnati: U.S. Department of Health, Education, and Welfare, National Institutes for Occupational Safety and Health, 104 pages.

NIOSH. 1973. *The Industrial Environment—Its Evaluation and Control*. Washington, D.C.: U.S. Department of Health, Education, and Welfare, National Institute for Occupational Safety and Health, 719 pages.

Pandolf, K. B. 1978. "Influence of Local and Central Factors in Dominating Rated Perceived Exertion During Physical Work." *Perceptual and Motor Skills, 46:* pp. 683–698.

Petersen, A. P. G., and E. E. Gross, Jr. 1967. *Handbook of Noise Measurement*. 7th Edition. Concord, Mass.: General Radio Company, 322 pages.

Randall, F. E., and J. J. Baer. 1951. *Survey of Body Size of Army Personnel, Male and Female Methodology*. Rpt. 122. Lawrence, Mass.: Quartermaster Climatic Research Laboratory, Research and Development Division.

Roebuck, J. A., Jr., K. H. E. Kroemer, and W. G. Thomson. 1975. *Engineering Anthropometry Methods*. New York: John Wiley & Sons, 459 pages.

Roozbazar, A. 1978. Workplaces for Short People, Tall People, or Both. *Industrial Engineering, 10:* pp. 18–21.

Sargeant, A. J., and C. T. M. Davies. 1973. "Perceived Exertion During Rhythmic Exercise Involving Different Muscle Masses." *Journal of Human Ergology, 2:* pp. 3–11.

Snook, S. H., C. H. Irvine, and S. F. Bass. 1970. "Maximum Weights and Work Loads Acceptable to Male Industrial Workers." *American Industrial Hygiene Association Journal, 31:* pp. 579–586.

Stevens, S. S. 1975. *Psychophysics—Introduction to Its Perceptual, Neural, and Social Prospects*. New York: John Wiley & Sons.

Torgeson, W. S. 1958. *Theory and Methods of Scaling*. New York: John Wiley & Sons.

White, R. M. 1961. *Anthropometry of Army Aviators*. TR EP-150. Natick, Mass.: U.S. Army Quartermaster Research and Engineering Center, Environmental Protection Research Division, 109 pages.

White, R. M. 1964. *Anthropometric Survey of the Armed Forces of the Republic of Vietnam*. Sponsored by the Advanced Research Projects Agency. Natick, Mass.: U.S. Army Natick Laboratories, 66 pages.

White, R. M., and E. Churchill. 1971. *The Body Size of Soldiers, U.S. Army Anthropometry–1966*. Tech. Rpt. 72-51-CE. Natick, Mass.: U.S. Army Natick Laboratories, 329 pages.

Woodson, W. E. 1981. *Human Factors Design Handbook*. New York: McGraw-Hill, pp. 980–984.

Yaglou, C. P., and D. Minard. 1957. "Control of Heat Casualties at Military Training Centers." *AMA Archives of Industrial Health 16:* pp. 302–316.

A SELECTIVE ANNOTATED BIBLIOGRAPHY ON THE SUBJECT OF HUMAN FACTORS

This bibliography is intended to provide an introductory guide for people who have recently become interested in the field of human factors. There has been no attempt to compile a complete list. A strong emphasis is placed on the industrial applications of human factors, especially the design of workplaces, production equipment, manufacturing systems, and environments. A list of books with information about human capabilities and of journals containing articles of interest to industrial human factors practitioners is also included. The bibliography is organized to correspond to the text sections.

1. General

Bailey, R. W. *Human Performance in Engineering, A Guide for System Designers.* Englewood Cliffs, N.J.: Prentice-Hall, 1982, 656 pages.
A textbook on human factors engineering from the Bell Laboratories, with special emphasis on information transfer and perception. The improvement of human performance through the presentation of information and design of equipment is covered from a psychologist's perspective. The human/computer interface and information-coding sections are especially thorough.

Edholm, O. G. *The Biology of Work.* New York: McGraw-Hill, 1967, 256 pages.
An easy-to-read introduction to the basic concepts of the effects of work on the body. Presents a broad but quick introduction to ergonomics and human factors.

Grandjean, E. *Fitting the Task to the Man: An Ergonomic Approach.* 3rd ed. London: Taylor & Francis Ltd., 1980, 379 pages.
Emphasis is placed on the factors that affect people at work. This book provides a summary of some important European ergonomics research that has not previously been available in English. The level of treatment is generally introductory.

Lehmann, G. *Praktische Arbeitsphysiologie.* 2nd ed. Stuttgart: Georg Thieme Verlag, 1962, 409 pages.
Classical text on industrial ergonomics/work physiology from the Max Planck Gesellschaft in West Germany. Particularly detailed in the measurement of job physical effort demands, the design of work/rest cycles to reduce fatigue, and environmental stressors in the workplace. (In German.)

McCormick, E. J., and M. S. Sanders. *Human Factors in Engineering and Design.* 5th ed. New York: McGraw-Hill, 1982, 512 pages.
A widely used basic text. The material covered includes many areas of application outside the design of work in industry, such as military and aerospace applications. The depth of coverage ranges from introductory to very detailed, with psychological material, such as information processing, predominant.

Murrell, K. F. H. *Human Performance in Industry.* New York: Reinhold, 1965, 496 pages.
Discusses ways in which ergonomics can be used to improve the design of work in industry. The coverage ranges from material of an introductory nature to a few topics that are treated in considerable depth. An excellent introduction to the field.

Parker, J. F., Jr., and V. R. West, eds. *Bioastronautics Data Book.* 2nd ed. NASA SP-3006. Washington, D.C.: U.S. Government Printing Office, 1973, 930 pages.
A NASA handbook prepared for the designers of aerospace equipment. The data are presented in a very condensed form. While this book may not be the best choice for a beginner, the large amount of data compiled makes it a worthwhile acquisition for the practicing human factors specialist.

Rubin, A. I., and J. Elder. *Building for People, Behavioral Research Approaches and Directions.* Washington, D.C.: U.S. Government Printing Office, National Bureau of Standards, U.S. Department of Commerce, 1980, 315 pages.
Prepared for practicing architects and students to acquaint them with human/environment data that should be incorporated into building design. Includes information about current architectural problems, methods to collect data to improve design, and suggestions about how such a human/environment data base should be further developed.

Scherrer, J., H. Monod, A. Wisner, P. Andlauer, A. Baisset, S. Bouisset, H. Desoille, J. M. Faverge, A. Dubois-Poulson, E. Grandjean, B. Metz, P. Montastruc, S. Pascaud, M. Pottier, and D. Rohr. *Physiologie du Travail (Ergonomie). Travail Physique Energetique.* Volume I, 387 pages. *Ambiances Physiques, Travail Psycho-Sensoriel.* Volume II, 362 pages. Paris: Masson and Company, 1967.
A comprehensive review of work physiology with much of the French literature included. Emphasis on the physiological responses to physical effort and environmental stressors, and sensory functions associated with industrial work. (In French.)

Shackel, B. *Applied Ergonomics Handbook.* Reprint of *Applied Ergonomics.* Volumes 1 and 2, Nos. 1–3. Surrey, England: IPC Science and Technology Press, Business Press Ltd., 1974, 122 pages.
A compilation of articles, appearing in the first two volumes of the journal *Applied Ergonomics,* that are intended to show the state of the art on industrial ergonomics in the early 1970s. Based on earlier booklets produced by the Ministry of Technology on *Ergonomics for Industry.* The articles cover workplace and equipment layout; environmental factors such as noise, light, and thermal conditions; systems design; safety, work organization, and design of work for the disabled.

Singleton, W. T., J. G. Fox, and D. Whitfield, eds. *Measurement of Man at Work*. New York: Van Nostrand Reinhold, 1971, 267 pages (also London: Taylor and Francis).
A compilation of 26 symposium papers. Most of the papers describe techniques for obtaining psychological and physiological measurements of people at work. Some of the studies reported were done in an industrial setting. Throughout the book the emphasis is on applications rather than on research.

Weiner, J., and H. Maule, eds. *Case Studies in Ergonomic Practice. Volume I: Human Factors in Work, Design, and Production*. London: Taylor and Francis, 1977, 150 pages.
A series of examples of the application of human factors principles to industrial production workplaces, equipment, and product design.

Woodson, W. E. *Human Factors Design Handbook*. New York: McGraw-Hill, 1981, 1072 pages.
This handbook focuses on the kinds of human factors data that are needed by equipment and workplace designers. The author includes information of value in industrial applications in an easy-to-read format with many illustrations. Topics covered include the design of the workspace, displays and controls, principles of vision and audition, and the use of body dimensions in design.

2. Workplace Design

Konz, S. *Work Design*. Columbus, Ohio: Grid Publishing Co., 1979, 592 pages.
Job and workplace design, from an engineer's perspective. Emphasis on improving work efficiency. The book is organized around design guidelines.

Roebuck, J. A., Jr., K. H. E. Kroemer, and W. G. Thomson. *Engineering Anthropometry Methods*. New York: John Wiley & Sons, 1975, 459 pages.
The application of scientific physical measurement methods to human subjects. Provides engineering design standards and techniques for the evaluation of engineering drawings, mock-ups, and manufactured products to ensure suitability of these products for the intended user population.

Tichauer, E. R. *The Biomechanical Basis of Ergonomics: Anatomy Applied to the Design of Work Situations*. New York: Wiley Interscience, 1978, 99 pages.
A good summary of the principles of biomechanics as applied to industrial jobs, typically in manual materials handling, and workplace and equipment design.

3. Equipment Design

Department of Defense. *Human Engineering Design Criteria for Military Systems, Equipment, and Facilities.* MIL-STD-1472B, December 31, 1974, 239 pages.
This standard establishes human engineering criteria, principles, and practices to be applied in the design and development of military systems, equipment, and facilities.

Grandjean, E., and E. Vigiliani, eds. *Ergonomic Aspects of Visual Display Terminals.* London: Taylor and Francis, 1980, 300 pages.
This book is drawn from papers presented at an international conference in Milan, Italy, in 1979. The topics covered include visual fatigue, visual display unit (VDU) workplace design, radiation exposure in VDU workplaces, and psychological stress associated with a VDU operator's work load.

Meister, D. *Human Factors: Theory and Practice.* New York: Wiley, 1971, 415 pages.
A general approach to the field of human factors. Meister discusses the human/machine interface, the analysis of human error, the principles that should govern applied research, an analysis of hardware design problems, and organization of a human factors group. The topics discussed reflect the aerospace industry background of the author.

Ramsey, H. R., and M. E. Atwood. *Human Factors in Computer Systems: A Review of the Literature.* AD-A075-679. Springfield, Va.: NTIS, Department of Commerce, Science Applications, Inc., 1979, 169 pages.
The results of a survey of 564 papers from human factors, computer, and behavioral science literature are presented. The areas surveyed include users, tasks, requirements analysis, interactive dialogue, output devices and techniques, input devices and techniques, and evaluation of system performance. Tables summarizing the salient points are included with indications of where further information can be found.

Swain, A., and H. Guttmann III. *Handbook of Human Reliability Analysis with Emphasis on Nuclear Power Plant Applications.* NUREG/CR/278 Washington, D.C.: U.S. Nuclear Regulatory Commission, 1980, 480 pages.
A summary of work on human reliability analysis, including principles of human behavior and ergonomics, mathematical models, and human error probabilities. These have been derived from other performance measures and by observation. The focus is on human reliability in nuclear power plant operations.

Van Cott, H. P., and R. G. Kinkade, eds. *Human Engineering Guide to Equipment Design.* Rev. ed. Washington, D.C.: U.S. Government Printing Office, 1972, 752 pages.
This handbook is intended to be used by the designers of military and aerospace equipment. Despite this orientation, it contains much basic information and data that can be used in industry. The data are discussed in greater depth than is usual with handbooks.

4. Information Transfer

Cornog, D. Y., and F. C. Rose, eds. *Legibility of Alphanumeric Characters and Other Symbols*. II. A Reference Handbook: National Bureau of Standards Miscellaneous Publication 262-2, Washington, D.C.: U.S. Government Printing Office, 1967, 460 pages.
Comprehensive review of 203 papers in the literature. There is a separate summary for each of the papers. The findings are cross-indexed under 78 categories that describe the conditions under which the experimental data were gathered. These include such factors as the type of lighting, the face design of the alphanumeric character, the layout of the characters, and the kind of experimental measure used.

Drury, C. G., and J. G. Fox, eds. *Human Reliability in Quality Control*. London: Halstead Press, 1975, 250 pages.
This book is a compilation of papers from a symposium at Buffalo, New York, in 1974. There are 21 papers grouped under three headings: "Models of Inspector Performance," "Factors Affecting Inspection Performance," and "Industrial Applications."

Harris, D. H., and F. B. Chaney. *Human Factors in Quality Assurance*. New York: John Wiley & Sons, 1969, 234 pages.
The application of human factors principles to the entire quality control system is discussed. Many specific applications are described.

5. Environment

Brouha, L. *Physiology in Industry*. 2nd ed. New York: Pergamon Press, 1967, 164 pages.
Basic introduction to industrial work physiology. Techniques of measuring the effects of the environment, especially heat, and the physical demands of the job are discussed. There are numerous examples of actual industrial applications.

Fanger, P. O. *Thermal Comfort*. New York: McGraw-Hill, 1973, 256 pages (also Copenhagen: Danish Technical Press).
Fanger reports on major studies done on people's comfort in various combinations of temperature, humidity, air velocity, and clothing. Very useful for determining how many people will be uncomfortable in a given thermal environment.

Hopkinson, R. G., and J. B. Collins. *The Ergonomics of Lighting*. London: MacDonald, 1970, 272 pages.
The authors summarize the research that lies behind current lighting practices. The relationship between visual function and the total luminous environment is stressed.

Kaufman, J. E., and J. F. Christensen, eds. *IES Lighting Handbook.* 5th ed. New York: Illuminating Engineering Society, 1972, 726 pages.
A compendium of information for the practicing lighting engineer containing chapters on basic physics of light and its measurement, visual processes, concepts of color, design of light fixtures and control of light, types of light sources, and chapters on specific applications, such as offices, industries, merchandising, and roadways.

Kryter, K. D. *The Effects of Noise on Man.* New York: Academic Press, 1970, 633 pages.
A detailed study of the effects of noise on humans. The functioning of the auditory system, subjective responses to noise, and physiological responses to noise are covered in depth.

Leithead, C. S., and A. R. Lind. *Heat Stress and Heat Disorders.* Philadelphia: F. A. Davis Company, 1964, 304 pages.
A comprehensive discussion of heat stress and acclimatization. It includes derivations of temperature indices, such as the effective temperature, plus graphs allowing estimation of the level of heat stress for various combinations of temperature, humidity, and wind velocity. There is also discussion of the physiological and pathological consequences of heat.

NIOSH. *The Industrial Environment—Its Evaluation and Control.* Washington, D.C.: U.S. Department of Health, Education, and Welfare, National Institute for Occupational Safety and Health, 1973, 719 pages.
A series of papers by specialists in ergonomics and industrial hygiene relating especially to environmental factors in the workplace and some aspects of job design.

Peterson, A. P. G., and E. E. Gross, Jr. *Handbook of Noise Measurement.* 7th ed. Concord, Mass.: General Radio, 1974, 322 pages.
Presents methodology for noise and vibration measurements. It clarifies the terminology and definitions used in these measurements, describes the measuring instruments and their use to aid the prospective user in selecting the proper equipment, and shows how these measurements can be interpreted to solve typical problems. There is also a chapter on human responses to noise and vibration.

6. Human Capacity

Belbin, R. M. *Training Methods for Older Workers.* Paris: OECD—McGraw-Hill, 1965, 72 pages.
Adult learning capacity, the problems of training and retraining, and the selection of training methods for older workers are discussed from a practical viewpoint.

Chapanis, A., ed. *Ethnic Variables in Human Factors Engineering*. Based on a symposium held at Oosterbeck, The Netherlands, June 19–23, 1972, under the auspices of the Advisory Group on Human Factors, NATO. Baltimore: Johns Hopkins University Press, 1975, 290 pages.

This book is drawn from papers from an international symposium that included the following topics: population differences in size, or anthropometrics and strength; movement stereotype differences; communications problems; human factors applications; and teaching in underdeveloped countries.

Damon, A., H. W. Stoudt, and R. A. McFarland. *The Human Body in Equipment Design*. Cambridge, Mass.: Harvard University Press, 1966, 360 pages.

An explanatory text discussing the proper application of anthropometric and strength data contained in the tables. Additional material covers other factors that affect equipment design.

Dirken, J. D. *Functional Age of Industrial Workers*. Groningen, The Netherlands, Wolters-Noordhoff Publishing, 1972, 251 pages.

A study of Dutch industrial workers relating their physiological, sensory, psychological, and social capacities to job demands and performance. The resulting model of functional age is a better predictor of performance than is chronological age.

Fitts, P. M., and M. L. Posner. *Human Performance*. Belmont, Calif.: Wadsworth Publishing Company, Inc., 1979, 162 pages.

Provides a comprehensive basic review of several aspects of human performance. Among these are learning and skilled performance, motivation, sensory capacities and perceptual processing, skills measurement, and perceptual-motor and language skills.

Graham, C. H., ed. *Vision and Visual Perception*. New York: John Wiley & Sons, 1965, 637 pages.

A comprehensive source of information on visual function. The capabilities and limitations of the human eye are discussed.

NASA. *Anthropometric Source Book. Volume I: Anthropometry for Designers*, 613 pages. *Volume II: A Handbook of Anthropometric Data*, 424 pages. *Volume III: Annotated Bibliography of Anthropometry*, 130 pages. Yellow Springs, Ohio: NASA Scientific and Technical Information Office, 1978.

Compilation of anthropometric data, primarily U.S. Air Force and Army men and women. Most complete and comprehensive source of these data.

Thompson, C. W. *Manual of Structural Kinesiology*. 8th ed. St. Louis: Mosby, 1977, 159 pages.

An introduction to kinesiology. The anatomy of the muscles and the movements of the body that they accomplish are clearly described. Numerous illustrations are included.

Welford, A. T. *Aging and Human Skill: A Report Centered on Work by the Nuffield Unit*. London: Oxford University Press, 1973, 300 pages (reprint of the 1958 edition, Greenwood).
A summary of the effects of age on performance. The types of tasks that are most likely to show an age effect are identified.

Welford, A. T. *Fundamentals of Skill*. London: Methuen, 1968, 426 pages.
A survey of the elements that affect performance on psychomotor tasks. Topics include decision making, short-term memory, stress, and work load.

7. Journals

American Industrial Hygiene Association Journal, published monthly, American Industrial Hygiene Association, 475 Wolf Ledges Parkway, Akron, Ohio 44311-1087.
The ergonomics group of AIHA frequently publishes work physiology, heat stress, and manual materials handling articles in this journal. The majority of articles relate to chemical and physical workplace exposures, however.

Applied Ergonomics, published quarterly, IPC House, Surrey, England.
The application of human factors to the solution of problems is emphasized. A very wide range of problem areas is covered. Theoretical research results and lengthy descriptions of experimental technique are avoided. The general style of the articles is such that they can be easily read by persons with little or no background in human factors.

British Journal of Industrial Medicine, published quarterly, British Medical Association House, Tavestock Square, London WCIH 9JR, England, and British Medical Journal, 1172 Commonwealth Avenue, Boston, Mass. 02134.
Articles on environmental stress are frequently included in this journal, especially heat, humidity, and cold stress and noise exposure. Many of the articles deal with chemical and physical environmental exposures.

Ergonomics, published bimonthly, Taylor & Francis Ltd., 10-14 Macklin Street, London WC2B 5NF, England.
The articles published include results of original research, evaluations of experimental techniques and methodologies, and application of human factors principles to practical problems. The emphasis is on industrial problems, but there are articles related to other areas such as sports, the military, and product design.

Ergonomics Abstracts, published quarterly, Taylor & Francis Ltd., 10-14 Macklin Street, London WC2B 5NF, England.
This service regularly scans more than 160 journals for articles related to ergonomics and human factors. Both English and foreign language sources are included. Several thousand abstracts are published and indexed under about 40 different topics each year. A comprehensive service.

European Journal of Applied Physiology, published quarterly, Springer-Verlag New York Inc., 175 Fifth Avenue, New York, N.Y.
This journal is a cross between *Ergonomics* and the *Journal of Applied Physiology*. Contains articles on work physiology, most of which are in English, some in French and German. A good source for information on ergonomics work in Germany that is written in English.

Human Factors Journal, published bimonthly, Human Factors Society, Santa Monica, Calif.
The types of articles published include the results of original research, literature surveys, and applications. The range of topics covered is broad and includes military and aerospace applications, industrial applications, environmental effects, and product and equipment design.

Journal of Human Ergology, published semiannually, Business Center for Academic Societies Japan, 4-16 Yayoi 2-chome, Bankyo-ku Tokyo 113, Japan.
This is the official journal of the Human Ergology Research Association of Japan. The articles are similar to those in *Ergonomics* and the *European Journal of Applied Physiology*, with a majority coming from Japanese ergonomists.

GLOSSARY OF HUMAN FACTORS TERMS

The terms listed here are commonly used in industrial human factors. This glossary is intended primarily to assist the reader in understanding the terms found in this volume. For a more comprehensive listing of ergonomics and human factors terms, see the *Ergonomics Glossary*, 1982 (*).

Abduction—movement of a limb away from the body's midline axis, such as elevating an elbow or raising an arm to the side.

Acromial—of or pertaining to the distal shoulder above the upper arm. The acromion provides a bony crest from which anthropometric measurements of the shoulder and arm are often made.

Adduction—movement of a limb toward the midline axis of the body, such as moving an arm across the front of the body.

Anthropometry—the study of people in terms of their physical dimensions. It includes the measurement of human body characteristics, such as size, breadth, girth, and distance between anatomical points, e.g., interpupillary distance. Anthropometry is also defined in terms of the techniques used to quantitatively express the form and dimensions of the body.

Biomechanics—the application of mechanical principles, such as levers and forces, to the analysis of body part structure and movement. This includes studies of range, strength, endurance, and speed of movements, and mechanical responses to such physical forces as acceleration and vibration.

Candelas per square meter (cd/m^2)—a measure of luminance (emitted or reflected light) obtained by means of a photometer. High values usually indicate a source of glare. In SI units 1 cd/m^2 = 0.29 footLamberts (fL).

Centimeter (cm)—metric measure of distance; 1 cm = 1/100 meter the standard measure of length in the SI system. One inch equals 2.54 cm, or 1 cm equals approx. 0.4 in.

Clo Unit—a measure of the thermal insulation provided by clothing. One clo is 0.16 degrees Celsius per watt per square meter of body surface area. The 0.16 becomes 0.18 if watts are changed to kilocalories. As clothing is added to the body, the clo value increases and it is more difficult to lose heat to the environment.

Compatibility—the consistency with which a response meets human expectations. For controls and displays, the consistency of the operator's movement of a control compared to its displayed response.

Contingency Analysis—the identification of non-routine situations to which a person may be exposed and that may entail special human performance requirements on a task.

Contrast—the brightness relationships of two non-specular adjacent surfaces viewed under the same illumination and in the same immediate surroundings.

Control—a mechanical or electronic device that directs the action of a mechanism or produces a change in the operation of a system or process.

Cursor—a moving element on a display. It is used to show the desired position or error on an instrument, and, on video displays, to mark row and column positions or to indicate where the next activity will occur.

dBA—decibels on the A-weighted scale, which de-emphasizes frequencies less than 1,000 Hz. A measurement of sound pressure level, most commonly used to assess the noise exposure of workers.

Decibel—a logarithmic scale often used to express quantities of sound, or electrical power, relative to a specified reference level.

Detectability—the quality of a signal, display, stimulus, or error that affects the probability of its presence being perceived.

Display—the presentation of information from a device or system in a form designed to be seen or heard by the human operator.

Dynamic Muscle Work—muscle contraction where muscle length changes during activity, resulting in motion around a joint. Most handling and assembly tasks are accomplished with dynamic work. Also see *static muscle work.*

Dynamic Visual Acuity—the measure of visual acuity used when the object being observed moves with respect to the operator.

Effort, Physical—the amount of muscle work performed on a job, often referred to as the physical work load. It is often defined by the number of objects handled per shift, their weight, the distance they are transported, and how long the task is performed.

Energy Expenditure—the power used during activity or rest. Expressed in watts or kilocalories per unit time, or in cubic centimeters of oxygen per kilogram of body weight per minute.

Ergonomics—the study of the design of work in relation to the physiological and psychological capabilities of people. One of several terms used to define similar fields of interest; others are human engineering, human factors, and human factors engineering. Ergonomics has been used predominantly outside of the U.S.A. The aim of the discipline is the evaluation and design of facilities, environments, jobs, training methods, and equipment to match the capabilities of users and workers, and thereby to reduce the potential for fatigue, error, or unsafe acts.

Fatigue—decreased ability to perform; feelings of being very tired after a work effort.

Font—an assortment of printer's type or characters comprised of all one size and style.

Foot (ft)—a measure of distance in the English system; one foot has 12 inches and is equivalent to about 30 centimeters.

Footcandle (fc)—in the English system of measurement, the measure of illuminance, or light falling on a surface. It is measured with an illumination meter laid on the surface of interest. One footcandle is equal to 10.8 lumens per square meter (lux) in the SI measurement system.

FootLambert (fL)—in the English system of measurement, a unit of luminance (emitted or reflected light). It is measured with a photometer pointed at the surface of interest. High values indicate possible glare sources in the workplace. One footLambert is equal to 3.43 candelas per square meter in the SI system.

Footrest—a support used primarily in sitting workplaces to help accommodate differences in the size of workers and relieve postural stresses.

Glare—a sensation produced by a luminance within the visual field that is sufficiently greater than the luminance to which the eyes are adapted. It can produce annoyance, discomfort, or distraction of the worker, or a reduction in visibility of the object of regard.

Gravity (g)—a measure of acceleration expressed as a rate of change of velocity per second. Vertical vibration of the body or limbs is often expressed in g units. Values below 0.3 g are of less concern than those above 1 g, which are associated with vibration illnesses.

Heart Rate—a physiological measure of the frequency with which the heart contracts. Expressed in beats per minute, it can be used as an estimate of job stress, work load, or environmental stress.

Heat Stress—the physiological load induced by working in a hot environment. Heat stress produces increased heart rate, body temperature, and sweat rate, and often increases an operator's feelings of fatigue. Severe heat stress can produce exhaustion or heat stroke.

Hertz (Hz)—the unit for frequency, in cycles per second, in the SI system. 1 Hz = 1 cycle per second. Noise levels are expressed in Hertz for specific frequency bands when an octave band analysis of a workplace or product is done.

Hue—the attribute of color determined primarily by the wavelength of light entering the eye. Spectral hues range from red through orange, yellow, green, and blue to violet.

Human Operator—a person who participates in some aspect of the operation or support of a system and its associated equipment and facilities.

Humerus—the long bone of the upper arm.

Inch (in.)—in the English system of measurement, a measure of distance, such as length, width, depth, height, or circumference. One inch is equal to 2.54 centimeters in the SI system.

Indicator—an instrument or device for displaying task-related information such as location, speed, pressure, or overload. Indicators may be mechanical, electrical, or electronic; examples of each are pointers, warning lights, and video display units.

Inferior—lowermost or below; the foot is *inferior* to the ankle.

Information—a quantitative property of a group of items that enables them to be categorized or classified. The amount of information in a group is measured by the average number of operations needed to categorize the items. Examples of operations are statements, decisions, and tests.

Input—the information or energy entered into a machine or system that is the quantity to be measured or otherwise operated upon. Also called input signal.

Interface—the common physical boundary between an operator and the equipment used, such as a control, display, seat, or workbench.

Job Analysis—a study to determine and identify duties, tasks, and functions in a job, together with the skills, knowledge, and responsibilities required of the worker. It is accomplished through measurement, observation, and interviews.

Job Design—the arrangement of tasks over a work shift, whether in terms of the distribution of light and heavy physical work or the arrangement of rest breaks in a mentally or perceptually demanding task, such as visual inspection. Good job design reduces the opportunities for fatigue and human error.

Kilogram (kg)—the unit for mass (kgm) and force (kgf) in the SI system of measurement. The weight of objects handled in the workplace and the force required to activate a control, such as a push button or lever, are expressed in these units. One kg is equal to 2.2 lbm and 9.8 Newtons.

Kinesthesis—a person's sense that informs the brain of the movements of the body or of its members and their location in space. This awareness of movement is accomplished through activation, usually by pressure or stretch, of special receptors in the muscular tissue, tendons, and joints.

Lateral—toward the side of the body away from the midline in a plane parallel to the midsagittal plane that divides the body into right and left halves.

Learning—any change in behavior or performance that occurs as a result of training, practice, or experience.

Legibility—the ease with which a label, document, or display can be read and understood. The design and size of characters, contrast, illumination, color of characters and background, and construction of textual information all affect legibility.

Load, Sensory—the number and variety of stimuli requiring operator response. For example, the load on the visual system is greater if several different classes of stimuli must be discriminated than if only one or a few types are present.

Loudness—the attribute of auditory sensation by which sounds may be ordered on a scale extending from soft to loud. The unit for measuring loudness is the sone.

Lux—a measure of illuminance, or light falling on a surface, in the SI measurement system. It is measured with an illumination meter that is set directly on the surface of interest. Low values may be problematic in difficult visual tasks such as detecting low contrast defects in a product. One lux unit is equal to 0.09 footcandles.

Maintainability—the design of hardware, software, and training to match the capabilities and characteristics of maintenance personnel. Guidelines for the design of instruments and fasteners, manual and visual access to equipment, efficient handling, ease of use of tools, identification of parts, and consideration of experience and knowledge are covered under maintainability.

Manual—operated or done by people rather than by machine.

Manual Dexterity—the ability to manipulate objects with the hands. Various degrees and types of dexterity have been identified through tests. There are three primary types: fine-finger, or precision dexterity; tweezer dexterity; and gross hand dexterity. The latter involves more of the physical hand and is much less precise than the other two.

Masking—the amount by which a sound's threshold of audibility is raised by the presence of another (masking) sound, as in "white noise." It is measured in decibels.

Medial—relating to the middle or center; nearer to the median or midsagittal plane, which divides the body into right and left halves.

Metabolism—the sum of all the physical and chemical processes by which living matter is produced and maintained; also the process by which energy is transformed for use by the body in muscular work and other processes.

Metacarpals—the long bones of the hand located between the wrist and the finger bones (phalanges).

Method of Magnitude Estimation—a psychophysical method used primarily to scale sensations or opinions. The subject assigns numbers to a set of stimuli that are proportional to some subjective dimension of the stimuli. This method allows a person to indicate how much a given item, such as a food product, differs from similar ones when several items are being compared.

Method of Paired Comparisons—a psychophysical method used primarily to scale responses. Pairs of stimuli are presented to a subject for comparison along some dimension.

Method of Rank Order—a psychophysical method used primarily to scale sensations. Stimuli are presented to a subject who orders them along some dimension. The order reflects where the stimuli line up in relation to one another, but does not indicate how far apart they are.

Midsagittal—the plane that divides the body vertically through the midline into right and left halves.

Milliamperes (mA)—0.001 of an ampere, the unit for electrical current in the SI measurement system. If electricity is contacted in the workplace, higher values of current increase the risk for ventricular fibrillation and its lethal results.

Monitor—to observe, listen in on, keep track of, or exercise surveillance over a process or activity; for example, to monitor radio signals, the quality of product in an assembly line, the progress of a chemical reaction, or the manufacturing steps in a production process.

Motor Skill—the ability to coordinate movement of hands, fingers, legs, or feet in a smoothly flowing sequence resulting in the performance of some act.

Multidimensional Scaling—a psychometric method used to describe stimuli that have more than a single attribute. Examples of such stimuli are foods which may be evaluated for flavor, appearance, and texture or photographs which may be described for color, contrast, exposure, and sharpness of detail.

Newton (N)—a measure of force in SI units. The force required to move a control, such as a lever, is expressed in newtons. One newton is equal to 10^5 dynes, approx. 0.102 kilogram-force, and approx. 0.225 pound-force.

Newton Meter (N · m)—a measure of torque (rotational force) in SI units. This measure is used to describe the force needed to rotate a control or turn a wheel.

Noise—unwanted signals that interfere with the detection of desired signals. Noise can be audible, such as voice communications, or visible, as in the case of radar or hard copy.

Noise Criterion (NC) Curves—any of several versions of criteria such as sound criteria (SC), noise criteria (NCA), or preferred noise criteria (PNC). These criteria are used to rate the acceptability of continuous indoor noise.

Operator Inputs—information received or sensed by the operator from instructions, displays, or the environment.

Operator Outputs—action taken by the operator based on some input, such as the activation of controls or verbal communication.

Operator Overload—the condition in which a person is required to perform more decision making, information handling, signal detection, or muscle work than he or she is able to handle effectively within a given time period.

Optimum Location Principle in Equipment Design—the principle of arrangement of displays and controls so that each is placed in its optimum location in relation to some criterion of use: convenience, accuracy, speed, or force applied.

Oxygen Consumption—the rate at which the body, or its tissues and cells, use oxygen. Expressed in liters per minute per unit of body or tissue weight.

Parallax—the difference in the apparent position of an object or pointer when a display is viewed from different angles.

Pascal (Pa)—1 Newton · meter^{-2} the unit of measure for pressure in the SI system. 1 Pa = 0.00015 psi.

Perception—the process for interpreting sensations. Also, the sensory awareness of external objects, qualities, or relations.

Perceptual-Motor Task—a task involving movement in response to a nonverbal stimulus, such as driving a car. The response is determined by sensory input that is organized by the operator.

Perceptual Skill—the detection and interpretation of information received through sensory channels.

Performance Decrement—a decrease in human proficiency that may be associated with operator overload, stress, or fatigue. It is characterized by increases in errors and misjudgments, omission of task elements, and reduced intensity of effort.

Performance Measures—objective and subjective evaluation methods to determine an individual's effectiveness. Productivity, job performance samples, proficiency and job knowledge tests, and evaluation checklists are used as objective measures. Peer, self, and supervisory ratings are subjective measures.

Peripheral Vision—the ability to see things to the side; primarily sensed by the rods of the eyes. For optimal perception, an object has to be focused on the cones in the retina's fovea. Peripheral vision is the major pathway for detecting moving defects in inspection tasks; it also plays a major role in providing vision in areas of low illumination.

Phalanges—the finger bones. The bones nearest the palm are the first phalanges, and the tips of the fingers are the third phalanges. The distal end of a phalanx is farthest away from the palm; the proximal end is nearest. Interphalangeal joints are located between the first and second and the second and third phalanges of the fingers.

Photopic Vision—vision associated with levels of illumination of about 30 cd/m^2 (candelas per square meter) or higher. It is characterized by the ability to distinguish colors and small details. It is also called foveal vision.

Popliteal—of or pertaining to the back of the knee, opposite the kneecap.

Population Stereotype—a behavioral sequence that is predictable, or the way most people expect something to be done. For example, rotating an electrical control in a clockwise direction is expected to increase the value of the setting; a design in which a clockwise rotation *decreases* the value would violate the population stereotype.

Pound (lb.)—a measure of mass (lbm) or force (lbf) in the English system; equal to approx. 0.45 kgm or 4.4 newtons.

Predictor Display—the means by which the operator is shown some measure of what will happen to the system in the future. These displays can be symbolic or pictorial, and they may predict system input, equipment output, or both. They provide advance information that allows the operator to anticipate the need for future action.

Psychometrics—the use of mathematical and statistical techniques in the measurement of psychological processes. In a psychometric study, a person may be asked to place a stimulus, such as a description of an object or situation, on a scale in relation to other similar stimuli. Quantitative measurements can be taken from such a scaling of opinion data. The technique is useful to help identify workplace conditions that are difficult to evaluate with physical or physiological measurements.

Psychomotor Ability—the action of a muscle resulting directly from a mental process, as in the coordinated manipulation of tools in assembly tasks.

Psychophysical Methods—standardized techniques for presenting stimulus material to a person for judging, or for recording the results of judgments. Originally developed to determine functional relationships between physical stimuli and correlated sensory responses, but used more widely now.

Quantitative Display—a display that provides numerical values, in contrast to one giving only descriptive information (a qualitative display).

Reliability, Human—the ability to perform with minimum errors. Reliability should decrease with increasing task complexity and with task-related and environmental stresses.

Response—(1) physiological—the muscular contraction, glandular secretion, or other activity of a person resulting from stimulation; (2) psychological—a behavioral action, usually motor or verbal, that follows an external or internal stimulus.

Root Mean Square (rms)—the most common value for measuring varying acoustical signals. The sound energy is integrated, and a single value is read off the meter. Most sound-level meters show rms values. They are also used to express electrical current values. It indicates the degree of variance of the signal.

Sagittal—any plane parallel to the midsagittal line that divides the body vertically into right and left halves.

Saturation—the extent to which a chromatic color differs from a gray of the same brightness. It is measured on an arbitrary scale from 0% (gray) to 100%.

Scotopic Vision—vision under very low levels of illumination such as occurs at night. It is characterized by lack of color vision and less discrimination of detail. The rods of the retina mediate scotopic, or night, vision.

Sensation—a subjective response aroused by stimulation of a sense organ.

Sensor—the nerve endings or sense organs that receive information from the environment, the body, or both.

Sequence-of-Use Principle in Equipment Design—the principle of arranging controls and displays so that those used in sequence are physically situated in order of their respective operation, as on a control console.

Shape Coding—varying the configuration of controls to make them distinctive. Shape coding is effective because the difference can be both seen and felt. The shape of a control should suggest its purpose, and it should be distinguishable not only with the bare hand, but also when wearing gloves.

Short-Term Memory—the storage of recently received information for a few seconds or minutes.

SI Units—abbreviation for Le Système International d'Unites (International System of Units). It is a coherent measurement system based on the metric unit of 10. An international committee recommended its use for worldwide scientific and commercial designations of measurements, such as distance, weight, force, illumination, noise, and volume.

Signal—an event describing some aspect of the work process that the operator should sense and respond to. It could be an auditory signal, such as an alarm bell, or a visual signal, such as a flashing light, indicating a product defect in an inspection task.

Signal Detection Theory—a psychophysical model describing human perception as a process in which decisions are made under conditions of risk and are based on uncertain sensory information. In inspection tasks, for example, the signal detection theory is used to identify the probability of detecting defects that exist versus the probability of missing them, or of rejecting good product.

Simulation—a set of test conditions which are designed to duplicate field-operating and usage environments. A good technique to help the designer of a workplace or a new piece of equipment anticipate human factors problems.

Spectrum Colors—the series of saturated colors that are normally evoked by photopic stimulation of the retina with light in the visible range.

Specular Reflection—scattering of light rays, as from a mirror, where the reflected radiation is not diffused. Also called regular, or simple, reflection.

Speech Interference Level (SIL)—a gross measure used for comparing the relative effectiveness of speech in the presence of noise. It is the simple numerical average of the noise decibel level in the three octave bands with centers at 500, 1,000, and 2,000 Hz.

Static Muscle Work—muscle contraction without motion; also known as isometric work. Standing is an example of static postural work; gripping or holding are examples of static manual work. Some muscles do static work while other muscles are doing dynamic work. An estimate of the static work can be made by multiplying the force of a static contraction by its duration.

Stature—the vertical distance from the top of the head to the floor. The subject stands erect and looks straight ahead.

Stimulus—internal or external energy that excites a receptor.

Stress—(1) deformation of a part of the body in response to increased force per unit area; (2) the effect of a physiological, psychological, or mental load which may produce fatigue and degrade a person's proficiency.

Surround Brightness—the brightness of an area immediately adjacent to the visual work area.

System—a composite of equipment, skills, and techniques, including all related facilities, equipment, material, services, and personnel, capable of performing and/or supporting an operational role.

System Analysis—identification of the dynamic, or functional, relationships between elements of an operational system.

System Engineering—the study and planning of a system where the relationships of various parts of the system are fully established before designs are committed. Also relates to the study of existing systems in which improvements may be sought.

Tabular Display—presentation of alphanumeric and other symbols in a row and column format.

Target Discrimination—the ability to detect a desired signal within a background of noise, or to distinguish between multiple signals.

Task—a group of related job elements performed within a work cycle and directed toward a goal. They can include discriminations, decisions, and motor activities required of a person to accomplish a given unit of work.

Task Analysis—an analytical process employed to determine the specific behaviors required of people when operating equipment or doing work. It involves measuring, over time, the detailed performance required of a person and a machine, their interactions, and the effects of environmental conditions and malfunctions. Within each task, behavioral steps are described in terms of the perception, decision-making, memory storage, and motor skills required, as well as the expected errors. The data may be used to establish criteria for equipment design and for personnel training.

Task Element—the smallest definable set of perceptions, decisions, and responses a person is required to perform in completing a task. An example of a task element is an operator responding to a signal on a display by actuating a control, and seeing that the response has produced the desired effect.

Tibial—of or pertaining to the shin bone of the lower leg.

Timbre—that attribute of auditory sensation which permits a listener to discriminate between two sounds of similar loudness and pitch but of different tonal quality.

Time-sharing—the division of an operator's perceptual, decision making, or response time among activities or tasks that must be performed at or about the same time. For example, operating a production machine often includes loading, monitoring, inspecting product, and making repairs and adjustments to it. These activities are time-shared according to the need to keep the equipment running.

Training—the instructions, planned circumstances, and directed activity by which a person acquires and/or strengthens new concepts, knowledge, skills, habits, or attitudes that will allow for the assigned performance of duties with maximum reliability, efficiency, uniformity, safety, and economy.

Transillumination—indirect illumination utilizing lighting from behind (backlighting) or along one side (edgelighting) on clear, fluorescent, or sandwich-type plastic materials. It may be used to light control panel switches or to enhance detection of quality deviations on transparent or translucent materials.

Transmittance—the percentage of light going through a material, such as plastic, divided by the amount of ambient light falling on that material. *Selective transmittance* is the passage of particular light wavelengths through a transparent or translucent material, such as a red filter.

Transverse—movement across the front of the body that is in a horizontal plane perpendicular to the body's midline axis that divides the body into right and left halves.

Trunk—the torso of the human body.

Validity of a Test—the degree to which a test measures what it was designed to measure. It is estimated using a coefficient of correlation between test scores and a criterion measure, such as actual on-the-job performance.

Viewing Angle—the angle formed by a line from the eye to the surface of the object being viewed.

Vigilance—an activity involving continuous visual or auditory watch; also the individual's ability to detect signals of varying frequencies during these observation periods.

Visual Acuity—the ability to see detail at various distances from the object of regard. It is the reciprocal of the visual angle, in minutes of arc, occupied by the smallest discriminable detail.

Visual Angle—the angle described by an object in the visual field at the point where it is focused in the eye. The size of the image on the retina is determined by the visual angle. Objects of different sizes or at different distances have the same image size on the retina if they subtend the same visual angle.

Visual Field—that space that can be seen when the head and eyes are motionless. Or, all visual stimuli that act upon the unmoving eye at a given moment.

Weber-Fechner Law—the psychophysical law that relates the sensation or response level of a sense organ to the intensity of the stimulus.

White Noise—noise that is a combination of many sound waves of different lengths that reinforce or cancel one another in a nonuniform, random fashion. The spectrum density is substantially independent of frequency range.

Windchill—a scale that indicates the combined effects of wind velocity and temperature and is used to express the severity of cold environments.

Work—an expression for human effort measured in physical units or in performance-output results. Also, a general description of a task.

Work Cycle—the total series of actions and events that characterize or describe an integral work assignment or a single operation.

Work Pace—the rate at which a task or activity is done. Work may be paced externally, as by machine rates or by other people on the assembly line, or it may be self-paced by the worker.

Work Physiology—the study of the body's response to physical and/or mental effort. Measurement of stress on the cardiovascular, respiratory, nervous, and musculoskeletal systems, in particular, are included

Work Space—the physical area in which a person performs work.

Work Space Layout—the design of the working area of a workstation, including provisions for seating, physical movement of people, operational maintenance, and adequate contacts between people and equipment and among people.

Work Study—the analysis of work methods, techniques, and procedures.

x-axis—the horizontal axis on a two-dimensional surface; also known as the abscissa. This axis contains values of the independent variable, such as time or frequency.

y-axis—the vertical axis on a two-dimensional surface; also known as the ordinate. This axis contains values of the dependent variable, such as intensity, or a physiological response such as heart rate.

(*)Anon (1982), *Ergonomics Glossary, Terms Commonly Used in Ergonomics*, International Publications Service, 114 East 32nd Street, New York, New York 10016.

INDEX